Methods in Molecular Biology™

Series Editor
John M. Walker
School of Life Sciences
University of Hertfordshire
Hatfield, Hertfordshire, AL10 9AB, UK

For further volumes:
http://www.springer.com/series/7651

RNA Folding

Methods and Protocols

Edited by

Christina Waldsich

Department of Biochemistry and Cell Biology, Max F. Perutz Laboratories, University of Vienna, Vienna, Austria

Humana Press

Editor
Christina Waldsich
Department of Biochemistry
and Cell Biology
Max F. Perutz Laboratories
University of Vienna
Vienna, Austria

ISSN 1064-3745 ISSN 1940-6029 (electronic)
ISBN 978-1-4939-6058-3 ISBN 978-1-62703-667-2 (eBook)
DOI 10.1007/978-1-62703-667-2
Springer New York Heidelberg Dordrecht London

© Springer Science+Business Media, LLC 2014
Softcover reprint of the hardcover 1st edition 2014
This work is subject to copyright. All rights are reserved by the Publisher, whether the whole or part of the material is concerned, specifically the rights of translation, reprinting, reuse of illustrations, recitation, broadcasting, reproduction on microfilms or in any other physical way, and transmission or information storage and retrieval, electronic adaptation, computer software, or by similar or dissimilar methodology now known or hereafter developed. Exempted from this legal reservation are brief excerpts in connection with reviews or scholarly analysis or material supplied specifically for the purpose of being entered and executed on a computer system, for exclusive use by the purchaser of the work. Duplication of this publication or parts thereof is permitted only under the provisions of the Copyright Law of the Publisher's location, in its current version, and permission for use must always be obtained from Springer. Permissions for use may be obtained through RightsLink at the Copyright Clearance Center. Violations are liable to prosecution under the respective Copyright Law.
The use of general descriptive names, registered names, trademarks, service marks, etc. in this publication does not imply, even in the absence of a specific statement, that such names are exempt from the relevant protective laws and regulations and therefore free for general use.
While the advice and information in this book are believed to be true and accurate at the date of publication, neither the authors nor the editors nor the publisher can accept any legal responsibility for any errors or omissions that may be made. The publisher makes no warranty, express or implied, with respect to the material contained herein.

Printed on acid-free paper

Humana Press is a brand of Springer
Springer is part of Springer Science+Business Media (www.springer.com)

Preface

Since the discovery of catalytic RNA in 1982, for which the Nobel Prize was later awarded, RNA research has continued to experience an increasing attention. During the past decade, the RNA community has been intrigued by the discovery of hundreds of novel functional, noncoding RNAs. As a consequence, it has become increasingly evident that RNA is the driving force in the most essential processes in the cell—*RNA Regulates Biology*. Although these RNAs are quite diverse and fulfill very different functions, they share a common theme: their strict dependence on acquiring specific 3D fold(s) in order to be functional and to accomplish their cellular tasks. The process of RNA folding describes how an RNA molecule undergoes the transition from an unfolded, disordered state to the native, functional conformation. So far, RNA structure and folding has been almost exclusively studied in vitro. The in vitro refolding conditions differ, however, significantly from the intracellular folding environment. In the cell factors such as the directionality of transcription and translation, *trans*-acting noncoding RNAs as well as proteins are likely to influence folding of RNA molecules. Hitherto, very little is known about RNA folding in vivo, which is in part due to the complex nature of the cellular environment, but also to the limited availability of suitable methodologies. In these respects, the compilation of methodologies herein lies at the heart of the topical research focus of in vitro and intracellular RNA structure formation. The main purpose of the book is to provide a comprehensive collection of experimental protocols, which are suitable to dissect RNA folding pathways and to characterize the structure of RNA folding intermediates at nucleotide or even atomic resolution. This book will be equally useful for experienced researchers as well as PhD students and postdoctoral fellows new to the field. Presented techniques include powerful tools with a long tradition in RNA research as well as more advanced, novel methods. As such, the methods span multiple disciplines, including molecular biology, biochemistry, biophysics, and computational biology. I believe that this special issue on exploring RNA folding is timely in light of the everlasting boom of RNA research. With the vast expansion of known RNA molecules, I wish to share my fascination and passion for RNA with new researchers joining our efforts to gain insights into the secrets of this astounding macromolecule. At last, I want to acknowledge the extraordinary group of people who contributed high-quality chapters to this special issue on modern-day research in RNA folding.

Vienna, Austria *Christina Waldsich*

Contents

Contributors

JOHN P. BIDA • *Department of Biochemistry, Stanford University, Stanford, CA, USA*
ELISA BIONDI • *Department of Molecular Microbiology and Immunology, University of Missouri, Columbia, MO, USA; Department of Biochemistry, Life Sciences Center, University of Missouri, Columbia, MO, USA; Foundation for Applied Molecular Evolution (FfAME), Gainesville, FL, USA*
DONALD H. BURKE • *Department of Molecular Microbiology and Immunology, University of Missouri, Columbia, MO, USA; Department of Biochemistry, Life Sciences Center, University of Missouri, Columbia, MO, USA*
PALLAVI K. CHOUDHARY • *Institute of Inorganic Chemistry, University of Zurich, Zurich, Switzerland*
PABLO CORDERO • *Department of Biomedical Informatics, Stanford University, Stanford, CA, USA*
MARIA COSTA • *Centre de Génétique Moléculaire du C.N.R.S., Gif-sur-Yvette, France*
RHIJU DAS • *Department of Biochemistry, Stanford University, Stanford, CA, USA; Department of Physics, Stanford University, Stanford, CA, USA*
FRANKA DEBELJAK • *Department of Biochemistry and Cell Biology, Max F. Perutz Laboratories, University of Vienna, Vienna, Austria*
SÉBASTIEN H. ESCHBACH • *Department of Biology, Faculty of Science, RNA Group, Université de Sherbrooke, Sherbrooke, QC, Canada*
OLGA FEDOROVA • *Howard Hughes Medical Institute, Yale University, New Haven, CT, USA; Department of Molecular, Cellular, and Developmental Biology, Yale University, New Haven, CT, USA*
BORIS FÜRTIG • *Institute for Organic Chemistry and Chemical Biology, Center for Biomolecular Magnetic Resonance, Johann Wolfgang Goethe-University, Frankfurt am Main, Germany*
SOFIA GALLO • *Institute of Inorganic Chemistry, University of Zurich, Zurich, Switzerland*
BRANT GRACIA • *Department of Chemistry and Biochemistry, Institute for Cellular and Molecular Biology, University of Texas at Austin, Austin, TX, USA*
ULRICH HAHN • *MIN Faculty, Chemistry Department, Institute for Biochemistry and Molecular Biology, Hamburg University, Hamburg, Germany*
STEFAN HANDL • *Department of Biochemistry and Cell Biology, Max F. Perutz Laboratories, University of Vienna, Vienna, Austria*
VERENA HEISIG • *Department of Biochemistry and Cell Biology, Max F. Perutz Laboratories, University of Vienna, Vienna, Austria*
IVO L. HOFACKER • *Institute for Theoretical Chemistry, University of Vienna, Vienna, Austria*
WIPAPAT KLADWANG • *Department of Biochemistry, Stanford University, Stanford, CA, USA*

PETR V. KONAREV • *Deutsches Elektronen-Synchrotron, European Molecular Biology Laboratory (EMBL) Outstation, Hamburg, Germany*
ANDREY S. KRASILNIKOV • *Department of Biochemistry and Molecular Biology, Center for RNA Molecular Biology, The Pennsylvania State University, University Park, PA, USA*
DANIEL A. LAFONTAINE • *Department of Biology, Faculty of Science, RNA Group, Université de Sherbrooke, Sherbrooke, QC, Canada*
ANDREAS LIEBEG • *Department of Biochemistry and Cell Biology, Max F. Perutz Laboratories, University of Vienna, Vienna, Austria*
RONNY LORENZ • *Institute for Theoretical Chemistry, University of Vienna, Vienna, Austria*
EILEEN MAGBANUA • *MIN Faculty, Chemistry Department, Institute for Biochemistry and Molecular Biology, Hamburg University, Hamburg, Germany*
BENOÎT MASQUIDA • *GMGM UMR 7156, IPCB, CNRS, Université de Strasbourg, Strasbourg, France*
MÉLANIE MEYER • *GMGM UMR 7156, IPCB, CNRS, Université de Strasbourg, Strasbourg, France*
SOMDEB MITRA • *Department of Chemistry, Columbia University, New York, NY, USA*
DARIO MONACHELLO • *Centre de Génétique Moléculaire du C.N.R.S., Gif-sur-Yvette, France*
BISHNU PAUDEL • *Department of Medicine, Section of Virology, Imperial College, London, UK*
DAVID RUEDA • *Department of Medicine, Section of Virology, Imperial College, London, UK*
RICK RUSSELL • *Department of Chemistry and Biochemistry, Institute for Cellular and Molecular Biology, University of Texas at Austin, Austin, TX, USA*
NORA SACHSENMAIER • *Department of Biochemistry and Cell Biology, Max F. Perutz Laboratories, University of Vienna, Vienna, Austria*
HARALD SCHWALBE • *Institute for Organic Chemistry and Chemical Biology, Center for Biomolecular Magnetic Resonance, Johann Wolfgang Goethe-University, Frankfurt am Main, Germany*
MATTHEW G. SEETIN • *Department of Biochemistry, Stanford University, Stanford, CA, USA*
KATHARINA SEMRAD • *Department of Biochemistry and Cell Biology, University of Vienna, Vienna, Austria*
ROLAND K.O. SIGEL • *Institute of Inorganic Chemistry, University of Zurich, Zurich, Switzerland*
DMITRI I. SVERGUN • *Deutsches Elektronen-Synchrotron, European Molecular Biology Laboratory (EMBL) Outstation, Hamburg, Germany*
CHRISTOPHER C. VANLANG • *Department of Chemical Engineering, Stanford University, Stanford, CA, USA*
CHRISTINA WALDSICH • *Department of Biochemistry and Cell Biology, Max F. Perutz Laboratories, University of Vienna, Vienna, Austria*
MICHAEL WILDAUER • *Department of Biochemistry and Cell Biology, Max F. Perutz Laboratories, University of Vienna, Vienna, Austria*
GEORGETA ZEMORA • *Department of Biochemistry and Cell Biology, Max F. Perutz Laboratories, University of Vienna, Vienna, Austria*
NORA ZINGLER • *Department of Molecular Genetics, University of Kaiserslautern, Kaiserslautern, Germany*

Chapter 1

Predicting RNA Structure: Advances and Limitations

Ivo L. Hofacker and Ronny Lorenz

Abstract

RNA secondary structures can be predicted using efficient algorithms. A widely used software package implementing a large number of computational methods is the `ViennaRNA Package`. This chapter describes how to use programs from the `ViennaRNA Package` to perform common tasks such as prediction of minimum free-energy structures, suboptimal structures, or base pairing probabilities, and generating secondary structure plots with reliability annotation. Moreover, we present recent methods to assess the folding kinetics of an RNA via 2D projections of the energy landscape, identification of local minima and energy barriers, or simulation of RNA folding as a Markov process.

Key words Energy directed folding, Minimum free-energy structure, Suboptimal structures, Structural reliability, Energy landscapes, Folding kinetics

1 Introduction

Computational approaches to RNA structure prediction can provide an easy alternative to costly experimental structure determination. More importantly, the two can often complement each other, e.g., when structure probing experiments are used to support a predicted structure. Ideally, we would like to be able to predict complete tertiary structures at atomic resolution. However, in spite of recent advances [1, 2], routine tertiary structure prediction is still out of reach. In this chapter, we therefore restrict ourselves to the prediction of *secondary structure*. We present the basic techniques for energy directed secondary structure prediction as well as somewhat more advanced methods to look at the *folding kinetics* of an RNA molecule. All computations will be done using the `ViennaRNA Package` and associated tools. While some of the tasks below could be performed with alternative implementations, such as David Mathews `RNAstructure` or Michael Zuker's `UNAFOLD`, we chose the ViennaRNA Package because (1) we are most familiar with it and (2) it provides a more comprehensive collection of computational methods.

Christina Waldsich (ed.), *RNA Folding: Methods and Protocols*, Methods in Molecular Biology, vol. 1086,
DOI 10.1007/978-1-62703-667-2_1, © Springer Science+Business Media, LLC 2014

RNA secondary structure prediction is a problem that has been tackled by several different approaches in the last three decades. First attempts to predict an optimal base pairing pattern of an RNA molecule were performed in the late 1970s/early 1980s. In 1978, Ruth Nussinov formulated the first computationally efficient dynamic programming algorithm [3] that solves the *Maximum circular matching* problem, i.e., it computes the structure with maximal number of base pairs.

In 1981 Zuker and Stiegler [4] used dynamic programming to compute minimum free-energy (MFE) structures in a loop based energy model and his approach still forms the basis for modern algorithms. The energy parameters needed are derived from measuring UV absorption in melting experiments [5, 6] and have been refined over the years, especially through the efforts of the group around Doug Turner. More recently, efforts have been made to complement measured parameters by statistical training [7]. The ViennaRNA Package provides three parameter sets for RNA folding, the Turner parameter set from 1999 to 2004 [5, 6], and the trained parameter set from Andronescu [7], as well as one set for folding single stranded DNA.

While the dynamic programming algorithms for MFE folding guarantees that the energetically best structure is found, inaccuracies of the energy parameters and limitations of the model, such as exclusion of pseudoknots, interaction with other molecules, or kinetic folding, all conspire to make predictions far from perfect. Benchmarks measuring the accuracy of prediction by counting the fraction of correct base pairs, report accuracies of about 40–70 % [6, 8] depending on the length of the RNAs and the definition of what constitutes a "correct" pair. While this may seem low at first glance, partially correct prediction can be quite useful and accuracy can be improved by including experimental knowledge or homology. Nevertheless, much of this chapter will focus on how to *deal* with the inherent inaccuracies of structure prediction.

Almost 10 years after Zuker's MFE algorithm John McCaskill published his work on how to use the MFE decomposition to efficiently calculate the partition function Z over all secondary structures of an RNA molecule [9]. This work finally made it possible to obtain insights into thermodynamic equilibrium properties like structure- and base pair probabilities.

Several more variations on RNA folding were introduced over the years, in particular in order to deal with the inherent inaccuracies of prediction. Rather than discuss all of these algorithms upfront, we will introduce them below using practical examples.

2 Materials

2.1 Hardware

A PC running Linux is generally the best choice for bioinformatical work, as this is the system most programs are developed on. Mac OS X is UNIX based and should therefore work well, but note that Macs usually come with no development tools preinstalled. Windows tends to be the least convenient choice, but we do provide pre-compiled ViennaRNA binaries for that platform as well (*see* **Note 1**). Of the tasks described below, only the methods relating to kinetic folding can require machines with large memory. For thermodynamic structure prediction even a typical laptop should be sufficient in terms of both speed and memory.

2.2 Software

1. `ViennaRNA Package` (`http://www.tbi.univie.ac.at/RNA`).
2. `barriers` (`http://www.tbi.univie.ac.at/RNA/Barriers`).
3. `treekin` (`http://www.tbi.univie.ac.at/RNA/Treekin`).
4. Text Editor (*not* a word processor).
5. Working `Perl` environment.
6. The `gri` plotting tool.
7. A `Postscript` Viewer (e.g., `gv`, `preview`, `gsview`).

2.3 Files

1. RNA sequence(s) in FASTA format.

3 Methods

3.1 General Remarks

The ViennaRNA Package [10] consists of command line tools, i.e., programs without a graphical user interface. This may seem unfamiliar for some readers in beginning, but allows to easily automate tasks (folding a million RNAs by clicking in menus is no fun). To try the examples below, open a `Terminal` and type in the commands (*see* **Note 2**). The example commands are typeset in fixed-width paragraphs highlighted by a `$` sign at the start of the line, representing the command line prompt. Type the text *after* the `$` sign into your terminal and hit the `return`-key to execute the command. The programs below read from *standard input* and write to *standard output*, meaning that the input can be typed interactively into the terminal and output can be read from the terminal. However, usually it is more convenient to use a file and redirect its content to the input (and possibly direct output into an output file). This is achieved by using the < and > symbols on the command line. It is even possible to chain programs together in a *pipe* using | symbol. The following command line illustrates the use of these symbols exemplarily:

```
$ program1 < infile | program2 > outfile
```

Here, two programs with the name program1 and program2 are executed. While program1 gets its input from infile the output it generates is passed as input to program2 via a *pipe*. The output of program2, again, is redirected into the file outfile.

3.2 Installation and Usage of the Required Tools

3.2.1 Installing the ViennaRNA Package from Source

The recommended way of installing the ViennaRNA Package is to download its latest (version 2.1.1 at the time of manuscript preparation) source code tarball from http://www.tbi.univie.ac.at/RNA, unpack, compile, and install it via the following commands:

```
$ tar xzvf ViennaRNA-2.1.1.tar
$ cd ViennaRNA-2.1.1
$ ./configure
$ make
$ sudo make install
```

In case you have no *sudo* privileges and/or you consider installing the ViennaRNA Package into a directory other than the default path (*see* **Note 3**). This procedure should work well for UNIX derived operating systems like Linux and MacOSX. Microsoft Windows users should consider downloading the precompiled binaries (*see* **Note 1**).

3.2.2 Installing barriers and treekin

Similar to the ViennaRNA Package installation above, the tools needed in the RNA folding kinetics section below follow the same procedure regarding their installation, i.e., the ./configure; make; sudo make install scheme (successive commands may be given on a single line by separating them with a ";"). Just download the source code tarballs of barriers and treekin from the respective Web sites and, again, unpack, compile, and install them:

```
$ tar xzvf Barriers-1.5.2.tar.gz
$ cd Barriers-1.5.2
$ ./configure; make; sudo make install
$ cd ..
$ tar xzvf Treekin-0.3.1.tar.gz
$ cd Treekin-0.3.1
$ ./configure; make; sudo make install
```

3.2.3 Getting Help on a Particular Program

All programs distributed with the ViennaRNA Package come with a detailed documentation provided by *man pages*. To access, for instance, the documentation of RNAfold, type

```
$ man RNAfold
```

To leave the man page, press the *q*-key. A brief help message is also available via the `--help` option, e.g.

```
$ RNAfold --help
```

3.2.4 Sequence and Structure Format

RNA sequence data may be provided in two different text file formats to be compatible to the programs within the `ViennaRNA Package`. First, sequences can be provided in `FASTA` format, i.e., a header line starting with '>' used as unique sequence identifier followed by an arbitrary number of lines specifying the sequence in letters of the alphabet $\{A, G, C, U, T\}$. The second, much more trivial, format relies on the RNA sequence only which has to be given as one single line.

Secondary structure input/output is always in so called dot-bracket (dot-parenthesis) notation. Here, the parenthesis characters '`(`' and '`)`' denote opening and closing of a base pair whereas the dot character '`.`' denotes unpaired nucleotides. See below for examples of input and output formats.

Here is an example input file in FASTA format, which we will use for the coming tasks.

```
>sequence1
CGUCAGCUGGGAUGCCAGCCUGCCCCGAAAGGGGCUUGGC
GUUUUGGUUGUUGAUUCAACGAUCAC
```

3.3 Thermodynamic Structure Prediction

3.3.1 Using RNAfold

Basic structure prediction for single sequences in the `ViennaRNA Package` is done by the `RNAfold` program. It implements algorithms to compute the minimum free-energy (MFE), the partition function and the base pair probabilities for an RNA sequence. In default mode it computes the MFE structure and creates a structure drawing.

```
$ RNAfold < sequence.fa
>sequence1
CGUCAGCUGGGAUGCCAGCCUGCCCCGAAAGGGGCUUGGCGUUUUGGUUGUUGAUUCAACGAUCAC
((((((((((....)))))..(((((....))))).)))))...(((((((((...))))))) )). (-30.60)
```

Here, the line starting with `$` contains the command the user has to type, the following three lines show the output, in particular the MFE structure in dot bracket notation and its free-energy in kcal/mol.

There are at least two reasons why one should perhaps do more than just MFE structure prediction: (1) An RNA molecule in thermodynamic equilibrium will fluctuate between different low energy structures rather than stay in the ground state; (2) Predictions may be inaccurate and MFE folding alone gives no indication how reliable the prediction is.

A convenient way to look at the equilibrium ensemble of RNA structures is to compute the partition function and pair probabilities. In `RNAfold` this is achieved simply by adding the `-p` option. Pair probabilities can be visualized in *dot plots*, an example of which can be seen in Fig. 1e. While interpreting a dot plot may require some getting used to, it provides information about *all* structural alternatives.

Even when we would like to generate a single representative structure, the MFE structure is only one of several possibilities. One attempt to define a structure that represents the whole equilibrium ensemble is the *centroid* structure. It is defined as the structure with the smallest distance to a Boltzmann weighted set of all structures. It turns out that the centroid structure can be constructed easily by using all base pairs with a probability $p_{ij}>0.5$. The centroid structure is automatically generated whenever we use the `-p` option.

Another alternative is the Maximum Expected Accuracy (MEA) structure. We can think of it as the structure that should contain the largest number of correct positions (correctly predicted pairs or correctly predicted unpaired nucleotides). More technically the MEA structure is the structure S maximizing:

$$EA = \sum 2\gamma p_{ij} + \sum p_i^u$$

where p_{ij} is the probability to form the pair (i,j) and p_i^u the probability that position i stays unpaired. γ is a parameter weighting the importance of base pairs versus unpaired sites.

Taken together we write:

```
$ RNAfold -p --MEA < sequence.fa > sequence.fa.out
```

To compute MFE structure, pair probabilities, centroid structure and MEA structure all in one go. The corresponding output can be seen in Fig. 1a–e.

The text output (Fig. 1a) now consists of seven lines, where the first three contain the FASTA header, the input sequence, and the MFE structure followed by its free-energy in kcal/mol. The next line depicts the pairing probabilities of the nucleotides in a pseudo dot-bracket notation, which uses the characters “`. , | { } ( )`” to denote that the pairing preference for each nucleotide (see the documentation of the RNAfold program that can be invoked by typing `man RNAfold`). The value at the end of this line shows the Gibbs free-energy

$$\Delta G = -RT\ln(Z)$$

in kcal/mol where Z is the partition function, R the gas constant, and T the absolute temperature. The next two lines show the centroid and MEA structure. The last line lists the Boltzmann probability of the MFE and the ensemble diversity. The latter indicates

```
$ RNAfold -p --MEA < sequence.fa
>sequence1
CGUCAGCUGGGAUGCCAGCCUGCCCCGAAAGGGGCUUGGCGUUUUGGUUGUUGAUUCAACGAUCAC
((((((((((....)))))..(((((....))))).)))))...(((((((((...))))))))). (-30.60)
{{,((((((({{{{||||,..(((((....)))))})))))}}}||||||,((,,,,,,,))))). [-31.38]
...(((((((((((((((...(((((....))))))))))))))))))))................ {-29.00 d=16.88}
...(((((((((((((((...(((((....)))))))))))))))))))).((((......)))). {-29.50 MEA=42.75}
 frequency of mfe structure in ensemble 0.280191; ensemble diversity 19.52
```

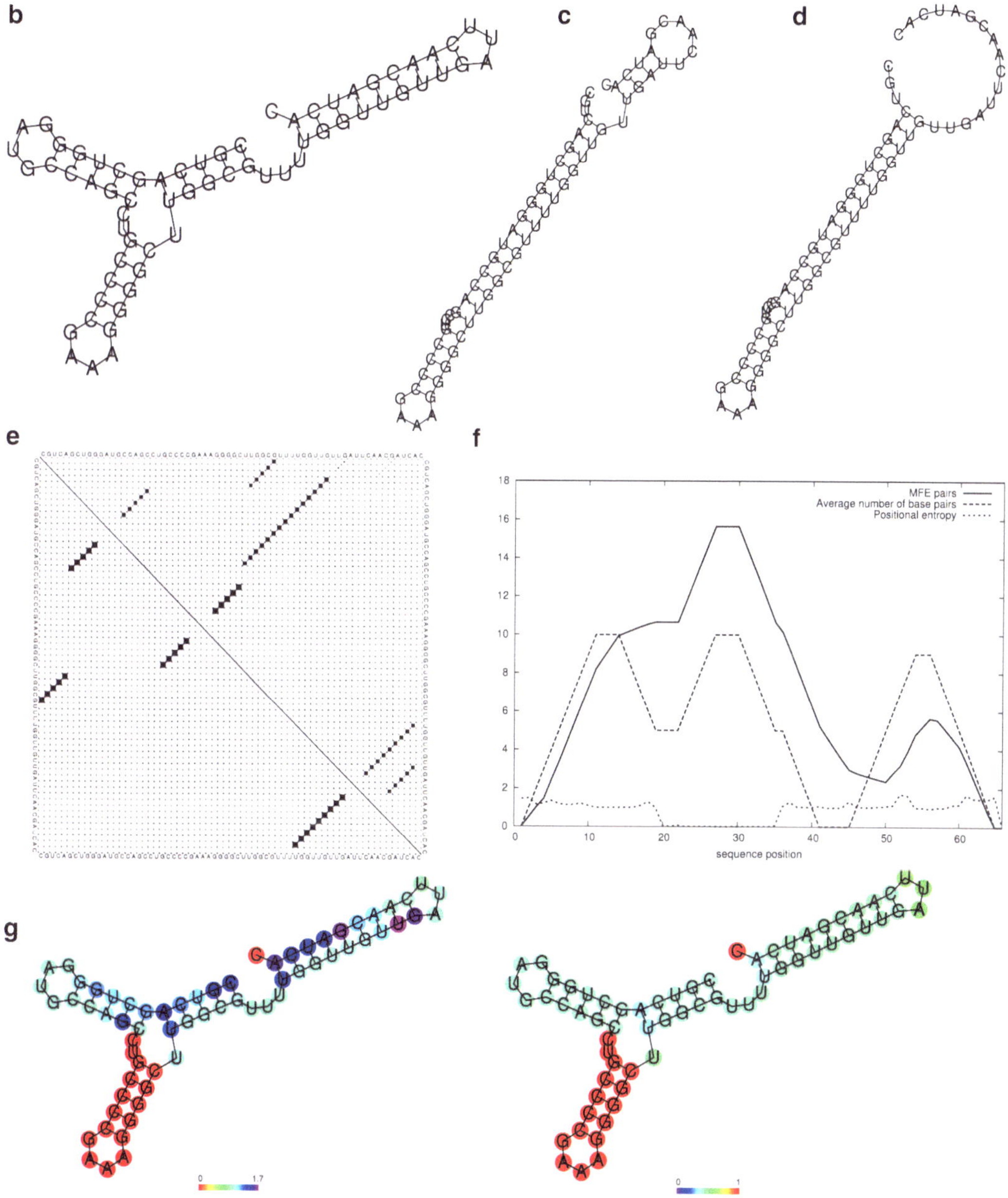

Fig. 1 Thermodynamic properties of an RNA obtained by `RNAfold`. (**a**) Executed commandline call and its text output. (**b**) Plot of the MEA structure. (**c**) Plot of the centroid structure. (**d**) Plot of the MEA structure. (**e**) Dot-plot showing the base pair probabilities for each base pair in the upper- and the MFE structure in the lower half of the matrix. The sizes of the black boxes scale with the square root of the respective base pair probability. (**f**) Mountain plot showing the base pairs of the MFE structure (*solid black line*), the average number of base pairs (*dashed line*) and the positional entropy (*dotted line*) according to the pairing probability. (**g**) Reliability plots as obtained by `relplot.pl`. *Left side* shows the positional entropy $S(i) = -\sum p_{ij} \log p_{ij}$. Right side displays the pair probabilities where paired nucleotides are colored according to their pair probability and unpaired nucleotides by their probability to be unpaired

whether the ensemble is dominated by a single structure (low diversity), or whether the RNA can form diverse structure with almost equal stability.

After running the `RNAfold` command, postscript drawings of the MFE structure (Fig. 1b) and the dot plot (Fig. 1e) can be found in files named `sequence1_ss.ps` and `sequence1_dp.ps` (*see* **Note 4**).

There are many more options to `RNAfold`, such as predicting structures for circular RNAs or for including RNA G-quadruplex formation, which we cannot all describe here due to space limitations. Run `RNAfold --help` to obtain a list of all available options.

3.3.2 Reliability of the Predictions

When using a single structure representative, one should at least have some indication how reliable the predicted structure is. The ensemble diversity mentioned above, provides a *global* reliability measure. In contrast, *local* reliability measures tell us which parts of a structure can be trusted. Such measure can be conveniently derived from the pair probabilities and added to structure drawings.

The `perl` script `relplot.pl` takes both postscript drawings produced by `RNAfold` with `-p` option, the secondary structure plot and the dot plot, and produces a new secondary structure plot with color coded reliability annotation. The annotation can show either the positional entropy $S(i) = -\sum p_{ij} \log p_{ij}$ for each nucleotide (default), or probabilities (pair probabilities for paired positions and probabilities for being unpaired otherwise). Both types of reliability plots can be seen in Fig. 1g, where the downward pointing arm is reliably predicted (low entropy/high probability) while the rest of the structure has poor reliability.

Finally, RNA structures can be represented by *mountain plots*, as in Fig. 1f. These are *x*–*y* plots with sequence position on the *x*-axis and the number of base pairs enclosing that position as mountain height on the y-axis. Thus, upward (downward) slopes represent the 5′ (3′) sides of helices, whereas plateaus represent loops. Mountain plots especially convenient for visually comparing long sequence. Running

```
$ mountain.pl sequence1_dp.ps
```

produces a set of *x*–*y* coordinates which can be plotted using any *x*–*y* plotting tool, such as `gnuplot` or `xmgrace`.

3.3.3 RNAsubopt

An overview over a larger set of representative structures can be obtained by inspecting suboptimal secondary structures. Even when the predicted MFE structure is wrong, highly accurate structures can usually be found within a small energy band from the MFE. Suboptimal folding is therefore an alternative approach to computing pair probabilities, and can provide insight into the diversity of structures that appear with high probability within the ensemble. The program `RNAsubopt` implements three different strategies of generating suboptimal secondary structures.

First, the default mode (`-e value` option) computes *all* structures within an energy band of size `value` around the MFE, i.e., it is an `exhaustive enumeration` of all structures S_i that exhibit a free-energy $E(S_i)$ with $E(S_i) < \text{MFE} + \text{value}$ [11]. Since in general the number of structures an RNA molecule is able to adopt grows exponentially with its sequence length, it is advisable to avoid too large energy ranges *value*.

Structurally diverse low free-energy states or another MFE structure (degenerate MFE structure) can be retrieved even by using a narrow energy bands of for example width 1–2 kcal/mol (`-e 2.0` option). In order to keep the number of output structures low, we will omit structures with isolated base pairs (base pairs not incorporated in a helix) using the `--noLP` option.

Taken the RNA sequence from the examples above, a proper call of `RNAsubopt` would be:

```
$ RNAsubopt -e 2.0 -s --noLP < sequence1.fa
> sequence1 [200]
CGUCAGCUGGGAUGCCAGCCUGCCCCGAAAGGGGCUUGGCGUUUUGGUUGUUGAUUCAACGAUCAC  -3060    200
((((((((((....)))))..(((((....))))).)))))...(((((((((...))))))))). -30.60
...(((((((((((((((...(((((....)))))))))))))))))))).(((((....))))). -30.50
((((((((((....)))))..(((((....))))).)))))....((((((((...)))))))).. -29.50
...(((((((((((((((...(((((....)))))))))))))))))))).((((......)))). -29.50
((((.(((((....)))))..(((((....)))))..))))...(((((((((...))))))))). -29.40
((.(((((((((((((((...(((((....))))))))))))))))))))).)).............. -29.40
...(((((((((((((((...(((((....))))))))))))))))))))..((((....)))).. -29.40
...((((((((((((((....(((((....))))).)))))))))))))).(((((....))))). -29.10
...(((((((((((((((...(((((....))))))))))))))))))))................ -29.00
((((((((((((((((((...(((((....))))))))))))))))))))........))).....  -28.90
```

Here, the `-s` option is used to sort the output by energy. The output itself consists of the FASTA header in the first line, the RNA sequence with MFE and energy band in dakal/mol (1 dakal/mol = 0.01 kcal/mol) in the second line. All following structures below the sequence resemble the set of suboptimals with their respective free-energy in kcal/mol. In this example one easily see that all structures share the central hairpin, but differ significantly at the 5′ and 3′ end. The output also contains some structures that are almost identical, such as number 1, 3, and 5.

In order to create a secondary structure plot for a given dot-bracket output one can utilize the `RNAplot` program (*see* **Note 5**).

The second mode, ***stochastic backtracking***, generates a random sample of secondary structures S_i according to their Boltzmann probability $p(S_i) = \frac{e^{-E(S_i)/RT}}{Z}$ within the ensemble [12]. `RNAsubopt` produces a list of secondary structures according to this approach using the `-p number` option where `number` specifies the number of structures to compute:

```
$ RNAsubopt -p 10 < sequence.fa
>sequence1
...(((((((((((((((...(((((....))))))))))))))))))))).(((((....))))).
(..(((((((((((((((...(((((....))))))))))))))))))))..).............
((((((((((....)))))..(((((....))))).)))))...(((((((((...))))))))).
((((((((((....)))))..(((((....))))).)))))...(((((((((...))))))))).
...(((((((((((((((...(((((....)))))))))))))))))))))..((((....))))..
((((((((((....)))))..(((((....))))).)))))....((((((((...))))))))..
.(((((((((....)))))..(((((....))))).))))....(((((((((...))))))))).
...(((((((((((((((...(((((....))))))))))))))))))))).(((((....))))).
((((((((((....)))))..(((((....))))).)))))...(((((((((...))))))))).
...(((((((((((((((...(((((....))))))))))))))))))))).((((......)))).
```

Since this is a *sampling* method, each run of `RNAsubopt` will produce a different set of structures. Also note that in contrast to the previous mode, the output does not contain the input RNA sequence but only the sampled structures.

Finally, a relatively small set of representative structures is generated by the third mode, activated by the `-z` option, which generates an energy sorted list of suboptimal structures according to the method of Zuker [13]. This particular algorithm generates for each possible pair the best structure containing that pair. Since the number of distinct base pairs for any sequence is limited by the sequence length n the number of structures obtained with this method will never exceed $n(n-1)/2$. Although this limitation in the number of representative structures results in a manageable amount of structures to visually inspect, it might miss some important low free-energy structures [11]. The following command computes the suboptimal structures according to Zuker's algorithm and the `head` program limits the output to the first ten structures:

```
$ RNAsubopt -z < sequence1.fa | head -12
>sequence1
CGUCAGCUGGGAUGCCAGCCUGCCCCGAAAGGGGCUUGGCGUUUUGGUUGUUGAUUCAACGAUCAC
((((((((((....)))))..(((((....))))).)))))...(((((((((...))))))))). [-30.60]
...(((((((((((((((...(((((....))))))))))))))))))))).(((((....))))). [-30.50]
((.(((((((((((((((...(((((....)))))))))))))))))))).)).............. [-29.40]
((((((((((((((((((...(((((....))))))))))))))))))))........))).....  [-28.90]
.(.(((((((((((((((...(((((....))))))))))))))))))))))(((((....))))). [-28.20]
((((((((((....))))...(((((....))))))))))))...(((((((((...))))))))). [-27.70]
...(((((((((((((((...(((((....)))))))))))))))))))))(((......)))....  [-27.70]
.(((((((((((((((((...(((((....))))))))))))))))))))...)))).......... [-27.00]
...(((((((((((((((...(((((....))))))))))))))))))))(.((((....))))). [-26.90]
((((((((((....)))).).(((((....))))).)))))...(((((((((...))))))))). [-26.80]
```

As visible in the last two outputs (exhaustive enumeration within 2.0 kcal/mol, and Zuker suboptimals), our example RNA sequence exhibits another, near optimum structure state that is quite different to its MFE state (second structure in the output). `RNAsubopt` does not automatically plot structures, this can, however, be accomplished using the `RNAplot` command, e.g.

```
$ RNAplot
CGUCAGCUGGGAUGCCAGCCUGCCCCGAAAGGGGCUUGGCGUUUUGGUUGUUGAUUCAACGAUCAC
...(((((((((((((((...(((((....))))))))))))))))))))).(((((....))))).
^D
```

In this example, we have simply typed in the sequence and structure using cut-and-paste, the `^D` (ctrl-D) terminates the program. The resulting structure plot can be seen in Fig. 2b.

3.4 Folding Kinetics

The dynamic programming approaches described so far assume that the RNA molecule is in thermodynamic equilibrium. Thus, they ignore folding kinetics, i.e., the dynamical process of how an RNA molecule explores structure space to adopt its native state. While statistical mechanics assures us that the RNA molecule will reach equilibrium eventually, some RNAs are known to be trapped in non-equilibrium states for long times, and will not reach equilibrium on a biologically relevant time scale. In such cases, only approaches that explicitly consider folding kinetics can give biologically meaningful results. Unfortunately, this requires significantly more computational effort.

3.4.1 Secondary Structure Free-Energy Landscapes

Kinetic folding can be thought of as moving around in an *energy landscape*. Each secondary structure an RNA molecule is able to form, is a point in the landscape, with the height of the point given by the free-energy of the structure. A neighborhood relation that is defined by a so called *move-set*, e.g., insertion/deletion of a base pair, specifies each point's relative position and completes the high-dimensional abstraction of the folding space. Local minima within the landscape then indicate states of attraction which can be potential kinetic traps, whereas saddle points connecting the local minima represent energy barriers between metastable states.

Energy landscapes are huge, however. For a 76 nt tRNA the landscape may have over 17 states. As we will see below, one therefore often performs a *coarse graining* that subsumes a set of (micro-) states into a macro-state.

3.4.2 `RNA2Dfold`

`RNA2Dfold` provides a classified dynamic programming that projects the high dimensional energy landscape of an RNA molecule into two dimensions [14], making it ideal for visualization purposes. For the projection it makes use of two user-defined reference structures and partitions the structures space according to the distances to the chosen reference structures. Thus, all structures with the same base pair distance to the references get lumped together. For each of the resulting distance classes, `RNA2Dfold` can compute the MFE

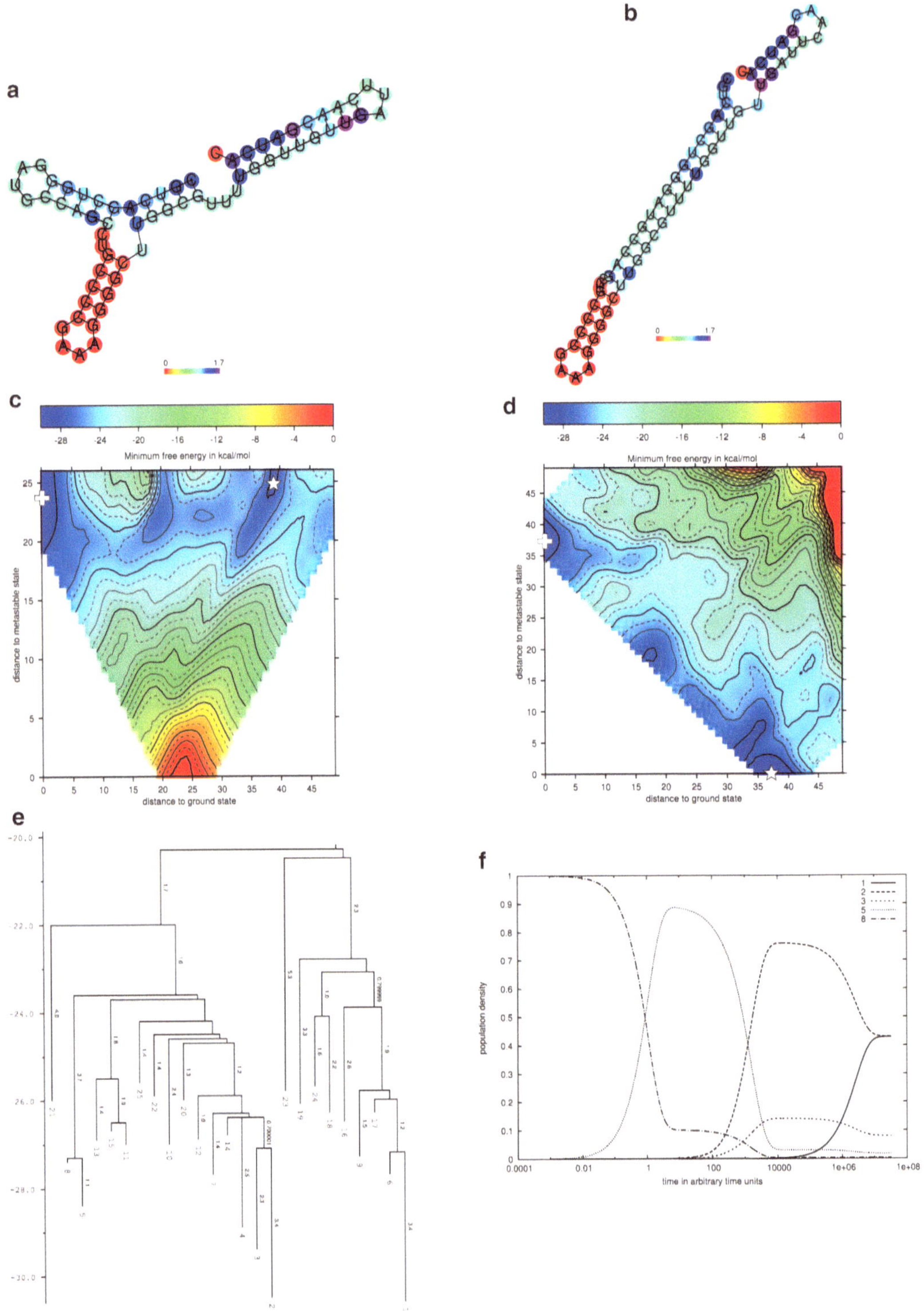

Fig. 2 Various plots which aim to understand RNA folding kinetics. The first two are secondary structure plots with positional entropy annotation for (**a**) the MFE structure and (**b**) the metastable structure. The next two plots show energy landscape projections of the Gibbs free-energies for all distance classes computed by `RNA2Dfold`. (**c**) The MFE structure and the unfolded state as references; (**d**) the MFE structure and the metastable state as references. In both plots the MFE structure and the metastable state are marked by a *white cross* and a *white star*, respectively. (**e**) Barrier tree showing the 25 lowest energy minima and the energy barriers separating them. (**f**) `treekin` simulation of the folding dynamics. The simulation is started with the complete population in the macro-state corresponding to minimum 8

structure and the partition function, or it can sample from the structure ensemble. Although the algorithm behind RNA2Dfold comes with a high asymptotic complexity, $O(n^7)$, it can be applied to biologically meaningful sequence lengths of up to 400 nt.

An input file for RNA2Dfold consists of three lines that specify the RNA sequence and the two reference structures in dot-bracket notation, respectively. After a successful run its output can be used easily to create a plot of the projected secondary structure landscape.

For the example below, assume that the reference structures are not yet known, but we have already computed the MFE structure using RNAfold. To find a good second reference structure, we will first compute a 2D landscape using the MFE structure and the unfolded state (*open chain*) as reference structures, i.e., the distance classes are defined by (a) the base pair distance to the MFE structure and (b) the number of base pairs. Thus, the appropriate input file sequence.1.2D with the three lines

```
CGUCAGCUGGGAUGCCAGCCUGCCCCGAAAGGGGCUUGGCGUUUUGGUUGUUGAUUCAACGAUCAC
((((((((((....)))))..(((((....))))).)))))...(((((((((...))))))))).
..................................................................
```

is used in the following RNA2Dfold call:

```
$ RNA2Dfold -p < sequence.1.2D > sequence.1.2D.out
```

The resulting output file sequence.1.2D.out consists of a table like format displaying (1) the distance to both references, (2) the probability of the appropriate distance class, (3) the probability of the MFE representative within the distance class, (4) the probability of the MFE representative in the ensemble of all structures, (5) free-energy of the MFE representative, (6) Gibbs free-energy of the distance class and (7) the secondary structure of the MFE representative in dot-bracket notation.

Since the output of RNA2Dfold quickly becomes large and visual inspection of the distance class MFEs or Gibbs free-energies in the textual output gets difficult, the ViennaRNA Package includes two scripts, 2Dlandscape_pf.gri and 2Dlandscape_mfe.gri, that allow for easy visualization of the MFE or partition function output of RNA2Dfold, respectively. Each of the above scripts uses the gri program to create a colored postscript plot with iso-contours. Here we use the partition function script to display the Gibbs free-energies of the distance classes:

```
$ 2Dlandscape_pf.gri sequence.1.2D.out
```

As visible in Fig. 2c there are two further low energy regions in a distance of about 20 and 40 base pairs to the MFE structure. After a closer investigation of the output file, the corresponding distance classes are at position 19, 23 and 39, 25 with an MFE representative of 28.45 kcal/mol and −30.52 kcal/mol, respectively. The relevant data lines of the output file are shown here:

```
k   l   MFE     MFE-structure
0   24  -30.60  ((((((((((....)))))..(((((....))))).)))))...(((((((((...))))))))).
19  23  -28.40  .........(((((((((...(((((....))))))))))))))(((((((((...))))))))).
39  25  -30.50  ...(((((((((((((((...(((((....)))))))))))))))))))).(((((....))))).
```

Since the distance class 39, 25 constitutes the lowest energy, structurally distant local minimum within the landscape, we replace the open chain with the representative of this distance class, and repeat the RNA2Dfold computation. The input file sequence.2.2D now contains:

```
CGUCAGCUGGGAUGCCAGCCUGCCCCGAAAGGGGCUUGGCGUUUUGGUUGUUGAUUCAACGAUCAC
((((((((((....)))))..(((((....))))).)))))...(((((((((...))))))))).
...(((((((((((((((...(((((....)))))))))))))))))))).(((((....))))).
```

Again, RNA2Dfold is called as in the example above and its output is then visualized with the gri script:

```
$ RNA2Dfold -p < sequence.2.2D > sequence.2.2D.out
$ 2Dlandscape_pf.gri sequence.2.2D.out
```

In the new projection, it is clearly visible that the two major local minima, which are 39 base pairs apart from each other, are well separated and thus are likely to be populated both with high probability. The resulting postscript plot is given in Fig. 2d. The 2D landscape immediately gives an estimate of the energy barrier separating the two minima. In the example we have to climb up to an energy of almost −20 kcal/mol to go from one reference state to the other.

3.4.3 barriers

Another concept of visualizing and classifying the high dimensional energy landscape in terms of macro states are *barrier trees*. Here, we are interested in finding all local minima in the landscape and measure the energy barriers separating them. The program barriers [15] takes a list of secondary structures ascending sorted by their respective free-energies. A flooding algorithm that coarse grains the energy landscape extracts local minima which are connected by saddle points. The resulting barrier tree then consists of (1) leaves representing local minima, (2) internal nodes representing the saddle points connecting the local minima and (3) edges connecting the nodes, where the length of an edge is proportional to the energy barrier separating two branches in the tree (Fig. 2e).

Since barriers is able to directly read output of RNAsubopt the computation of barrier trees for RNA sequences is quite simple:

```
$ RNAsubopt -s --noLP -e 20 < sequence.fa | \
barriers -G RNA-noLP --bsize --max=25 > sequence.bar
```

The above command calls `RNAsubopt` that produces a sorted list (`-s` option) of secondary structures without isolated base pairs (`--noLP` option) within an energy range of 20 kcal/mol around the MFE (`-e 20` option) (*see* **Note 6**). This list is directly passed to `barriers` where the option `-G RNA-noLP` tells `barriers` that the input consists of RNA secondary structures without isolated base pairs and the `--bsize` adds information on the size of the gradient basins. Using the optional parameter `--max=25` restrains the output to at most 25 local minima. All output is written to the file `sequence.bar`. Additional to the text output in the output file `barriers` creates a postscript file named `tree.ps` which contains an actual drawing of the barrier tree (Fig. 2e).

3.4.4 RNA Folding Kinetics as a Markov process

`RNA2Dfold` and `barriers` give qualitative information about the folding kinetics by providing an, albeit static, picture of the energy landscape. Ideally, however, we would like to observe the folding process as a function of time. This can be achieved by modeling folding as a Markov process.

For the Markov process we will need transition rates between neighboring conformations. A common Ansatz for the rates is to use the well known Metropolis rule:

$$r_{x\to y} = \Gamma \cdot \begin{cases} e^{-\frac{E(y)-E(x)}{RT}} & ,if\ E(x) < E(y) \\ 1 & ,otherwise \end{cases}$$

Here $r_{x\to y}$ is the transition rate between neighboring structures x and y, the rate is set to 0 for states that are not neighbors.

We now want to compute the probability $p_i(t)$ that our RNA is in structure i at time t. Given the matrix R of all transition rates, this can be expressed in a so called Master equation

$$\frac{d}{dt}\vec{p}(t) = R\vec{p}(t) \text{ with formal solution } \vec{p}(t) = e^{t\cdot R} \cdot \vec{p}(0)$$

where $\vec{p} = (p_1, \ldots, p_n)$ is the vector of all state probabilities, $\vec{p}(0)$ is the initial population density. Note that $\sum p_i = 1$ applies for $\vec{p}(t)$ at any time t.

In theory the formal solution given above allows to compute the refolding of an RNA molecule. However, it is unfeasible in practice due to the astronomically large structure space. The barrier tree, however, provides a coarse graining that reduces the size of the state space to a manageable number. Instead of considering every possible secondary structure, we only deal with transitions between macro-states, each represented by a local minimum.

The `barriers` program is in fact capable of computing the transition rate matrix between macro-states while computing the barrier tree. This rate matrix in turn can be used to solve the master equation and therefore simulate the refolding process of an RNA given an initial conformation. Note, however, that `barriers` is still based on exhaustive enumeration of suboptimal structures, which in practice limits its usefulness to RNAs less than about 100 nt in length.

3.4.5 treekin

The treekin program implements an efficient solver for Markov processes [16]. Given a matrix of transition rates in the system and an initial population density vector the program solves the master equation of the system for a specified number of time steps. The resulting output can then be plotted to reveal the (re-)folding behavior of the RNA molecule.

First, RNAsubopt is run for the enumeration of secondary structures. In this example, the number of suboptimal structures amounts to about 1.2 million structures. In any case keep in mind that the number of secondary structures compatible with an RNA sequence grows exponentially with the length of the sequence. Thus, it is advisable to increase the energy range in small steps only.

```
$ RNAsubopt -s --noLP -e 20 < sequence.fa >\
sequence.sub
```

This output is then used as input for the barriers program. To generate the rate matrix for macro-states we add the --rates option.

```
$ barriers -G RNA-noLP --bsize --rates < sequence.\
sub > sequence.bar
```

As before, the sequence.bar output file contains of a list of local minima sorted ascending by their energy. Each of the minima and its associated macro-state has a unique number and represents a state in the following treekin simulation. The rate matrix is stored in an additional output file called rates.out.

We are now ready to compute the time-dependent folding using treekin. For the example we start with an RNA initially in local minimum number 8 from the sequence.bar file, and simulate the dynamics over 11 decades from time $t = 10^{-3}$ to $t = 10^{8}$.

```
$ treekin -m I --t0 1e-3 --t8 1e8 --p0 8=1 -f\
rates.out < sequence.bar > treekin.out
```

The data treekin.out can now be visualized using your favorite x–y plotting tool, preferably using a logarithmic time scale. For xmgrace a suitable command for plotting would be:

```
$ xmgrace -log x -nxy treekin.out
```

The result is shown in Fig. 2f. We see that the RNA first refolds from the starting structure 8 to structure 5, which achieves almost 90 % probability at $t \approx 10$; at $t \approx 10{,}000$ the dominant structure is number 2 (the metastable structure shown in Fig. 2b); after about 10^7 time steps, we reach equilibrium where the MFE (structure number 1, Fig. 2a) and the metastable structure (structure number 2) are almost equally likely.

3.5 Using the ViennaRNA Webservers

For the analysis of a small set of RNA sequences, where no automated processing is needed, most tools of the `ViennaRNA Package` are available as online tools from the `ViennaRNA Webservers` at `http://rna.tbi.univie.ac.at` [17]. The `RNAfold webserver`, for instance, allows to compute the MFE and partition function properties, such as base pair probabilities and reliability annotations. All graphical output can be downloaded as vector graphics (*EPS* or *PDF*), but can also be converted to a *GIF* or *PNG* bitmap picture. Additionally, all the command line calls used in the calculations are displayed, such that a user is able to reproduce the data on his/her own computer.

At this time, there is no webserver that provides access to `RNAsubopt`. This is mainly due to the exhaustive enumeration of secondary structures that makes it prone to produce an enormously large amount of data. However, the `barriers webserver` which is capable of processing RNA sequences with a length of up to 100 nt uses `RNAsubopt` as a backbone to produce the data needed for the barrier tree computation. It also allows for folding kinetics analysis via `treekin` which is displayed in a population density versus time plot.

4 Notes

1. *Operating systems.* For the examples discussed in this chapter we recommend using Linux since it should come with all the required tools by default. A build environment such as *Xcode* has to be installed first in order to compile the discussed programs on OS X for Apple computers. For users of Microsoft Windows an executable installer of the `ViennaRNA Package`, the `barriers` program and `treekin` can be downloaded from the appropriate webpages. However, after installation the *PATH* environment variable should be adjusted to include the respective installation directories.
2. *Command line terminal.* Most users of Linux might already be familiar with the concept of a command line terminal. However, OS X and Windows also provide a so called *terminal emulator* even if the majority of users of these operating systems possibly have never got in touch with it. The default *terminal emulator* in OS X is the `Terminal.app` while Microsoft Windows provides the `cmd.exe`. Both of them can be used for all the examples provided.
3. *Installation into a nonstandard directory.* For Linux or OS X users who have no *sudo* privileges and/or want to install the programs used above into a different location, e.g., into their home directory, the `configure` script provides an easy way to accomplish just that.

```
$ ./configure --prefix=/your/path/
```

Here, the commandline option `--prefix` instructs the `configure` script to use `/your/path/` as installation target. However, when installing into a nonstandard directory the user has to manually add the installation path to the *PATH* environment variable for example by adding the line

```
PATH=/your/path:${PATH}
```

at the end of the *bash* configuration file `~/.bashrc`.

4. *Visualizing secondary structure plots.* All graphical output produced by the programs of the `ViennaRNA Package` comes in the printer friendly postscript format. `RNAfold`, e.g., produces a secondary structure plot of the MFE structure, named according to the FASTA header with the suffix `_ss.ps` or, if no such header is provided, `rna.ps` (Fig. 1a). The same accounts for the dot-plot which displays the base pair probabilities. This file is produced by `RNAfold` using the `-p` option with the filename having either the suffix `_dp.ps` or a filename of `dot.ps` (Fig. 1b). Display the resulting plots using one of the recommended postscript viewers such as `gv` for Linux, `preview` for OS X or `gsview` for Windows, respectively:

```
$ gv sequence1_ss.ps
```

5. *Plotting secondary structures.* To visualize the secondary structures of for example the MEA structure or the centroid structure use the program `RNAplot`. This program takes an RNA sequence followed by its structure in dot-bracket notation as input and produces a postscript structure plot. A list of all possible output formats is available by calling `RNAplot` with the `--help` option.
6. *Exhaustive enumeration of secondary structures.* As mentioned in the main text, the number of secondary structures an RNA molecule is able to form grows exponentially with its sequence length. Therefore, only a slight increase in the energy range used with `RNAsubopt` could result in an enormous amount of additional structures. When using `RNAsubopt` as input for the `barriers` program a set of about ten million structures should usually suffice. Increasing the energy range by a step wise decreasing amount of free-energy can be used as a conservative strategy to asymptotically converge to this number of structures.

Acknowledgements

This work was supported in part by the Austrian Science Foundation FWF project "SFB F43 RNA regulation of the transcriptome."

References

1. Das R, Karanicolas J, Baker D (2010) Atomic accuracy in predicting and designing noncanonical RNA structure. Nat Methods 7:291–294
2. Reinharz V, Major F, Waldispuhl J (2012) Towards 3D structure prediction of large RNA molecules: an integer programming framework to insert local 3D motifs in RNA secondary structure. Bioinformatics 28:i207–i214
3. Nussinov R, Pieczenik G, Griggs JR et al (1978) Algorithms for loop matchings. SIAM J Appl Math 35:68–82
4. Zuker M, Stiegler P (1981) Optimal computer folding of large RNA sequences using thermodynamics and auxiliary information. Nucleic Acids Res 9:133–148
5. Mathews DH, Disney MD, Childs JL et al (2004) Incorporating chemical modification constraints into a dynamic programming algorithm for prediction of RNA secondary structure. Proc Natl Acad Sci USA 101: 7287–7292
6. Mathews DH, Sabina J, Zuker M et al (1999) Expanded sequence dependence of thermodynamic parameters improves prediction of RNA secondary structure. J Mol Biol 288:911–940
7. Andronescu M, Condon A, Hoos HH et al (2007) Efficient parameter estimation for RNA secondary structure prediction. Bioinformatics 23:i19–i28
8. Doshi KJ, Cannone JJ, Cobaugh CW et al (2004) Evaluation of the suitability of free-energy minimization using nearest-neighbor energy parameters for RNA secondary structure prediction. BMC Bioinformatics 5:105
9. McCaskill JS (1990) The equilibrium partition function and base pair binding probabilities for RNA secondary structure. Biopolymers 29:1105–1119
10. Lorenz R, Bernhart SH, Höner zu Siederdissen C et al (2011) ViennaRNA package 2.0. Algorithms Mol Biol 6:26
11. Wuchty S, Fontana W, Hofacker IL et al (1999) Complete suboptimal folding of RNA and the stability of secondary structures. Biopolymers 49:145–165
12. Ding Y, Lawrence CE (2003) A statistical sampling algorithm for RNA secondary structure prediction. Nucleic Acids Res 31:7280–7301
13. Zuker M (1989) On finding all suboptimal foldings of an RNA molecule. Science 244:48–52
14. Lorenz R, Flamm C, Hofacker IL (2009) 2D projections of RNA folding landscapes. In: Grosse I, Neumann S, Posch S, Schreiber F, Stadler PF (eds) German conference on bioinformatics, Bonn. Lecture notes in informatics. Gesellschaft f. Informatik, pp 11–20
15. Flamm C, Hofacker IL, Stadler PF et al (2002) Barrier trees of degenerate landscapes. Zeitschrift für Physikalische Chemie 216:155
16. Wolfinger MT, Svrcek-Seiler WA, Flamm C et al (2004) Efficient computation of RNA folding dynamics. J Phys A Math Gen 37:4731
17. Gruber AR, Lorenz R, Bernhart SH et al (2008) The Vienna RNA websuite. Nucleic Acids Res 36:W70–W74

Chapter 2

Cis-Acting 5' Hammerhead Ribozyme Optimization for In Vitro Transcription of Highly Structured RNAs

Mélanie Meyer and Benoît Masquida

Abstract

RNA-mediated biological processes usually require precise definition of 5′ and 3′ ends. RNA ends obtained by in vitro transcription using T7 RNA polymerase are often heterogeneous in length and sequence. An efficient strategy to overcome these drawbacks consists of inserting an RNA with known boundaries in between two ribozymes, usually a 5′ hammerhead and a 3′ hepatitis delta virus ribozymes, that cleave off the desired RNA. In practice, folding of the three RNAs challenges each other, potentially preventing thorough processing. Folding and cleavage of the 5′ hammerhead ribozyme relies on a sequence of nucleotides belonging to the central RNA making it more sensitive than the usual 3′ hepatitis delta virus ribozyme. The intrinsic stability of the central RNA may thus prevent correct processing of the full transcript. Here, we present a method in which incorporation of a full-length hammerhead ribozyme with a specific tertiary interaction prevents alternative folding with the lariat capping GIR1 ribozyme and enables complete cleavage in the course of the transcription. This strategy may be transposable for in vitro transcription of any highly structured RNA.

Key words Ribozymes, RNA folding, Lariat capping GIR1 ribozyme, In vitro transcription, RNA tertiary interactions

1 Introduction

To study the massive amount of RNAs identified by transcriptomic analysis, it is important to identify their boundaries in order to grab the "right" molecules. The history of ribozymes well illustrates this idea. At the beginning, the hammerhead ribozyme (HMH) was characterized in short versions that prevented determining the full extent of the catalytic properties of the natural versions [1]. Thus, biochemical studies showing that hammerhead ribozymes were orders of magnitude more efficient to fold and cleave under low magnesium concentrations if a tertiary interaction between stem I and the loop from stem II was occurring [2], came years later as a surprise (Fig. 1). This finding was further demonstrated by the crystal structure of the *Schistosoma* hammerhead ribozyme in which

Christina Waldsich (ed.), *RNA Folding: Methods and Protocols*, Methods in Molecular Biology, vol. 1086,
DOI 10.1007/978-1-62703-667-2_2, © Springer Science+Business Media, LLC 2014

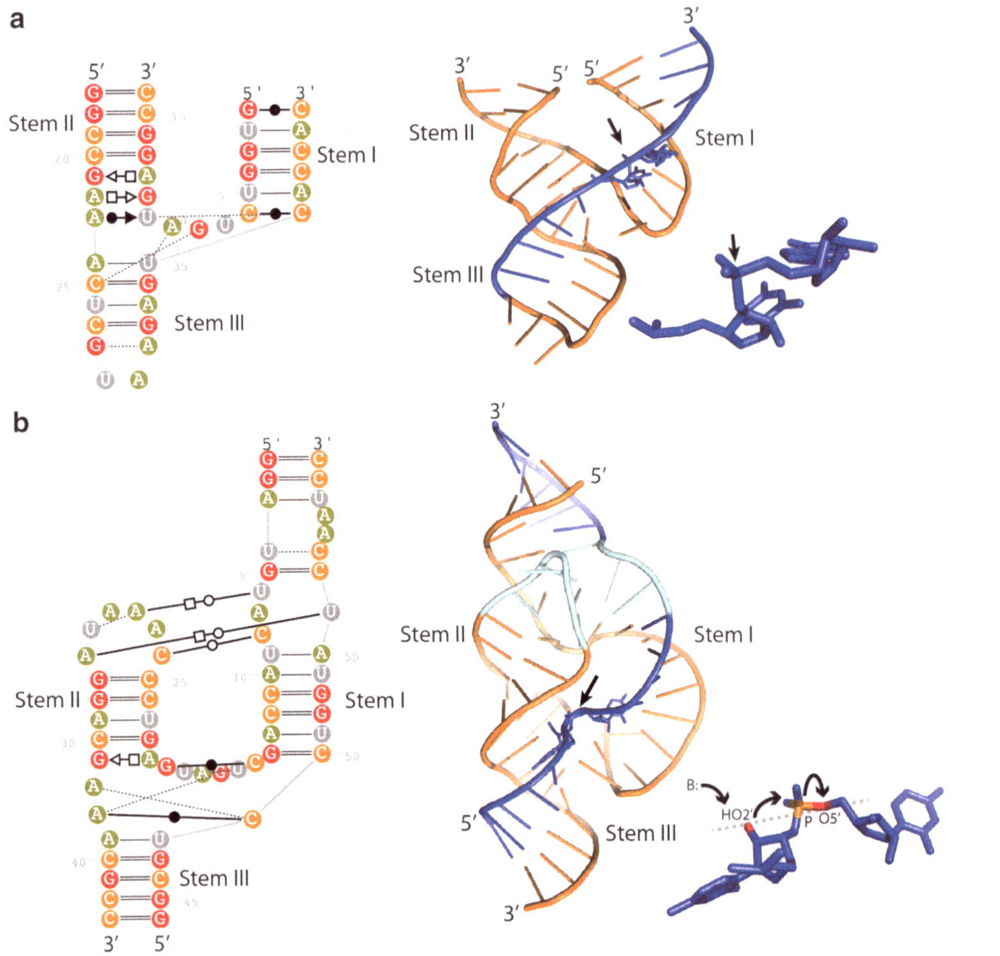

Fig. 1 Comparative secondary and three-dimensional structures of the hammerhead ribozyme cores from the minimal (**a**) versus the full-length (**b**) versions. Secondary structures (represented using the program Assemble [22]) indicate the sequence with nucleotides colored by type. For clarity, stem I is depicted with the 5′ strand on the *left*. Schemes in following figures depict the 5′ strand of stem I on the *right*. The crystal structures of the minimal [5] and full-length [3] hammerhead ribozymes present an *orange core* and a *blue substrate*. Nucleotides critical for tertiary interactions taking place in the full-length hammerhead are *colored in cyan* (**b**). The cleavage site is indicated with an *arrow*. The dinucleotides around the scissile bond are enlarged and displayed as *sticks*. The structures show the different conformations taking place in the two-ribozyme versions. A *dashed gray line* shows the alignment of the three colored atoms (02′, P, 05′) involved in the in-line attack mechanism that takes place in the efficient version of the hammerhead ribozyme

setting the correct tertiary interaction leads to a shape of the catalytic site in which the chemical groups responsible for catalysis are ready for in-line attack [3]. This was not the case in crystal structures of the shorter versions [4, 5] (Fig. 1). Other examples like studies on the hairpin [6] and Varkud satellite [7] ribozymes also show that the originally defined molecules were not including all the determinants required for their full efficiencies, and further, the characteristic modular architecture of RNA molecules [8]. The loss of structural integrity of important modules thus affects the overall RNA architecture, which in the end perturbs the function.

The potential occurrence of functional modules within the ends of RNA molecules justifies that they should be well defined for most studies. Unfortunately, the most widely used enzyme for in vitro production of RNA, the T7 RNA polymerase (T7RNAP), is prone to incorporate residues not encoded by the template at both the 5′ [9] and 3′ ends [10]. These uncoded residues eventually jeopardize experimental strategies such as crystallographic studies, dye-labeling for FRET studies, or in general any technique based on RNA ligation [11]. Moreover, any sequence immediately downstream from the promoter cannot be synthesized, since T7RNAP requires a specific G-rich starting sequence [12].

The errors made by T7RNAP can be concentrated in the tails of ribozymes located 5′ and 3′ of the RNA of interest that will be cleaved off [13]. This strategy can also be adapted to various systems including the production of RNAs with unfavorable 5′ sequences [14, 15], and/or to the use of ribozymes in *trans* [16]. On the 5′ side, a hammerhead ribozyme is usually chosen (Fig. 2). Only few nucleotides with any sequence downstream from the cleavage site are required for hybridization to fold the active ribozyme. On the 3′ side, the hepatitis delta virus (HδV) ribozyme is preferred because it is entirely located 3′ from the cleavage site and cleaves after any nucleotide, which leaves total freedom to the user. After cleavage, the central RNA is left with 5′ OH and 2′,3′ cyclic phosphate groups.

As long as the central RNA adopts a simple structure, this strategy does not need specific adaptation. However, if the structure of the central RNA is complex, it may interfere with proper folding of the flanking ribozymes and prevent efficient processing. While a short incubation at a temperature favoring structural rearrangements (~60 °C) usually restores cleavage by the 3′ HδV ribozyme, it is not the case for the 5′ hammerhead. This situation happened when we tried to express a lariat capping GIR1 ribozyme [17] with well-defined ends (Fig. 2). Nearly no cleavage was observed with a hammerhead version bearing a rather short stem I (5 bp). We assumed that the used truncated hammerhead had suboptimal activity because it could not form the critical tertiary interactions taking place between stem I and the loop of stem II. Our purpose was to design 5′ hammerhead constructs so that their substrate (stem I) resulted from hybridization to a longer sequence of the central lariat capping GIR1 ribozyme (LCrz) while preserving unpaired nucleotides, which interact with the facing loop from stem II (Fig. 3). This tertiary interaction, thereby preserved in the fusion between the hammerhead and the lariat capping GIR1 ribozymes, accelerates folding and catalysis and allows decreasing the required magnesium concentration [2].

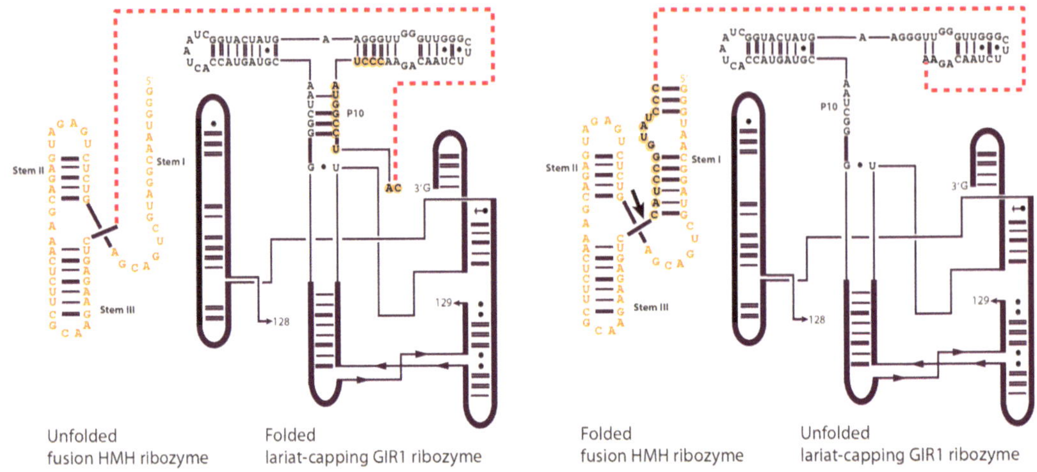

Fig. 2 Mutually exclusive theoretical secondary structures of the pre-RNA ribozyme transcript. If the pre-transcript presents an unfolded inactive hammerhead and a folded lariat-capping GIR1 ribozymes, the situation is unfavorable (*left panel*). However, the opposite situation (*right panel*) favors maximal cleavage during the in vitro transcription reaction. The bulge in stem I potentially forms a more stable structure with the loop closing stem II than if stem I is closed by a couple of additional A–U pairs. An *arrow* indicates the cleavage site

Design of the fusion hammerhead constructs using the program RNAfold [18] and their comparison to wild-type sequences of full-length hammerhead ribozymes was carried out. Selected hammerhead sequences with a stem I integrating the 5′ end of the lariat capping GIR1 ribozyme [17] were assayed for efficient cleavage in the course of in vitro transcription reactions (Fig. 4). The direct involvement of the bulge in stem I was investigated by determining the observed kinetic constants of the successful candidates (Fig. 5, [19]). Cloning of efficient fusion hammerhead ribozymes upstream from the lariat capping GIR1 ribozyme, followed by the HδV ribozyme, was performed in plasmid pUC19, further transcribed in vitro as linearized templates (Fig. 6). Northern blot analysis showed that the optimized hammerhead ribozymes were very efficient in the full context since they cleave to completion in the course of transcription, thus validating our design (Fig. 7). This method, validated on two different sequences of the lariat capping GIR1 ribozyme, is a practical addition to the molecular biologist's toolbox that addresses defining accurate ends to highly structured RNAs.

2 Materials

Prepare all solutions using Millipore or DEPC water. All buffers are filtered using a 0.22 μm sterile filter and kept at room temperature unless otherwise specified.

2.1 In Silico Design of Fusion Hammerhead Ribozyme Sequences

1. RNAfold [18] Web server can be used http://rna.tbi.univie.ac.at/cgi-bin/RNAfold.cgi.

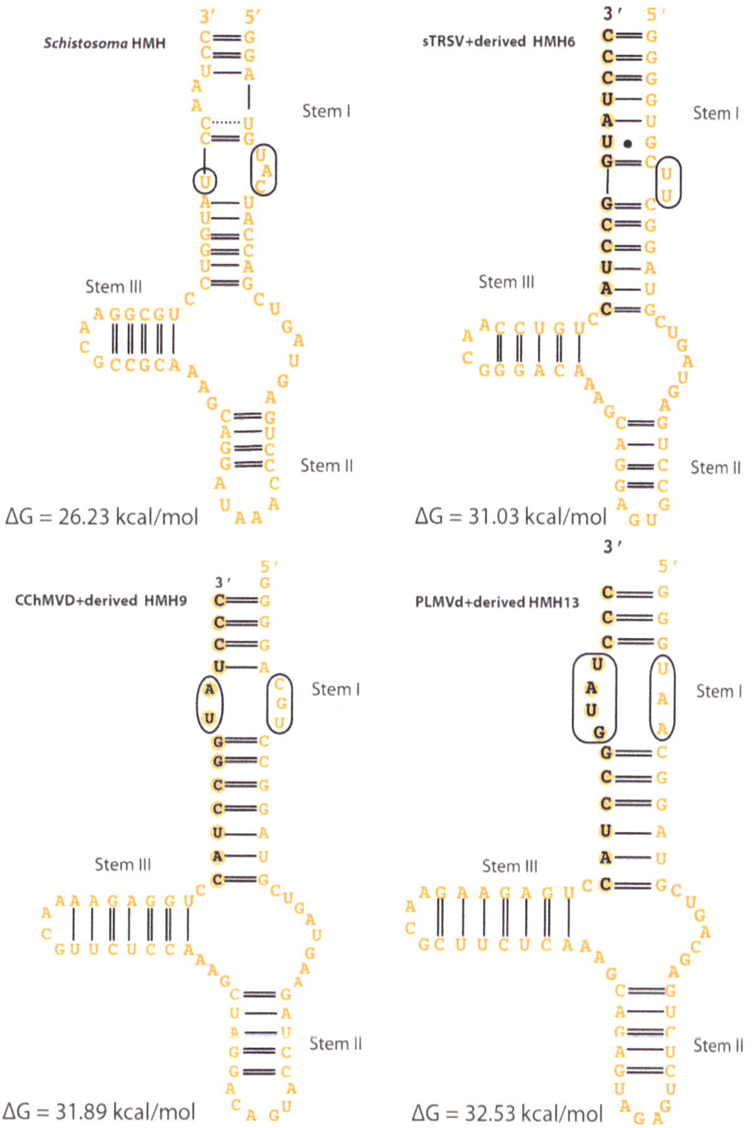

Fig. 3 Secondary structure diagrams of the fusion hammerhead ribozyme candidates (outputs from RNAfold) show a correct catalytic core structure and unpaired nucleotides in stem I. The secondary structure of the hammerhead ribozyme from *Schistosoma* is shown as a reference model with nucleotides from the bulge in stem I interacting with the *loop* from stem II *circled*. Some of the *circled* unpaired nucleotides in candidates HMH6, 9, and 13 potentially interact with the *loop* from stem II

Sequence of the full-length wild-type hammerhead ribozyme from *Schistosoma*: 5′ GGAUGUACUACCAGCUGAUGA GUCCCAAAUAGGACGAAAC GCCGCAAGGCGUCC UGGUAUCCAAUCC 3′.

Sequence of the full-length fusion hammerhead 6 ribozyme derived from sTRSV+: 5′ GGGGUGCUUCGGAUGC

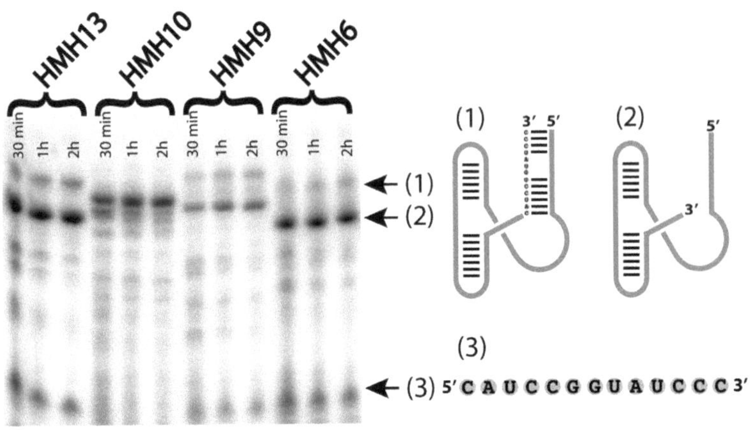

Fig. 4 Body-labeled RNA pattern from a standard transcription reaction of the full-length hammerhead ribozymes fused to the 13 nucleotides at the 5′ end of the lariat-capping GIR1 ribozyme. This step is meant to check whether ribozymes with the modified sequence, but lacking the main part of the lariat-capping GIR1 ribozyme, cleave in the course of the transcription reaction. The main RNA species numbered 1–3 are visible on the gel, (1) uncleaved hammerhead, (2) cleaved hammerhead core, (3) cleaved substrate (13 nt). HMH10, used as a negative control, carries a mutation ΔC15 with respect to the *Schistosoma* hammerhead ribozyme numbering (*see* Fig. 1b)

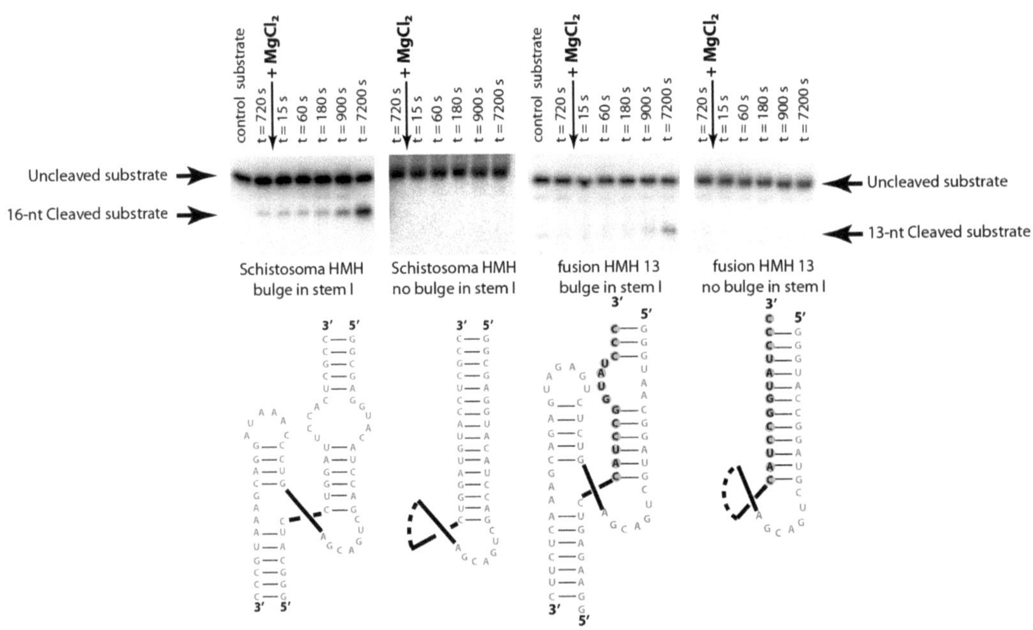

Fig. 5 Example of kinetic analysis of *trans* fusion hammerhead lariat capping GIR1 ribozymes. The model hammerhead from *Schistosoma* (*left panel*) is compared to HMH13 (*right panel*). A 50-fold excess of the in vitro transcribed hammerhead ribozyme core was added to the 3′ ^{32}P-labeled RNA substrate to achieve single turnover conditions. Two kinds of cores (and substrates in the case of the *Schistosoma* ribozyme) were used. The first kind preserved the bulge in stem I whereas the second kind forms a duplex with perfect complementarity. Cleavage is observed only when the bulge is present. The fraction of cleaved substrate was plotted versus time. Data were fitted to the equation $F_{cl} = F_{max}(1 - e^{-kt}) + F_0$ to determine k_{obs} values. The *Schistosoma* ribozyme presents a k_{obs} similar to HMH13, $k_{obsSch} = 0.013 \pm 4.7 \times 10^{-3}$ min^{-1} and $k_{obsHMH13} = 0.016 \pm 8.3 \times 10^{-3}$ min^{-1} showing that our designed hammerheads are as efficient as the *Schistosoma* model ribozyme at 0.1 mM $MgCl_2$

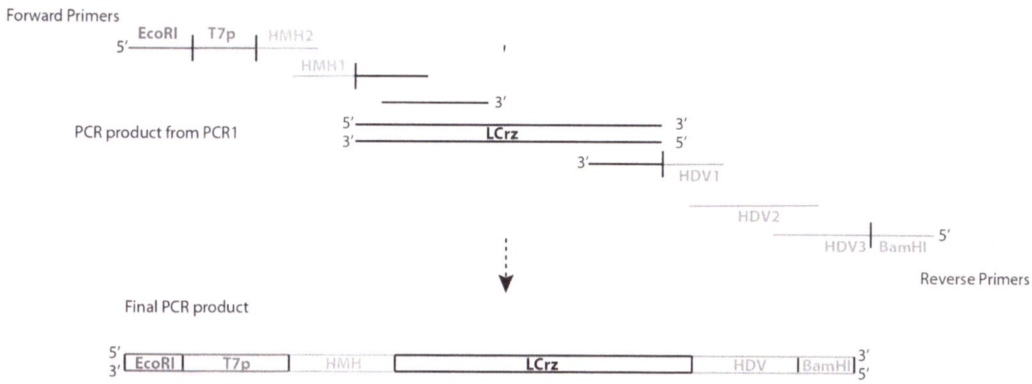

Fig. 6 Description of the PCR strategy of the DNA template encoding the ternary ribozyme system. A plasmid containing a 5′ truncated lariat-capping GIR1 ribozyme (LCrz) was used as a template for the first PCR amplification. Two further PCR steps created the flanking hammerhead and hepatitis δ virus (HδV) ribozymes, a T7 promoter and restriction sites for cloning (*Eco*RI, *Bam*HI)

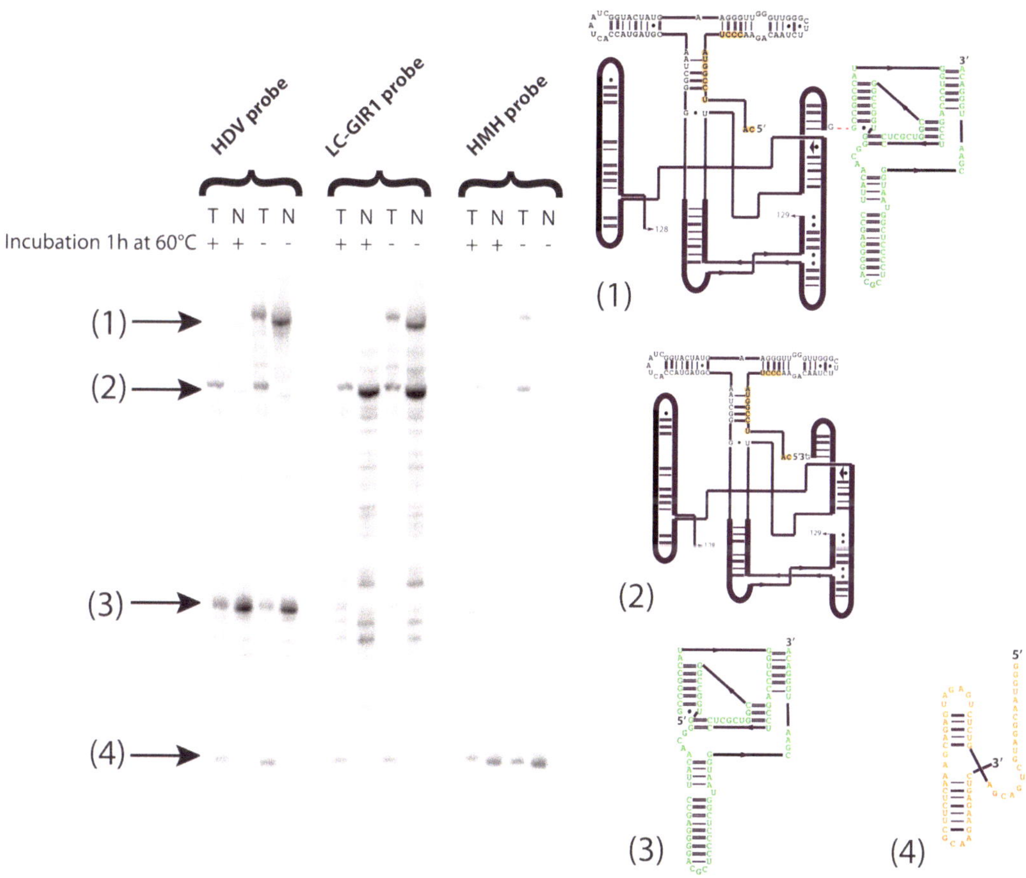

Fig. 7 Northern blot analysis of a transcription reaction of the ternary ribozyme construct. Body-labeled (T) and cold (N) transcription reactions were performed for 2 h and fractionated side by side by PAGE. Northern blot analysis was performed using a set of probes specific for (i) HδV, (ii) LCrz, and (iii) the hammerhead ribozymes. To complete HδV cleavage reactions, samples were incubated 1 h at 60 °C. Four main RNA products are visible on the membranes. (1) corresponds to the lariat-capping GIR1 ribozyme tethered to the HδV ribozyme. (2) is the correctly processed lariat-capping GIR1 ribozyme. (3) and (4) correspond to the two cleaved flanking ribozymes. It is important to note that cleavage by the hammerhead ribozyme goes to completion during the transcription reaction implying that the ternary ribozyme transcript cannot be detected

UGAUGAGUCCGUGAGGACGAAACAGGGCAA CCUGUCCAUCCGGUAUCCC 3′.

Sequence of the full-length fusion hammerhead 9 ribozyme derived from CChMVD+: 5′ GGGGACGUCCGGAUGC UGAUGAAGAUCCAUGACAGGAUCGAAAC CUCUUGCAAAAGAGGUCCAUCCGGUA UCCC 3′.

Sequence of the full-length ΔC15 fusion hammerhead 10 ribozyme derived from sLTSV−: 5′ GGGAGGUGCGG AUGAUGAGUCCGAAAGGACGAAACAGU AGGAAUACUGUCCAUCCGGUAUCCC 3′.

Sequence of the full-length fusion hammerhead 13 ribozyme derived from PLMVd+: 5′ GGGUAACGGAUGCUGAC GAGUCUCUGAGAUGAGACGAAACUCUUCG CAAGAAGAGUCCAUCCGGUAUCCC 3′.

2.2 In Vitro Determination of Fusion Hammerhead Ribozymes Cleavage Efficiencies

2.2.1 Co-transcriptional Cleavage of Fusion Hammerhead Ribozymes

1. DNA oligos are used as template for amplification by PCR creating the dsDNA template required for the in vitro transcription of the full-length hammerhead ribozymes from *Schistosoma* and full-length fusion hammerhead ribozymes HMH6, HMH9, HMH10, and HMH13.

   Forward (+) and reverse (−) strands for each hammerhead ribozyme:

   HMHSchistosoma+:5′TAATACGACTCACTATAGGATGTA CTACCAG CTGATGAGTCCCAAATAGGACGAAACGCCG CAAGGCGTCCTGGTATCCAATCC 3′.

   HMHSchistosoma−: 5′GGATTGGATACCAGGACGC CTTGC GGCGTTTCGTCCTATTTGGGACTCATCAGCT GGTAGTACATCCTATAGTGAGTCGTATTA 3′.

   HMH6+: 5′ TAATACGACTCACTATAGGGGTGCTTCGG ATGCTGATGAGTCCGTGAGGACGAAACAGGGCAA CCTGTCCATCCGGTATCCC 3′.

   HMH6−: 5′ GGGATACCGGATGGACAGGTTGCCCTGT TTCGTCCTCACGGACTCATCAGCATCCG AAGCACCCCTATAGTGAGTCGTATTA 3′.

   HMH9+: 5′ TAATACGACTCACTATAGGGGACGT CCG GATGCTGATGAAGATCCATGACAGGATCGAAA CCTCTTGCAAAAGAGGTCCATCCGGTATCCC 3′.

   HMH9−: 5′ GGGATACCGGATGGACCTCTTTTGCAAGA GGTTTCGATCCTGTCATGGATCTTCATC AGCATCCGGACGTCCCCTATAGTGAGTCGTATTA 3′.

   HMH10+: 5′ TAATACGACTCACTATAGGGAGGTGCGG ATGATGAGTCCGAAAGGACGAAACAGTAGG AATACTGTCCATCCGGTATCCC 3′.

HMH10–: 5′ GGGATACCGGATGGACAGTATTCCTATG TTTCGTCCTTTCGGACTCATCATCCGCACCTCCCTA TA GTGAGTCGTATTA 3′.

HMH13+:5′TAATACGACTCACTATAGGGTaaCGGATGC TGACGAGTCTCTGAGATGAGACGAAACTCTTCGC AAGAAGAGTCCAT CCGGTATCCC 3′.

HMH13–: 5′ GGGATACCGGATGGACTCTTCTTGCGAA GAGTTTCGTCTCATCTCAGAGACTCGTCAGCA TCCGTTACCCTATAGTGAGTCGTATTA 3′.

2. DNA primers used for PCR amplification of the dsDNA template:

Primer 1: 5′ TAATACGACTCACTATA 3′.

Primer 2 for fusion hammerhead ribozymes: 5′ GGGATACCGG ATGGAC 3′.

Primer 2 for hammerhead from *Schistosoma*: 5′ GGATTGGATA CCAG 3′.

3. Phusion High-Fidelity DNA polymerase (2,000 U/mL) stored at −20 °C.
4. 5× Phusion High-Fidelity DNA polymerase HF reaction buffer stored at −20 °C.
5. 25 mM dNTPs mix stock solution is stored at −20 °C.
6. 6× glycerol blue loading buffer: 50 % glycerol, 0.1 % xylene cyanol, 0.1 % bromophenol blue.
7. 2 % (w/v) agarose gel: agarose for routine use is dissolved in 100 mL of 1× TBE.
8. Steady red [α-^{32}P]-UTP at a concentration of 10 mCi/mL and a specific activity of 3,000 Ci/mmol is stored at 4 °C.
9. Micro Bio-Spin 6 chromatography columns (Bio-Rad), or comparable product.
10. His-tag-T7 RNA polymerase is produced in the laboratory from plasmid pT7-911Q [20].
11. 5× TMSDT buffer: 200 mM Tris–HCl, pH 8.1, 110 mM $MgCl_2$, 5 mM spermidine, 25 mM DTT, 0.05 % Triton stored at −20 °C.
12. Alkaline phosphatase is stored at a concentration of 1 U/μL in 1× TMSDT at 4 °C.
13. 100 mM ATP, 100 mM UTP, 100 mM CTP, and 100 mM GTP stock solution are stored at −20 °C.
14. 2× loading buffer: 8 M urea, 0.025 % xylene cyanol, 0.025 % bromophenol blue.
15. 10× Tris-borate-EDTA buffer (TBE) to run PAGE: 0.89 M Tris base, 0.89 M boric acid, 20 mM EDTA, pH 8.0.

16. Denaturing PAGE: 10 % acrylamide/bisacrylamide (19:1), 8 M urea in 1× TBE.
17. Gel size: 35 cm width × 42 cm length × 0.08 cm thickness.
18. Apparatus to vacuum dry gels.
19. 35 × 43 cm imaging plate (FujiFilm), or comparable product.
20. Bio-imager (FujiFilm), or comparable equipment.

2.2.2 Kinetic Analysis of Fusion Hammerhead Ribozymes with Trans substrates

1. Substrate RNAs were purchased (e.g., at Dharmacon, Thermo Scientific).
2. Shaker MixMate (Eppendorf), or comparable equipment.
3. 3 M KCl stock solution.
4. T4 RNA ligase (5 U/μL) is stored at −20 °C.
5. 10× T4 RNA ligase reaction buffer: 500 mM Tris–HCl, pH 7.8, 100 mM $MgCl_2$, 100 mM DTT, 10 mM ATP stored at −20 °C.
6. Steady red ^{32}pCp at a concentration of 10 mCi/mL and a specific activity of 3,000 Ci/mmol is stored at 4 °C.
7. X-ray autoradiograms (e.g., Super RX from Fuji Medical).
8. 10× kinetic buffer: 500 mM Tris–HCl, pH 8.0, 1 M NaCl, 1 mM EDTA, pH 8.0.
9. 1 M $MgCl_2$ stock solution.
10. 2× formamide-EDTA blue: 80 % formamide, 50 mM EDTA, pH 8.0, 0.02 % bromophenol blue.
11. Science Lab 2003 Image Gauge 4.2 (FujiFilm), or comparable quantification software.
12. KaleidaGraph 4.02, or comparable software for curve fitting.

2.3 Cloning of Ternary Ribozyme Constructs

1. DNA oligos used to create the insert corresponding to the ternary ribozyme constructs to be cloned:

   Forward primers:

   EcoRI_T7p_HMH2: 5′ GACGGCCAGTGAATTCTAA TACGACTCACTATAGGGTAACGGATGCTGACGAGT CTCTGAGA 3′.

   HMH1_LCrz: 5′ CTGACGAGTCTCTGAGATGA GACGA AACTCTTCGCAAGAAGAGTCCATCCGGTATCCCAA GA 3′.

   Reverse primers:

   BamHI_HδV3: 5′ GGATCCTCTAGAGTCGACCTTGTCC CATTCGCCATTACCGAGGGGACGGTCCCCTCGG AATGTT 3′.

   HδV2: 5′ GTCCCCTCGGAATGTTGCCCACCGGCCGCC AGCGAGGAGGCTGGGACCATGGCCGGC 3′.

HδV1_LCrz: 5′ GGACCATGGCCGGCGATTGTCTT GGGATACCGGATGCTTCCTTTCGG AACGACT 3′.

2. Gel-extraction kit.
3. pUC19 vector stored at a concentration of 1 mg/mL at −20 °C.
4. Cloning kit.
5. Miniprep spin kit.

2.4 Northern Blot Analysis of the Cleavage Pattern of the Ternary Ribozyme Transcripts

1. DNA probes for Northern blot:

   HMH: 5′ GACTCTTCTTGCGAAGAGTTTC 3′.

   LCrz: 5′ CATGATACTTCCCAACCCAACC 3′.

   HδV: 5′ CTCGGAATGTTGCCCACCGGCC 3′.
2. Steady red [γ-^{32}P]-ATP at a concentration of 10 mCi/mL and a specific activity of 3,000 Ci/mmol is stored at 4 °C.
3. T4 polynucleotide kinase (PNK; 10 U/μL). Store at −20 °C.
4. 10× PNK buffer: 500 mM Tris–HCl, pH 7.6, 100 mM $MgCl_2$, 50 mM DTT, 1 mM spermidine. Store at −20 °C.
5. Nylon membrane for Northern blot analysis.
6. Stratalinker UV cross-linker (Stratagene).
7. Hybridization bottle: 35 × 300 mm.
8. 6× SSPE: 3 M NaCl, 0.2 M NaH_2PO_4, and 0.02 M EDTA. pH is adjusted to 7.4 with NaOH.
9. Pre-hybridization buffer: 1× SSPE, 2.5 % SDS, 2.5 % Denhart. Pre-hybridization buffer is stored at 30 °C to avoid SDS precipitation.
10. 50× Denhart's solution: 1 % (w/v) ficoll 400, 1 % (w/v) polyvinylpyrrolidone, 1 % (w/v) bovine serum albumin (fraction V). Denhart's solution should be freshly made.
11. Hybridization incubator.

3 Methods

3.1 In Silico Design of Fusion Hammerhead Ribozyme Sequences

The aim of this step is to design fusion hammerhead ribozymes presenting a hammerhead catalytic core fused to a substrate stem corresponding to the 5′ sequence of the lariat capping GIR1 ribozyme. The ribozyme strand presents both ends rooted in stem I, while loops close stems II and III. Stems II and III and the single-stranded regions form the catalytic core are directly copied from genuine hammerhead sequences presenting tertiary contacts between stem I and the loop of stem II (sTRSV+, CChMVD+, PLMVD+, [2]). Stem I should be approximately a dozen nucleotides long with a 3′ strand being the 5′ end of the central RNA, the lariat-capping GIR1 ribozyme in this case [17]. Consequently the

5′ strand of stem I is designed to be partially complementary to the latter and starts with three G residues for transcriptional efficiency (Fig. 2). Moreover, a bulge intervening six nucleotides after the cleavage site has to be preserved in stem I. The purpose of this bulge is to avoid formation of a perfect duplex which would destroy the tertiary contacts with the loop of stem II (Fig. 1) that are crucial for efficient folding and cleavage of the ribozyme (Fig. 3, [19]). In the sequences resulting from the fusion between hammerhead ribozymes and the lariat-capping GIR1 ribozyme, the presence of (1) a correct three-way junction forming the catalytic core and of (2) the critical unpaired nucleotides in stem I was systematically checked using RNAfold [18] (*see* **Note 1**).

3.2 In Vitro Determination of Fusion Hammerhead Ribozymes Cleavage Efficiencies

Successful sequences were transcribed from commercial template DNA oligos. The extent of cleavage was monitored in the course of the transcription reactions (Fig. 4). More detailed kinetic analysis of the most efficient transcripts, HMH13, were performed under single turnover conditions using fusion hammerhead ribozymes consisting of distinct catalytic and substrate strands (Fig. 5).

3.2.1 Co-transcriptional Cleavage of Fusion Hammerhead Ribozymes

To produce RNAs corresponding to candidate sequences selected in silico by in vitro transcription [12, 21], DNA templates preceded by a Class III T7 promoter (5′ TAA TAC GAC TCA CTA TA 3′) should be produced (Fig. 4).

1. Design full-length template DNA oligos, forward and reverse strands that will be used as dsDNA templates for in vitro transcription of the candidates sequences selected in silico (*see* **Note 2**).
2. Design primers for PCR amplification of the template DNA oligos (*see* **Note 3**).
3. Prepare 50 μL of PCR reaction containing: 300 ng of each DNA template, 0.5 μL of the Phusion High-Fidelity DNA polymerase (2,000 U/mL), 10 μL of the 5× Phusion HF buffer, 1.2 μM of forward primer, 1.2 μM of reverse primer, 400 μM dNTPs mix, water to 50 μL.
4. Run the PCR reaction in a thermal cycler following the set up displayed in Table 1.
5. Check the efficiency of amplification by mixing 5 μL of the PCR reaction to 1 μL 6× glycerol blue loading buffer and load onto a 2 % (w/v) agarose gel.
6. After ethidium bromide staining, observe the DNA migration pattern under UV.
7. Clean the PCR reaction using the Micro Bio-Spin columns.
8. Set up 50 μL in vitro transcription reactions containing: 1 μg of the cleaned PCR reaction, 10 μL of 5× TMSDT buffer, 5 mM ATP, 5 mM GTP, 5 mM CTP, 2.5 mM UTP, 25 μCi

Table 1
PCR program

Step	Temperature (°C)	Time	Number of cycles
Initial denaturing	95	3 min	1
Denaturing	95	30 s	30
Annealing	45[a]/50[b]	1 min	
Elongation	72	1 min	
End	4	10 min	1

[a]Annealing temperature for PCR of **step 4** in Subheading 3.2.1
[b]Annealing temperature for PCR of **step 3** in Subheading 3.3

[α-^{32}P]-UTP (10 mCi/mL, 3,000 Ci/mmol), 0.2 μL of alkaline phosphatase (1 U/μL) (*see* **Note 4**), T7 RNA polymerase as indicated by your provider (*see* **Note 5**), water to 50 μL.

9. Incubate at 37 °C for 2 h.
10. Remove a 12 μL aliquot of each transcription reaction after 30 min, 1 h and 2 h.
11. Stop the transcription by mixing the sample to an equal volume of 2× loading buffer.
12. Clean the transcription reaction using the Micro Bio-Spin columns (*see* **Note 6**).
13. Fractionate each sample by 10 % denaturing PAGE. Run the gel for 2 h at 60 W (*see* **Note 7**).
14. Vacuum dry the gel onto Whatman 3MM paper for 4 h at 80 °C or alternatively seal it in a plastic film and store it at −80 °C.
15. Expose the gel to an imaging plate in a cassette for 10 min (*see* **Note 8**).
16. Read the imaging plate using a Bio-imager.

3.2.2 Kinetic Analysis of Fusion Hammerhead Ribozymes with Trans substrates

This step can be performed as in [19] to check whether the bulge in stem I is functional or to determine an optimal magnesium concentration for cleavage (*see* **Note 9**). The hammerhead cores are obtained by in vitro transcription of dsDNA templates while substrates are obtained by RNA solid phase chemical synthesis (Fig. 5).

1. Design two sets of fusion hammerhead ribozymes and their substrates as separate RNA molecules. From this step the catalytic cores of the fusion hammerhead ribozymes will be handled separately from their substrate domains.
2. Purify the complementary DNA oligos that will form the dsDNA templates used for in vitro transcription of the different ribozymes' cores by 10 % denaturing PAGE.

3. Mix 20 μg of the forward and the reverse pure template DNA oligos and add KCl up to a final concentration of 500 mM. To hybridize the template DNA oligos heat the sample for 1 min at 95 °C and allow it to cool to room temperature for 10 min.
4. Set up 200 μL transcription reactions containing 40 μg of dsDNA template, 40 μL of 5× TMSDT buffer, 5 mM ATP, 5 mM GTP, 5 mM CTP, 5 mM UTP, 0.8 μL of alkaline phosphatase (1 U/μL) (*see* **Note 4**), T7 RNA polymerase as indicated by your provider (*see* **Note 5**), water to 200 μL.
5. Purify transcription reactions by 10 % denaturing PAGE.
6. Identify RNA bands corresponding to the fusion hammerhead core by UV shadowing.
7. Cut the band of interest out of the gel and place it into a 1.5 mL tube.
8. Immerse the gel band in about 800 μL of water and place the tube in a shaker overnight at 4 °C.
9. Decant the eluate and transfer it to a fresh tube.
10. Recover the fusion hammerhead cores by ethanol precipitation of the eluate.
11. Set up a 3′ radio-labeling reaction for the RNA substrates: 500 ng of RNA substrate, 40 μCi ^{32}pCp (10 mCi/mL, 3,000 Ci/mmol), 10 % DMSO, 1 μL T4 RNA ligase (5 U/μL), 3 μL 10× T4 RNA ligase buffer, water to 30 μL.
12. Incubate at 16 °C overnight.
13. Clean the labeling reactions using the Micro Bio-Spin columns (*see* **Note 6**).
14. Purify the labeled RNA substrates by 20 % denaturing PAGE at 50 W for 2 h.
15. Localize the positions of the labeled RNAs using an X-ray sensitive film.
16. Repeat **steps 7–10**.
17. Set up a reaction for analyzing the kinetics of the hammerhead cleavage under single turnover conditions as follows: 0.5 μM of purified fusion hammerhead core RNA, 10 nM of the purified 3′ radio-labeled substrate RNA, 1.5 μL of 10× kinetic buffer, water to 13 μL.
18. Incubate the reaction at 70 °C for 2 min and then at 25 °C for 10 min.
19. Take a 2 μL aliquot of the reaction.
20. Add $MgCl_2$ to a final concentration of 0.1 mM.
21. Incubate the reaction at 25 °C. Take a 2 μL aliquot of the reaction after 15 s, 1 min, 3 min, 15 min, 2 h.

22. Quench the reaction by mixing the aliquot with 2 µL 2× formamide-EDTA blue.
23. Place samples in ethanol cooled down by dry ice.
24. Separate the cleaved from uncleaved substrates by 12 % denaturing PAGE at 50 W for 2 h (*see* **Note 7**).
25. Vacuum dry the gel onto Whatman 3 MM paper for 4 h at 80 °C or alternatively seal it in a plastic film and store it at −80 °C.
26. Expose the gel to an imaging plate overnight in a cassette (*see* **Note 10**).
27. Reveal the radioactivity pattern on the imaging plate using a Bio-imager.
28. Quantify the amount of cleaved substrate using appropriate software.
29. Plot the fraction of substrate cleaved versus time and determine k_{obs} by fitting the data to the equation, $F_{cl} = F_{max}(1 - e^{-kt}) + F_0$ using an appropriate software like KaleidaGraph (*see* **Note 11**).

3.3 Cloning of Ternary Ribozyme Constructs

The most efficient fusion hammerhead ribozyme is cloned upstream from the lariat capping GIR1 ribozyme followed by an HδV ribozyme ending with a restriction endonuclease site used for linearization to permit run off transcription. Since cloning is the best choice for storing matrices in the laboratory's library, it is advised to anticipate the cloning step at the very beginning of the sequence design by adding a 5′ sequence either corresponding to the desired restriction enzyme recognition sequence or to the recombination sequence specified in the cloning kit.

All lariat capping GIR1 RNA constructs from *Didymium iridis* (DiGIR1) were transcribed from plasmid DNA templates obtained by cloning of an insert generated by three successive PCR amplifications. The 5′ PCR primers were designed to contain an *Eco*RI restriction site, a T7 promoter, the sequence of the co-transcriptional self-cleaving hammerhead ribozyme, and the 5′ sequence of the DiGIR1 ribozyme. The 3′ PCR primers included a *Bam*HI restriction site, a posttranscriptional self-cleaving HδV ribozyme, and the 3′ sequence of the lariat capping DiGIR1 ribozyme (Fig. 6).

1. Design forward and reverse primers for the successive PCRs (*see* **Note 12**).
2. Prepare 50 µL of PCR reaction containing: 50 pg of template plasmid DNA, 0.5 µL of the Phusion High-Fidelity DNA Polymerase (2,000 U/mL), 10 µL of the 5× Phusion HF buffer, 1.2 µM of forward primer, 1.2 µM of reverse primer, 400 µM dNTP mixture, water to 50 µL (*see* **Note 13**).

3. Run the PCR reaction in a thermal cycler following the set up displayed in Table 1 (*see* **Note 14**).
4. Mix the PCR reaction to 10 μL 6× glycerol blue loading buffer and load the entire PCR reaction onto a 2 % (w/v) agarose gel.
5. After ethidium bromide staining, observe the DNA migration pattern under UV.
6. Cut the DNA band of interest out of the gel.
7. Recover the DNA insert in 30 μL of water with a gel-extraction kit (*see* **Note 15**).
8. Insert the purified PCR product into a pUC19 plasmid using a cloning kit.
9. Directly transform XL-1 blue *Escherichia coli* competent cells with the recombinant plasmid product (*see* **Note 16**).
10. Grow the transformed XL-1 blue *E. coli* competent cells on LB/Ampicillin (100 μg/mL) plates overnight at 37 °C.
11. Grow 3 mL of LB/Ampicillin (100 μg/mL) cultures from individual colonies.
12. Extract DNA using a miniprep spin kit.
13. Sequence the extracted DNA using the M13 sequencing primers.
14. Prepare glycerol stocks of the clones containing the desired sequences by mixing 1 mL of saturated bacterial culture to 600 μL of sterilized 50 % glycerol solution in 2 mL screw cap tubes. Store the glycerol stocks at –80 °C.

3.4 In Vitro Transcription of the Ternary Ribozyme RNA Using Linearized Plasmids as Templates

The transcription reaction of the plasmid template is performed as for the PCR template except that the plasmid is linearized prior to the reaction and that the mass added is greater, since the molarity of the template is important.

1. For in vitro transcription, linearize the plasmid with *Bam*HI restriction endonuclease following the enzyme provider instructions.
2. Set up 50 μL in vitro transcription reactions containing: 2.5 μg of linearized plasmid DNA, 10 μL of 5× TMSDT buffer, 5 mM ATP, 5 mM GTP, 5 mM CTP, 5 mM UTP, 0.2 μL of alkaline phosphatase (1 U/μL) (*see* **Note 4**), T7 RNA polymerase as indicated by your provider (*see* **Note 5**), water to 50 μL.
3. In parallel set up 50 μL in vitro transcription reactions containing: 2.5 μg of linearized plasmid DNA, 10 μL of 5× TMSDT buffer, 5 mM ATP, 5 mM GTP, 5 mM CTP, 2.5 mM UTP, 25 μCi [α-^{32}P]-UTP (10 mCi/mL, 3,000 Ci/mmol), 0.2 μL of alkaline phosphatase (1 U/μL) (*see* **Note 4**), T7 RNA polymerase as indicated by your provider (*see* **Note 5**), water to 50 μL.

4. Incubate at 37 °C for 2 h.
5. Incubate half of each transcription at 60 °C for 1 h.
6. Stop transcription reaction by mixing it to an equal volume of 2× formamide-EDTA blue loading buffer.

3.5 Northern Blot Analysis of the Cleavage Pattern of the Ternary Ribozyme RNA (See Fig. 7)

1. Design three DNA probes that will be used to hybridize specifically either to the hammerhead ribozyme or to the central RNA or to the HδV ribozyme (*see* **Note 17**).
2. Label each DNA probe by mixing: 1 μM of DNA oligo, 2.5 μL of [γ-^{32}P]-ATP (10 mCi/mL, 3,000 Ci/mmol), 1 μL of T4 polynucleotide kinase (10 U/μL), 5 μL 10× PNK buffer, water to 50 μL.
3. Incubate 30 min at 37 °C, then 10 min at 65 °C.
4. Clean the labeling reactions using the Micro Bio-Spin columns (*see* **Note 6**).
5. Fractionate samples from in vitro transcription reactions of the ternary ribozyme RNA by 8 % denaturing PAGE. Load in four consecutive lanes the 60 °C pre-incubated or not samples in order to be able to compare the cold to the corresponding body-labeled transcription reactions. Repeat the same operation two times. Each set of four lanes will be used for a different DNA probe. Run for 3 h at 60 W (*see* **Note 7**).
6. Transfer the RNA bands from the gel onto positively charged nylon membrane using a vacuum gel dryer for 4 h at 80 °C.
7. Wash the membrane in water to remove the dried gel particles (*see* **Note 18**).
8. Cross-link the RNA onto the membrane using a UV cross-linker.
9. Cut the membrane to separate each set of four lanes. From this step you will be dealing with three membranes in parallel.
10. Insert delicately each membrane into a hybridization bottle (*see* **Note 19**).
11. Add 20 mL of pre-hybridization buffer.
12. Incubate at 60 °C for 30 min in a rolling hybridization oven.
13. Add the total amount of one labeled DNA primer per hybridization bottle.
14. Incubate at 37 °C over night in the rolling hybridization oven.
15. Wash each membrane with 20 mL of 6× SSPE at 37 °C for 5 min in the rolling hybridization oven. Repeat this step three times.
16. Wash each membrane with 20 mL of 6× SSPE at 42 °C for 5 min in the rolling hybridization oven.
17. Seal each membrane in a plastic film.

18. Expose an imaging plate to the plastic sealed membrane for 3 h in a cassette (*see* **Notes 8** and **10**).
19. Reveal the radioactivity pattern on the imaging plate using a Bio-imager.

4 Notes

1. The sequence of the adapted substrate starts after the cleavage site and preserves several unpaired residues in stem I of the ribozyme core to favor tertiary contacts with the loop of stem II. The best situation is when the UAC sequence forms a bulge. When this not possible, avoid building an extended helix by inserting at least non-complementary residues.
2. Do not forget to insert the T7 promoter sequence upstream of the DNA templates.
3. These primers can be the same for all fusion hammerhead ribozymes candidates if you use as primer 1 the sequence of the T7 promoter and as primer 2 the sequence of the central RNA.
4. It is not compulsory to add alkaline phosphatase but it is known to increase in vitro transcription yields.
5. We produce our own T7 RNA polymerase, which reduces cost considerably.
6. It is not mandatory to clean transcription reaction although it is preferable to remove not incorporated radio-labeled nucleotides to avoid contamination of the electrophoresis buffer.
7. To have better defined bands, gels are pre-run for 1 h at 60 W.
8. The radioactivity pattern of the gel can as well be obtained on X-ray autoradiogram.
9. This step is not mandatory and was carried out as a proof of principle. It can be performed if the ribozyme construct is inactive to further check for efficiency of a set of hammerhead constructs carrying different kinds of bulges in stem I.
10. Exposure time of the imaging plate depends on the amount of ^{32}P-labeled material on the membrane or gel.
11. F_{cl} is the fraction of cleaved substrate, F_0 is the initial amount of substrate cleaved before addition of $MgCl_2$, F_{max} is the maximal amount of substrate cleaved, t is the time in minute and k_{obs} is the observed rate of cleavage.
12. For optimal yield, forward and reverse primers should have close melting temperature.
13. Keep your samples on ice while preparing the PCR reaction.
14. Insert your samples in the thermal cycler when the temperature of the sample holder is at 90 °C minimum.

15. To avoid any troubles with next steps of the cloning process, 10 % of 3 M ammonium acetate, pH 5.2, is added to the recovered DNA and precipitated with 3 volumes of pure ethanol. DNA is then resuspended in 30 μL of water.
16. Very efficient chemically competent *E. coli* cells can be prepared if all the material is maintained ice-cold throughout the whole preparation process.
17. The three primers should have close melting temperature and the melting temperature must be above 42 °C.
18. Be delicate not to scratch the membrane.
19. Be careful that the side of the membrane on which the RNA is cross-linked is facing the inside of the hybridization bottle. There should not be any membrane overlap in the hybridization bottle.

Acknowledgements

This work was supported by the Centre National de la Recherche Scientifique and the University of Strasbourg through a PhD grant to M.M. We are grateful to Laure Schaeffer for technical assistance.

References

1. Fedor MJ, Uhlenbeck OC (1990) Substrate sequence effects on "hammerhead" RNA catalytic efficiency. Proc Natl Acad Sci USA 87:1668–1672
2. Khvorova A, Lescoute A, Westhof E et al (2003) Sequence elements outside the hammerhead ribozyme catalytic core enable intracellular activity. Nat Struct Biol 10:708–712
3. Martick M, Scott WG (2006) Tertiary contacts distant from the active site prime a ribozyme for catalysis. Cell 126:309–320
4. Pley HW, Flaherty KM, McKay DB (1994) Three-dimensional structure of a hammerhead ribozyme. Nature 372:68–74
5. Scott WG, Finch JT, Klug A (1995) The crystal structure of an all-RNA hammerhead ribozyme: a proposed mechanism for RNA catalytic cleavage. Cell 81:991–1002
6. Rupert PB, Ferre-D'Amare AR (2001) Crystal structure of a hairpin ribozyme-inhibitor complex with implications for catalysis. Nature 410:780–786
7. Lacroix-Labonte J, Girard N, Lemieux S et al (2012) Helix-length compensation studies reveal the adaptability of the VS ribozyme architecture. Nucleic Acids Res 40: 2284–2293
8. Masquida B, Beckert B, Jossinet F (2010) Exploring RNA structure by integrative molecular modelling. N Biotechnol 27:170–183
9. Helm M, Brule H, Giege R et al (1999) More mistakes by T7 RNA polymerase at the 5′ ends of *in vitro*-transcribed RNAs. RNA 5:618–621
10. Milligan JF, Uhlenbeck OC (1989) Synthesis of small RNAs using T7 RNA polymerase. Methods Enzymol 180:51–62
11. Solomatin S, Herschlag D (2009) Methods of site-specific labeling of RNA with fluorescent dyes. Methods Enzymol 469:47–68
12. Beckert B, Masquida B (2011) Synthesis of RNA by *in vitro* transcription. Methods Mol Biol 703:29–41
13. Price SR, Ito N, Oubridge C et al (1995) Crystallization of RNA-protein complexes I. Methods for the large-scale preparation of RNA suitable for crystallographic studies. J Mol Biol 249:398–408
14. Fechter P, Rudinger J, Giege R et al (1998) Ribozyme processed tRNA transcripts with

unfriendly internal promoter for T7 RNA polymerase: production and activity. FEBS Lett 436:99–103
15. Mörl M, Lizano E, Willkomm DK et al (2005) Production of RNAs with homogenous 5′ and 3′ ends. In: Hartmann RK, Bindereif A, Westhof E (eds) Handbook of RNA biochemistry, Wiley VCH Verlag Gmbh & Co, Weinheim, Germany, vol 1. pp 22–35
16. Ferré-D'Amaré AR, Doudna JA (1996) Use of *cis*- and *trans*-ribozymes to remove 5′ and 3′ heterogeneities from milligrams of *in vitro* transcribed RNA. Nucleic Acids Res 24:977–978
17. Nielsen H, Westhof E, Johansen S (2005) An mRNA is capped by a 2′, 5′ lariat catalyzed by a group I-like ribozyme. Science 309: 1584–1587
18. Hofacker IL, Fontana W, Stadler PF et al (1994) Fast folding and comparison of RNA secondary structures. Monatshefte für Chemie/Chemical Monthly 125:167–188
19. Canny MD, Jucker FM, Kellogg E et al (2004) Fast cleavage kinetics of a natural hammerhead ribozyme. J Am Chem Soc 126: 10848–10849
20. Ichetovkin IE, Abramochkin G, Shrader TE (1997) Substrate recognition by the leucyl/phenylalanyl-tRNA-protein transferase. Conservation within the enzyme family and localization to the trypsin-resistant domain. J Biol Chem 272:33009–33014
21. Milligan JF, Groebe DR, Witherell GW et al (1987) Oligoribonucleotide synthesis using T7 RNA polymerase and synthetic DNA templates. Nucleic Acids Res 15:8783–8798
22. Jossinet F, Ludwig TE, Westhof E (2010) Assemble: an interactive graphical tool to analyze and build RNA architectures at the 2D and 3D levels. Bioinformatics 26:2057–2059

Chapter 3

RNA Structural Analysis by Enzymatic Digestion

Elisa Biondi and Donald H. Burke

Abstract

Enzymatic probing is a rapid, straightforward method for determining which regions of a folded RNA are structurally constrained. It can be carried out using very small amounts of material, and is especially suitable for short RNAs. Here we report a protocol that we have found to be useful and readily adaptable to the evaluation of RNAs up to 150–200 nucleotides in length. Considerations for optimization are also included. In brief, the method includes folding end-labeled RNA into its native conformation, partial digestion with structure-sensitive nucleases, and identification of the cleavage sites by electrophoretic separation of the cleavage fragments.

Key words RNA, Secondary structure, RNase, Enzymatic probing

1 Introduction

The biological functions of RNA are largely determined by their secondary and higher-order structures. In addition, several human diseases are thought to be caused by RNA misfolding that arises from single nucleotide polymorphisms [1]. Thus there is extensive interest in RNA structure. While atomic resolution structure from NMR and crystallographic analysis would be desirable for a full understanding of these RNAs, secondary structure maps are much more readily obtained and can be highly informative. Many experimental tools are now available for determining RNA secondary structure. These methods are especially powerful when multiple strategies are used in combination and when the experimental output is paired with informatics approaches such as comparative sequence analysis and computational structure prediction (for example, *see* refs. [2, 3]).

Most experimental methods for RNA structure determination are based on covalent adduction or cleavage of the RNA chain by reagents whose reactivity is governed by local RNA structure. For example, in-line probing (ILP) [4] samples the frequency with

Christina Waldsich (ed.), *RNA Folding: Methods and Protocols*, Methods in Molecular Biology, vol. 1086,
DOI 10.1007/978-1-62703-667-2_3, © Springer Science+Business Media, LLC 2014

which the 2′ O, P and 5′ O are aligned to approximately 180° for nucleophilic attack; while this is not strictly equivalent to monitoring single-strandedness, the in-line geometry is disfavored by A form helices and by several other stable RNA conformations, and cleavage sites are often equated with RNA flexibility. Metal ions such as Mg^{2+}, Tb^{3+}, and Pb^{2+} enhance reactivity of the 2′ oxygen by acidifying metal-bound water molecules and increasing deprotonation of the ribose hydroxyl, greatly accelerating ILP-like probing methods [5]. Similarly, 2′ alkylation methods such as SHAPE (selective 2′ hydroxyl alkylation analyzed by primer extension, [6]) monitor nucleophilicity of the 2′ O, which is enhanced in flexible regions because of increased dynamic sampling of conformations in which the 2′ OH approaches the neighboring phosphate.

Once the structure-specific damage has been inflicted on the RNA, several approaches can be applied to determine the locations at which the damage occurred, and thereby arrive at structural inferences. For 5′ or 3′ terminally radio-labeled RNA, cleavage sites are readily identified by direct readout of fragment sizes from denaturing gel electrophoresis; this approach is especially useful for relatively small transcripts, as described herein. Sites of cleavage and sites of covalent adduction can both be determined by primer extension methods through either run off cDNA synthesis (for cleavage) or by reverse transcription pausing (for bulky adduction). The modifications are then read either by measuring cDNA size or by high-throughput sequencing of the fragment population, in analogy with recently developed ribosomal profiling analyses methods [7].

We have made extensive use of enzymatic cleavage methods to analyze secondary structures of ribozymes [8] and aptamers [3]. These methods are rapid, straightforward and can be carried out using very small amounts of material. Importantly, this approach is especially suitable for short RNAs because it does not require appending flanking nucleotides for annealing cDNA primer or for eliminating noise near the RNA 5′ end, both of which are required for SHAPE analysis of small RNAs [9].

The enzymatic digestion method is based on the sequence and structural preferences of natural nucleases (nuclease specificities and reactivities are reviewed in [10]). In brief, the method is to allow the RNA to fold into its native conformation, digest briefly with nuclease, and then identify the cleavage sites by electrophoretic separation of the cleavage fragments. Nuclease T1 cleaves on the 3′ side of guanosine residues within unstructured regions of the RNA (Table 1). Because it is active over a wide range of temperatures and denaturant concentrations, T1 can be applied to both the native and unfolded RNA. Nucleases S1 and V1 have no known sequence preference, but S1 preferentially cleaves unstructured nucleic acid (RNA or DNA) while V1 preferentially cleaves double-stranded RNA (Table 1).

Table 1
Summary of cleavage products and specificities

Reagent	Cleavage products[a]	Concentrations/ temperatures/times[b]	Specificity and notes[c]
Alkaline ladder	2′,3′ cyclic phosphate, 5′ OH	pH 9.0/90 °C/ 5–15 min	Cleaves after every ribose
T1 nuclease ladder	2′,3′ cyclic phosphate, 5′ OH	1 U/55 °C/5–15 min	Cleaves after single-stranded G; active in EDTA
T1 nuclease	2′,3′ cyclic phosphate, 5′ OH	0.05–1.0 U/T_{nat}/ 1–5 min	Cleaves after single-stranded G; active in EDTA
S1 nuclease	3′ OH; 5′ phosphate	19–95 U/T_{nat}/ 5–15 min	Cleaves after single-stranded nucleotides; Inactivated by EDTA; Optimal at low pH and 1 mM Zn^{2+}
V1 nuclease	3′ OH; 5′ phosphate	0.0005–0.1 U/T_{nat}/ 5–15 min	Cleaves after double-stranded nucleotides and some other structures; inactivated by EDTA

[a] See text for details on how to avoid errors of data interpretation associated with differences in electrophoretic gel migration caused by differences in the cleavage products
[b] Additional details are given in **steps 7–9** in Subheading 3.4. "T_{nat}" refers to the temperature at which the RNA is folded into its native conformation
[c] Structural probing must be performed under native conditions in which the RNA is properly folded. Use caution in adjusting probing conditions to favor the preferences of the nuclease, as this raises the risk that the RNA will no longer be in its native conformation

A critical element for interpreting electrophoretic data of enzymatic digestions is the nature of the cleavage products. As with alkaline digestions, T1 nuclease leaves a 2′,3′ cyclic phosphate, and the products of the two reactions can be compared directly (Table 1). In contrast, both S1 and V1 reactions leave 3′ OH cleavage products. While these differences have no effect on readout by primer extension methods, S1 or V1 digestion products of 5′ radio-labeled RNA carry less negative charge than the corresponding T1 or alkaline digestion products and migrate more slowly on polyacrylamide gels, as though the S1 and V1 products were about one nucleotide longer than their actual size [11, 12].

Here we report a useful and easy protocol to probe RNAs up to 150–200 nucleotides in length, or within this distance of the 5′ or 3′ terminus of longer molecules. The method is described for use in combination with gel electrophoretic analysis, although it can be readily adapted to analysis by capillary electrophoresis or by high-throughput sequencing of cleavage fragment endpoints.

2 Materials

Prepare each solution and compound using clean, sterile, RNase-free labware and Milli-Q water with resistivity of 18 MΩ. We have not found it necessary to use DEPC-treated water in our RNA manipulations, as long as everything is cleanly maintained.

2.1 Dephosphorylation of Transcripts and 5′ Labeling

1. Calf intestinal alkaline phosphatase (CIP; 10 U/μL). Store at −20 °C.
2. 10× CIP buffer: 500 mM Tris–HCl, pH 7.9, 1 M NaCl, 100 mM $MgCl_2$, 10 mM dithiothreitol. Store at −20 °C.
3. Phenol–chloroform–isoamyl alcohol solution, (25:24:1), pH 5.2 ± 0.3, saturated with an aqueous layer. Store at 4 °C.
4. 5 M ammonium acetate, pH 5.5. Store at room temperature.
5. Glycogen (10–50 mg/mL). Store at −20 °C.
6. 98 % EtOH. Store at room temperature.
7. Cooling tabletop centrifuge for 1.5 mL tubes.
8. T4 polynucleotide kinase (PNK; 10 U/μL). Store at −20 °C.
9. 10× PNK buffer: 700 mM Tris–HCl, pH 7.6, 100 mM $MgCl_2$, 50 mM dithiothreitol. Store at −20 °C.
10. [γ-^{32}P]-ATP, 3.3 μM, 3,000 Ci/mmol. Store at −20 °C or at 4 °C if "easy-tide."
11. Heat block with heated lid, or programmable thermocycler.
12. −20 °C or −80 °C freezer.
13. SpeedVac.
14. Spectrophotometer.

2.2 PAGE Preparation and Gel-Extraction of Labeled Molecules

1. Gel-loading buffer: 95 % formamide, 30 mM EDTA, bromophenol blue and xylene cyanol in traces. Store at room temperature.
2. Colorless gel-loading buffer: 10 M urea, 15 mM EDTA. Store at room temperature.
3. 40 % acrylamide/bisacrylamide (19:1) solution. Store at 4 °C.
4. Urea. Store at room temperature.
5. 10 % (w/v) ammonium persulfate (APS). Store at 4 °C.
6. N,N,N′,N′-tetramethylethylene-1,2-diamine (TEMED). Store at 4 °C.
7. 10× Tris-borate-EDTA (TBE), pH 7.5–8.0: 0.89 M Tris, 0.89 M boric acid, 20 mM EDTA. Store at room temperature.
8. Vertical electrophoresis equipment (medium and sequencing sizes).

9. Saran wrap.
10. X-ray film and film developer, and/or phosphorimager and screen.
11. Fluorescent stickers if available.
12. Light box if available.
13. Clean razor blade and tweezers.
14. Filter spin columns (We use Pierce 0.8 mL centrifuge columns, with polyethylene filters and 0.03 mm pore size).
15. Elution buffer (EB): 0.5 M ammonium acetate, pH 5.5, 10 mM EDTA, pH 8.0. Store at room temperature.
16. Rotator. Store at room temperature.
17. 4 °C refrigerator.

2.3 Preparation of Ladders

1. Na_2CO_3 buffer: 0.5 M Na_2CO_3, pH 9.0 or above, 10 mM EDTA, pH 8.0. Store at room temperature.
2. 0.25 M sodium-citrate, pH 5.0. Store at room temperature.
3. RNase T1 (1 U/μL). Store at −20 °C.
4. Dry-ice plus ethanol bath.
5. Heat block with heated lid, or programmable thermocycler.

2.4 RNase Digestions

1. 5′ (or 3′) labeled RNA. Store at −20 °C.
2. Native folding buffer appropriate to the RNA that is being studied.
3. Programmable thermocycler with ramp rate of 0.1 °C/s.
4. RNase T1 (1 U/μL). Store at −20 °C.
5. RNase S1 (95 U/μL). Store at −20 °C.
6. RNase V1 (0.1 U/μL). Store at −20 °C.
7. Heat block with heated lid, or programmable thermocycler.
8. Dry-ice plus ethanol bath.

2.5 Gel Analysis

1. Vertical gel apparatus.
2. Silicon-based glass coater. Store at room temperature.
3. Gel dryer or used X-ray film.
4. Whatman 3MM paper.
5. Saran wrap.
6. Gel imaging system (e.g., FujiFilm FLA-9000).
7. Gel analysis software (normally coming with the gel imaging system, use for example FujiFilm ImageQuant).

3 Methods

3.1 Dephosphorylation of Transcripts and 5′ Labeling

When the molecule to be analyzed is the product of in vitro transcription, it is necessary to treat it with a phosphatase to remove the 5′ phosphates, in order to be able to label it with a new, radioactive phosphate (*see* **Note 1**). For chemically synthesized RNA, this pretreatment is not needed because these RNAs have homogeneous 5′ OH.

3.1.1 Dephosphorylation of Transcripts

1. Treat 100–200 pmol of RNA with 1 μL CIP (10 U/μL) in 40 μL reaction in 1× CIP buffer, 1 h at 37 °C in a heat block or dry bath with heated lid to prevent evaporation.
2. Stop the reaction by adding 260 μL H_2O and 300 μL phenol–chloroform–isoamyl alcohol mixture (*see* **Note 2**), vortex well, and pass the supernatant to a new 1.5 mL tube.
3. Add 33 μL 5 M ammonium acetate, pH 5.5 and mix well.
4. Add 1 μL glycogen (20 mg/mL), mix well.
5. Add 2.5 volumes (approximately 750 μL) 98 % EtOH and leave at −20 °C for 1 h or more, or at −80 °C for 20 min.
6. Spin 20 min at 17,000 × *g* (or maximum achievable speed) at 4 °C.
7. Remove the ethanol and dry the pellet in a SpeedVac for 30 min.
8. Resuspend pellet in 10–20 μL H_2O and read concentration at a spectrophotometer.

3.1.2 5′ Labeling of RNA

1. Treat 30–50 pmol of dephosphorylated RNA with 1 μL PNK (10 U/μL) in 20 μL reaction volume, 1× PNK buffer, and 8–10 μL [γ-^{32}P]-ATP at 3.33 or 1.67 μM, for 30 min at 37 °C.
2. Stop the reaction with three volumes of denaturing gel-loading buffer and load directly on preparative gel for extraction.

3.2 Gel-Extraction of Labeled Molecules

1. Prepare a medium-sized denaturing polyacrylamide gel (about 16 cm (w) × 22 cm (h)), 1–1.2 mm tick, with fairly wide wells (about 1.5–3 cm). The percentage of the gel will depend on the length of the molecule: 8–10 % will be fine for molecules of 60 nt and above, while 12–15 % can be used for shorter molecules. Note, however, that higher gel percentages reduce the efficiency of elution.
2. Load the 5′ labeled RNA into one or two wells (keep the gel-to-RNA ratio as low as possible, without overloading the wells).
3. Run the gel at 15–25 W until optimal separation is achieved, which will depend on RNA length and gel percentage.

4. Disassemble the gel, wrap it with Saran wrap, and expose it to X-ray film or to a phosphorimager screen. We routinely use fluorescent stickers as position labels.
5. Develop the film or screen, align the gel to the image of the stickers, and cut the RNA band(s).
6. Put each band in one spin column with 500 μL elution buffer (EB), crush the gel inside the column (for example, with clean metal tweezers), taking care not to damage the filter membrane.
7. Put the columns into a rotator and rotate at 4 °C overnight (*see* **Note 3**).
8. Spin down eluted RNA in a new 1.5 mL tube with a low-speed benchtop minispinner, such as a picofuge.
9. Wash the leftover gel in the column with another 100–200 μL EB, and add it to the previous aliquot.
10. Proceed to ethanol precipitation as in Subheading 3.1.1. There is no need for addition of extra salt, because ammonium acetate at 0.5 M was already present in the elution buffer.
11. Dry the pellet and resuspend in 20 μL H_2O. Labeled RNA is now ready and has a concentration between 0.5 and 2 μM (depending on the efficiency of the gel-extraction procedure, which varies with RNA type and working hands).

3.3 Preparation of Ladders

Depending on the molecule tested, some optimization in terms of incubation times and amount of enzyme to be used is needed, as optimal digestions vary with length, degree of structure and sequence composition. We found that the following is a good starting point for most RNAs between 30 and 140 nt in length.

3.3.1 Alkaline Digestion

This reaction will cleave the molecule after each nucleotide, producing a homogeneous ladder from nucleotide 1 and up.

1. Prepare three separate tubes, each with 1 μL 5′ ^{32}P-labeled RNA (between 50,000 and 150,000 cpm/μL), 1 μL Na_2CO_3 buffer, and 8 μL H_2O, yielding a total volume of 10 μL (*see* **Note 4**). Treat the buffer as if it was an enzyme by adding it immediately before starting the reaction.
2. Put the tubes at 90 °C, one for 5 min, one for 10 min, and the other for 15 min.
3. Stop the reactions with 10 μL colorless gel-loading buffer (*see* **Note 5**) mixing well, and by snap cooling in a bath of dry ice plus ethanol. Store the samples at −20 °C.
4. Load only 10 μL of this ladder on the gel.

3.3.2 Denaturing T1 Digestion

The T1 digestion in denaturing conditions will cut the molecule 3′ of each G residue (*see* **Note 6**), regardless of its structural state in the native condition. This, together with the alkaline digestion, will allow to precisely assign the length of any band obtained from the structural analysis (BOTH ladders are needed together, only one of them will not be enough).

1. Prepare three tubes, each with the following: 1 μL 5′ ^{32}P-labeled RNA (between 50,000 and 150,000 cpm/μL), 1 μL 0.25 M Na-citrate, pH 5.0, 1 μL RNase T1 (1 U/μL), 7 μL colorless gel-loading buffer, for a total volume of 10 μL.
2. Put the tubes at 55 °C, one for 5 min, one for 10 min, and the other for 15 min.
3. Stop the reactions with 10 μL colorless gel-loading buffer mixing well, and follow by snap cooling in a bath of dry ice plus ethanol. Store the samples at −20 °C.
4. Load only 10 μL of this ladder on the gel.

3.4 RNase Digestions

RNase digestions for structural analysis need to be performed on RNA molecules under conditions in which they are in their active form ("native" conditions). To achieve this conformation, in most cases the RNA needs to be unfolded and slowly refolded in optimal conditions. These conditions depend completely on the system to be analyzed, but in most cases they include 3–20 mM $MgCl_2$ or another divalent cation (*see* **Note 7**), and 20–100 mM buffering agent at pH 7.0–8.0 (such as HEPES or Tris–HCl). Once the RNA achieved its optimal conformation, enzymatic digestion can be performed.

Again, some optimizations in terms of incubation times and units of enzyme to use will be needed. A good general procedure is the following:

1. Prepare enough material for nine reactions: three time points and three dilutions for each of the three enzymes.
2. Heat the RNA (50,000–150,000 cpm per reaction) in H_2O for 5 min at 85–95 °C. At 8 μL final volume of RNA per reaction, this corresponds to 72 μL total.
3. Let it cool down to room temperature at 0.1 °C/s in a thermocycler.
4. Add 9 μL 10× folding buffer of choice when the temperature reaches 60–70 °C. (Buffer concentration is approximately 1.1× at this point, but will be diluted to 1.0× when enzyme is added).
5. Move to ice when the temperature reaches 20–25 °C.
6. Aliquot 9 μL in each of the nine tubes, for each enzyme (27 tubes total).
7. Equilibrate the RNA at the temperature of the RNA "native" conditions (often 25 °C or 37 °C) for 1–2 min.

8. Start the reactions by adding 1 μL of enzyme at different concentrations: for T1 RNase: 1 U, 0.1 U (1:10), 0.05 U (1:20). For S1 RNase: 95 U, 47.5 U (1:2), 19 U (1:5). For V1 RNase: 0.1 U, 1×10^{-3} U (1:100), 5×10^{-4} (1:200).
9. Stop the reactions with 10 μL of colorless gel-loading buffer and snap cooling on dry ice + ethanol at each concentration at three time points: for T1 RNase: 1 min, 2 min, 5 min. For S1 RNase: 5 min, 10 min, 15 min. For V1 RNase: 5 min, 10 min, 15 min.
10. Load 10 μL of each time point on the gel.

3.5 PAGE and Gel Analysis

1. Prepare 10–20 % acrylamide/bisacrylamide gel solution, 7 M urea, in 1× TBE according to the length of the molecule to be analyzed. For 70–150 nucleotides long molecules we suggest 10–12 % gels, while 15–20 % should be used for smaller molecules.
2. Spread some silicon-based glass-coating to the inner surface of the glass plates to avoid bubbles, but make sure to remove excess coating (*see* **Notes 8** and **9**).
3. Cast the gel plates. For RNA structural analysis we use 46 × 18 cm sequencing gel, 0.4 mm thick.
4. To 60 mL of gel solution, add 60 μL TEMED and 600 μL 10 % (w/v) APS.
5. Let it polymerize 1–2 h.
6. After removing the comb, wash the wells very well with a syringe or a pipette, taking care to remove polyacrylamide debris.
7. Pre-run the gel for 45–60 min at 30–60 W.
8. Wash the urea from the wells again before loading the samples. Because the analysis samples have no dyes, some dye-containing gel-loading buffer can be loaded in the side-most wells to better follow the run.
9. Run at 30–60 W until the bromophenol blue (bfb) is at a suitable distance from the bottom of the gel. Depending on the concentration of the gel, to obtain single nucleotide resolution the bfb needs to be between 10 cm (lower % gels) and 20 cm (higher %) from the bottom.
10. When the run is completed, disassemble the plates. The gel can be either dried in a gel dryer, or directly exposed to a phosphorimager screen keeping the cassette at −20 °C or −80 °C to avoid band diffusion.
11. Using a gel dryer: pass the gel from the glass plate to a piece of saran wrap, then cover the other side of the gel with Whatman 3MM paper, making sure to avoid bubbles. Place the sandwich on a gel dryer with the Saran wrap facing up,

apply vacuum, and dry for 30 min at 80 °C. The gel is then ready to be exposed to a phosphorimager screen.

12. In case of a direct exposure of the gel, apply a used (developed) X-ray film to the gel when it is still on one of the glass plates. Start from the bottom and keep the film curved backwards to avoid creating bubbles between the film and the gel. After positioning all the film on the gel, level any residual bubbles with a hard flat surface (like the side of a tube rack). At this point the film and gel can be peeled back off the glass plate together; the gel will stick to the film and will come off the glass very easily (*see* **Note 10**). Expose the gel to the phosphor screen and keep the cassette at −20 °C or −80 °C.
13. Develop the exposed plate in a phosphorimager.
14. Analyze results with the appropriate image software provided with the phosphorimager (*see* **Note 11**).
15. The susceptibility data obtained from this image can now be compared to the starting 2D model(s) for the molecule.

4 Notes

1. In this chapter we report the preparation and probing for 5′ labeled RNA, but essentially the same procedure can be followed after labeling the RNA on the 3′-end. This can be achieved in several ways: for example by using a DNA polymerase [13] or terminal transferase and a labeled [α-^{32}P]-dNTP, or [α-^{32}P]-pCp (cytidine-3′,5′-bis-phosphate) and RNA ligase. For a more comprehensive analysis, and especially when the molecules are long enough to preclude optimal gel resolution of bands on both ends of the RNA, it is suggested that the probing is performed both on 5′ and 3′ labeled RNAs, as we did for our 126 nt kinase ribozymes [8].
2. Note that this protocol uses calf intestinal phosphatase (CIP) to remove phosphates from the 5′ termini of RNA. This enzyme is extremely robust and can only be deactivated by organic extraction. We find it cleaner to use this method, but to speed up the process, one could use Antarctic alkaline phosphatase or shrimp alkaline phosphatase instead of CIP and then deactivate the enzyme after the reaction with a simple treatment of 10 min at 70 °C.
3. To speed up the gel-extraction process, it is possible to subject the crushed gel submerged in the elution buffer to a few cycles of freezing and thawing, followed by separation without the overnight elution step. Notice though, that it is better to perform this procedure in a 1.5 mL tube rather than straight into the spin column, because its filter could be damaged by the

freezing, causing some polyacrylamide to end up in the final RNA. Traces of polyacrylamide in the sample make small RNAs much more resistant to enzymatic digestion and can enhance non-enzymatic degradation in a totally unspecific and irreproducible fashion.

4. Notice that, contrary to the majority of methods in literature, this protocol does not make use of any kind of carrier RNA. The enzyme units, times, and temperatures are optimized for the exact pmoles and nucleotide length of the RNA under examination. This method has the advantage of using very small amounts of components, thus avoiding the need for organic extraction and/or ethanol precipitation of samples before loading aliquots on gel.
5. Because of the high concentration of urea in the colorless gel-loading buffer, it is important to check for urea crystallization prior to each use. In case of crystal formation, briefly warm up the solution (for example, by microwaving the tube for 5–8 s) to allow the urea to redissolve.
6. For certain molecules the optimization of denaturing T1 digestion for the production of a G-ladder can also be performed by adjusting enzyme concentration. A good starting point for this optimization is to test 1:10 and 1:100 dilutions of the enzyme at the same time range reported for the non-diluted enzyme. Dilutions can be made in water right before each use and cannot be stored afterwards.
7. Especially for longer RNAs, it is not recommended to add a buffer containing divalent cations during the denaturation process, because the RNA might get degraded when exposed for prolonged times to the cations at high temperatures, in particular Mg^{2+}. The buffer can be added during refolding, when the temperature reaches 60–70 °C or lower.
8. When coating glass plates with silicon-based substances, it is important to make sure that the excess is removed, or the gel will slide out of the plates from the top, and end up floating in the buffer on the upper chamber. To remove excess silicon, we make a final wipe of the plates with 100 % EtOH or MeOH.
9. Siliconized plates can be used two or three times without the need of reapplying the coating, as long as the plates are not washed with strong detergents. Notice that freshly siliconized plates tend to adsorb radioactivity and will thus need to be decontaminated, but this is easily achievable by wiping the surface with a Kimwipe and 70 % (v/v) EtOH.
10. When choosing the drying or X-ray film method for exposing a gel to the phosphorimager screen, it needs to be taken into account that dried gels can be stored for a long time in a radiation-safe drawer or container, while film-sticking gels are

faster to set up, but do not allow for long-term storage of the gel, because the bands will diffuse.

11. Depending on the format of the image, one advanced software is available specifically for RNA footprinting analysis (SAFA, Semi-Automated Footprinting Analysis, https://simtk.org/home/safa, [14]), but data can be retrieved with any image analyzer, as long as the peak profiling for each band is correctly aligned with the others, and corrected for background.

Acknowledgements

This work was supported by a grant from the National Science Foundation Chemistry of Life Processes program [grant number CHE-1057506].

References

1. Halvorsen M, Martin JS, Broadaway S et al (2010) Disease-associated mutations that alter the RNA structural ensemble. PLoS Genet 6:e1001074
2. Ditzler MA, Lange MJ, Bose D et al (2013) High-throughput sequence analysis reveals structural diversity and improved potency among RNA inhibitors of HIV reverse transcriptase. Nucleic Acids Res 41:1873–1884
3. Whatley AS, Ditzler MA, Lange MJ et al (2013) Potent inhibition of HIV-1 reverse transcriptase and replication by nonpseudoknot, "UCAA-motif" RNA aptamers. Mol Ther Nucleic Acids 2:e71
4. Soukup GA, Breaker RR (1999) Relationship between internucleotide linkage geometry and the stability of RNA. RNA 5:1308–1325
5. Walter NG, Yang N, Burke JM (2000) Probing non-selective cation binding in the hairpin ribozyme with Tb(III). J Mol Biol 298:539–555
6. Wilkinson KA, Merino EJ, Weeks KM (2006) Selective 2′-hydroxyl acylation analyzed by primer extension (SHAPE): quantitative RNA structure analysis at single nucleotide resolution. Nat Protoc 1:1610–1616
7. Ingolia NT, Brar GA, Rouskin S et al (2012) The ribosome profiling strategy for monitoring translation *in vivo* by deep sequencing of ribosome-protected mRNA fragments. Nat Protoc 7:1534–1550
8. Biondi E, Nickens DG, Warren S et al (2010) Convergent donor and acceptor substrate utilization among kinase ribozymes. Nucleic Acids Res 38:6785–6795
9. Merino EJ, Wilkinson KA, Coughlan JL et al (2005) RNA structure analysis at single nucleotide resolution by selective 2′-hydroxyl acylation and primer extension (SHAPE). J Am Chem Soc 127:4223–4231
10. Knapp G (1989) Enzymatic approaches to probing of RNA secondary and tertiary structure. Methods Enzymol 180:192–212
11. Biondi E, Maxwell AW, Burke DH (2012) A small ribozyme with dual-site kinase activity. Nucleic Acids Res 40:7528–7540
12. Brown TS, Bevilacqua PC (2005) Method for assigning double-stranded RNA structures. Biotechniques 38:368–372
13. Huang Z, Szostak JW (1996) A simple method for 3′-labeling of RNA. Nucleic Acids Res 24:4360–4361
14. Laederach A, Das R, Vicens Q et al (2008) Semiautomated and rapid quantification of nucleic acid footprinting and structure mapping experiments. Nat Protoc 3:1395–1401

Chapter 4

The Mutate-and-Map Protocol for Inferring Base Pairs in Structured RNA

Pablo Cordero, Wipapat Kladwang, Christopher C. VanLang, and Rhiju Das

Abstract

Chemical mapping is a widespread technique for structural analysis of nucleic acids in which a molecule's reactivity to different probes is quantified at single nucleotide resolution and used to constrain structural modeling. This experimental framework has been extensively revisited in the past decade with new strategies for high-throughput readouts, chemical modification, and rapid data analysis. Recently, we have coupled the technique to high-throughput mutagenesis. Point mutations of a base paired nucleotide can lead to exposure of not only that nucleotide but also its interaction partner. Systematically carrying out the mutation and mapping for the entire system gives an experimental approximation of the molecule's "contact map." Here, we give our in-house protocol for this "mutate-and-map" (M^2) strategy, based on 96-well capillary electrophoresis, and we provide practical tips on interpreting the data to infer nucleic acid structure.

Key words RNA structure, Chemical mapping, Capillary sequencing, Systematic mutation

1 Introduction

Elucidating the structures of nucleic acids is essential to understanding the mechanisms that govern their many biological roles, ranging from packing and transmission of genetic information to regulation of gene expression (e.g., refs. [1–3]). Many biophysical, biochemical, and phylogenetic methods have been developed to probe nucleic acid folds [4–6], but rapid determination of a functional RNA's secondary and tertiary structures remains generally challenging. Amongst available methods, chemical and enzymatic mapping (or "footprinting") approaches are particularly facile. First, the molecule is covalently modified by the mapping reagent at nucleotides that are exposed or unstructured. The modifications can then be detected in various ways, including reverse transcription: here,

Pablo Cordero and Wipapat Kladwang contributed equally to this work.

Christina Waldsich (ed.), *RNA Folding: Methods and Protocols*, Methods in Molecular Biology, vol. 1086,
DOI 10.1007/978-1-62703-667-2_4, © Springer Science+Business Media, LLC 2014

the adducts formed by modification will stop the reverse transcriptase, generating complementary DNA (cDNA) of specific lengths that highlight the molecule's reactive sites and that can be read out by gel or capillary electrophoresis. Chemical mapping can provide a range of useful information: DNA and RNA can both be interrogated along their Watson–Crick base edges through alkylation or formation of other adducts [7–9]. Additionally, RNA can be probed through 2′ hydroxyl acylation or metal-ion-catalyzed 2′ hydroxyl in-line cleavage at phosphates to reveal "dynamic" nucleotides [10–12], or to determine accessibility of the phosphate [13] or more complex features such as participation in helices [14–17]. DNA/RNA duplexes or DNA structures can be similarly analyzed [18], although this chapter focuses on RNA cases where the method has been most thoroughly tested.

Chemical mapping is applicable to essentially any RNA, including large nucleic acid systems in vivo or in vitro [19–21]. While a useful source of information, the one-dimensional nature of the resulting data (one reactivity value per nucleotide) limits the conclusions that can be deduced from these experiments, as several structural hypotheses can be consistent with the detected reactivity data [9, 22, 23]. To help address these ambiguities, we developed a "mutate-and-map" (M^2) strategy whereby each nucleotide of the RNA is systematically mutated and its interaction partners are discovered by their "release" and increase in chemical accessibility [18, 24, 25]. Chemical mapping of single point mutants reveal the molecule's approximate "contact map" and can be efficiently performed using current molecular biology tools. Automated analysis protocols can then quantitate these data and inform structural models. The data appear particularly constraining at the level of secondary structure. Furthermore, in several cases, unambiguous information related to tertiary contacts or pseudoknots formed by the RNA are obtained. In some mutants, the single mutations produce substantial perturbations to the structure, but these are straightforward to detect as perturbations in the chemical reactivity profiles, and, indeed, reflect the propensity of the wild-type RNA to switch its structure, a property of functional interest. Overall, the M^2 method provides rich data sets with information content that generally exceeds prior chemical or enzymatic approaches to RNA structure.

The M^2 experimental protocol begins by assembling DNA templates corresponding to the RNA mutants of interest, using PCR assembly [24]. The DNA is then purified and in vitro transcribed. The resulting RNA is folded or, for DNA/RNA or other multicomponent systems, assembled into the desired complex, and then subjected to standard chemical mapping protocols using fluorescently labeled primers and capillary electrophoresis. Each purification step is carried out with magnetic beads and multichannel pipettors in 96-well format, enabling the probing of hundreds of mutants in parallel. The requirement for synthesis via PCR assembly limits the length range of the current procedure to approximately 300 nucleotides. However, after a correct PCR assembly is found, the complete protocol can be performed on the timescale of 2 days: each major step in the procedure (PCR assembly and DNA purification, transcription and RNA purification, chemical mapping and reverse transcription, and capillary electrophoresis) takes on the order of 3–5 hours. Quantitation of the data and its use in secondary structure inference is similarly rapid. More sophisticated analyses of the data as well as accelerations from next-generation synthesis and sequencing methods have been piloted in our laboratory, but the focus of this chapter will be on our stable in-house protocol, which has been applied to dozens of RNA sequences at the time of writing.

2 Materials

2.1 General Materials and Equipment

All of the procedures described below are performed in a 96-well plate format. Each sample plate contains 96 mutants; so if the sequence to be probed has 119 nucleotides, then the wild-type construct plus all single nucleotide mutants will occupy 120 wells, or one and a quarter sample plates. Multichannel pipettors are used to reduce the pipetting effort required in each step.

1. At least 15 boxes of extra-long P10 tips per sample plate; three to four boxes of P100 refill tips per plate; at least one box each of P20 filter tips, and P200 filter tips for premixes.
2. Several 96-well V-bottom plates, at least four per sample plate. In many steps, an "auxiliary" plate will be used to pre-aliquot premixes and reduce pipetting effort.
3. Plastic plate seal films, at least three per sample plate used.
4. 100 mL 70 % ethanol per sample plate (*see* **Note 1**).
5. 96-post magnetic stands (VP Scientific), one per sample plate.
6. 0.5 mL dNTP mixture (10 mM each dATP, dCTP, dTTP, dGTP, prepared from 100 mM stocks of each dNTP—stored at −20 °C) per sample plate.
7. 2 mL of Agencourt AMPure XP beads (Beckman Coulter—stored at 4 °C) for nucleic acid purification.
8. Two 96-well format dry incubators (Hybex) set to 42 °C and 90 °C before starting the chemical mapping part of the experiment. If these types of incubators are not available, a water bath can be used.
9. 50 mL RNAse-free sterile water. We use Barnstead MilliQ purified water.

10. A spectrophotometer-based system to probe UV light absorbance (e.g., a NanoDrop system).
11. Access to an ABI capillary electrophoresis sequencer. We use an ABI3100 sequencer (16 capillary) for local tests and send samples to an external company (Elim Biosciences) for higher throughput runs on 96-capillary ABI 3730 sequencers.

2.2 Sequence Assembly and Fluorescent Primers

The M^2 method requires a fluorescent primer that will be used to reverse transcribe the RNA to detect modification sites and a working set of DNA primers from which to assemble the desired sequence as well as primer variants per mutant.

1. Fluorescent primer: The primer should include a fluorophore, such as FAM (6-fluorescein amidite), attached to the 5′ end that can be read by the capillary sequencer. We include a poly(A) stretch in the 5′ end of 20 nucleotides, which enables rapid and quantitative purification by poly(dT) beads. This stretch is followed by the reverse complement of the primer binding region of the RNA (*see* **Note 2** and Table 1 for the sequence of the fluorescent primer we use). This primer can be ordered from any oligo-synthesizing company.
2. Sequence to assemble: This sequence is generally composed of a T7 promoter in the 5′ end, followed by a buffer region starting with two guanines to allow for transcription, the sequence of the RNA of interest, a short 3′ buffer region, and the primer binding region (Fig. 1, Table 1 and *see* **Note 2** for the primer binding region sequence we use). Check that adding these sequences does not potentially alter the structure of the RNA of interest by inspecting the minimum energy secondary structures of the RNA with and without the additional sequences as predicted by a secondary structure prediction program, such as RNAstructure [26]. Lack of perturbation of the sequence of interest can also be tested empirically by using an alternate set of primers and buffers and primer binding sites. For details on how to calculate an assembly for a given sequence, *see* **Note 3**.
3. The plate layouts generated in the step above are filled only in wells where the mutant primers differ from the wild-type primers (*see*, e.g., green wells in Figs. 1 and 2). All other wells should be filled with the appropriate wild-type primer (for example, the white wells in Plate 1-Primer 3 and Plate 2-Primer 3 of Fig. 2c should be filled with the wild-type primer 3). *See* below for further details on PCR assembly of the primers.

2.3 Mutant DNA Assembly Components

1. 1,200 μL of 5× HF (High Fidelity) buffer and 120 μL of Phusion polymerase (2,000 U/mL) per sample plate (stored at −20 °C).

Table 1
DNA oligos required to assemble the MedLoop motif (as in Fig. 1) and P4–P6 domain of the Tetrahymena ribozyme (as in Fig. 2) as well as three single nucleotide mutants of each constructs. The fluorescent primer used in the protocol for reverse transcription is shown in the last row. Sequence color codes are the same as in Figs. 1 and 3; mutations are colored in orange. All sequences are shown from 5′ to 3′. Abbreviations for modification are those used by Integrated DNA Technologies (IDT): /5–6FAM/, 5′-6-fluorescein. For the P4–P6 domain, to avoid confusion, mutant positions are given with no offset (i.e., this numbering does not correspond with the nucleotide numbering of the full group I intron of the Tetrahymena ribozyme)

MedLoop WT	TTCTAATACGACTCACTATAGGAACGACGAACCGAAAACCGAAGAAATGAAGAAAGGTTTTCGGTACCGACCTG GTTGTTGTTGTTGTTTCTTTGGTTTTCAGGTCGGTACCGAAAACCTTTCTTCAT
MedLoop A01U	TTCTAATACGACTCACTATAGGAACGACGATCCGAAAACCGAAGAAATGAAGAAAGGTTTTCGGTACCGACCTG GTTGTTGTTGTTGTTTCTTTGGTTTTCAGGTCGGTACCGAAAACCTTTCTTCAT
MedLoop A23U	TTCTAATACGACTCACTATAGGAACGACGAACCGAAAACCGAAGAAATGAAGTAAGGTTTTCGGTACCGACCTG GTTGTTGTTGTTGTTTCTTTGGTTTTCAGGTCGGTACCGAAAACCTTACTTCAT
MedLoop U31A	TTCTAATACGACTCACTATAGGAACGACGAACCGAAAACCGAAGAAATGAAGAAAGGTTTACGGTACCGACCTG GTTGTTGTTGTTGTTTCTTTGGTTTTCAGGTCGGTACCGTAAACCTTTCTTCAT
P4–P6 domain WT	TTCTAATACGACTCACTATAGGCCAAAACAACGGAATTGCGGGAAAGGGGTCAACAGCCG GGCCATCTCAAAGTTTCCCCTGAGACTTGGTACTGAACGGCTGTTGACCCCTTTCCCG GGGAAACTTTGAGATGGCCTTGCAAAGGGTATGGTAATAAGCTGACGGAC GTTGACTTAGGACTTGGCTGCGTGGTTAGGACCATGTCCGTCAGCTTATTACCATAC CAGCCAAGTCCTAAGTCAACAGATCTTCTGTTGATATGGATGCAGTT GTTGTTGTTGTTGTTTCTTTGGTTTGGTTTTGAACTGCATCCATATCAACAGAAG
P4–P6 domain C32G	TTCTAATACGACTCACTATAGGCCAAAACAACGGAATTGCGGGAAAGGGGTGAACAGCCG GGCCATCTCAAAGTTTCCCCTGAGACTTGGTACTGAACGGCTGTTCACCCCTTTCCCG GGGAAACTTTGAGATGGCCTTGCAAAGGGTATGGTAATAAGCTGACGGAC GTTGACTTAGGACTTGGCTGCGTGGTTAGGACCATGTCCGTCAGCTTATTACCATAC CAGCCAAGTCCTAAGTCAACAGATCTTCTGTTGATATGGATGCAGTT GTTGTTGTTGTTGTTTCTTTGGTTTGGTTTTGAACTGCATCCATATCAACAGAAG

(continued)

Table 1
(continued)

P4–P6 domain U64A	TTCTAATACGACTCACTATAGGCCAAAACAACGGAATTGCGGGAAAGGGGTCAACAGCCG GGCCATCTCAAAGATTCCCCTGAGACTTGCTACTGAACGGCTGTTGACCCCTTTCCCG GGGAATCTTTGAGATGGCCTTGCAAAGGGTATGGTAATAAGCTGACGGAC GTTGACTTAGGACTTGGCTGCGTGGTTAGGACCATGTCCGTCAGCTTATTACCATAC CAGCCAAGTCCTAAGTCAACAGATCTTCTGTTGATATGGATGCAGTT GTTGTTGTTGTTGTTTCTTTGGTTTGGTTTTGAACTGCATCCATATCAACAGAAG
P4–P6 domain U170A	TTCTAATACGACTCACTATAGGCCAAAACAACGGAATTGCGGGAAAGGGGTCAACAGCCG GGCCATCTCAAAGTTTCCCCTGAGACTTGGTACTGAACGGCTGTTGACCCCTTTCCCG GGGAAACTTTGAGATGGCCTTGCAAAGGGTATGGTAATAAGCTGACGGAC GTTGACTTAGGACTTGGCTGCGTGGTTAGGACCATGTCCGTCAGCTTATTACCATAC CAGCCAAGTCCTAAGTCAACAGATCTTCTGTTGATATGGATGCAGTA GTTGTTGTTGTTGTTTCTTTGGTTTGGTTTTGTACTGCATCCATATCAACAGAAG
Fluorescent primer for reverse transcription	/5-6FAM/AAAAAAAAAAAAAAAAAAAAGTTGTTGTTGTTGTTTCTTT

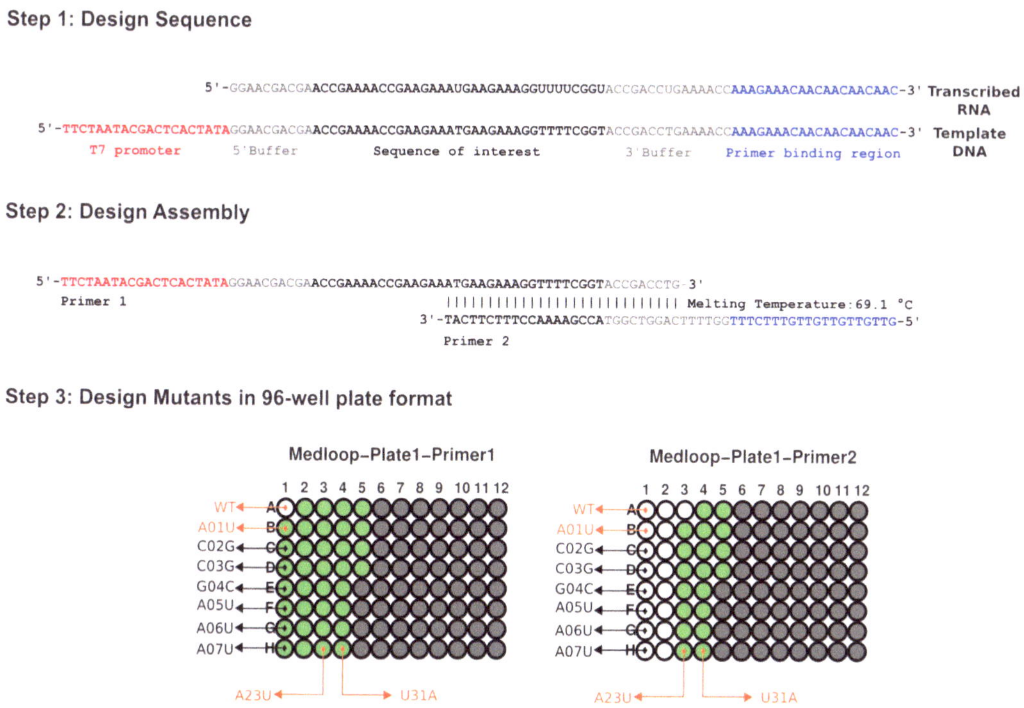

Fig. 1 Steps for designing the sequences necessary for building and probing RNA with the M² strategy. Here, the steps to design the assembly of the MedLoop RNA are shown. Step 1: Designing the sequence includes adding a T7 promoter, a 5′ buffer (normally beginning with two guanines to allow transcription of the RNA), the sequence of interest, a short 3′ buffer, and a sequence where the fluorescent primer will bind. Step 2: Primers required for the PCR assembly of the complete sequence can be obtained from software packages such as NA_thermo (*see* **Note 3**). Step 3: Example DNA primer plate layout for a library of single nucleotide mutants and wild-type of a two primer assembly of the MedLoop RNA. *Green wells* mark where each primer varies due to each single nucleotide mutation while *white wells* use the respective wild-type primers; *gray wells* are not used for any mutant and are left empty. Combining all primer plates per well into one plate using the appropriate concentrations and performing PCR will assemble all mutants simultaneously. *Arrows* to the first 8 wells (A1 to H1) indicate the mutant contained in that well (e.g., A01U, the mutant whose first base, an adenine, is mutated to a uracil is contained in well B1). Oligos used to assemble mutants marked in *orange* are given in Table 1

2. Standard reagents needed for agarose gel electrophoresis. We use Tris-borate-EDTA buffer (TBE) with 0.5 μg/mL ethidium bromide (stored at room temperature), and 96-well gel casting systems.

2.4 RNA Transcription

1. 10× transcription buffer: 400 mM Tris–HCl, pH 8.1, 250 mM $MgCl_2$, 35 mM spermidine, 0.1 % Triton X-100. Filter the buffer using a sterile 60 mL syringe through a 0.2 μm filter. 10 mL of 10× transcription buffer can be used for at least thirty 96-well sample plates, and can be stored frozen at −20 °C for at least 6 months.

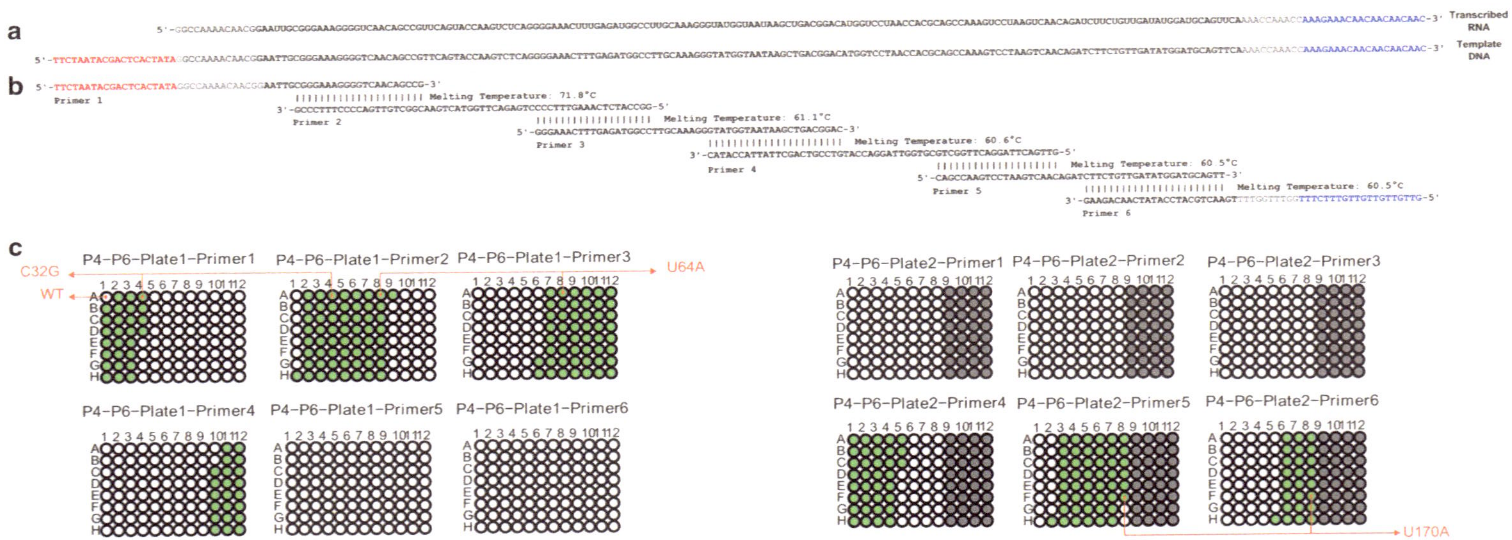

Fig. 2 Mutant libraries for larger RNAs, such as the P4–P6 domain of the *Tetrahymena* group I intron ribozyme shown here, require several plates. In this case, the RNAs of interest are assembled through six primers. Synthesizing all the mutants requires almost two full 96-well plates. (**a**) The sequence, with T7 promoter, buffer regions, and primer binding region included. (**b**) The six-primer assembly of the RNA. (**c**) Plate layouts for assembling the mutant library are shown with the same color coding as in Fig. 1. Oligos used to assemble mutants marked in *orange* are given in Table 1

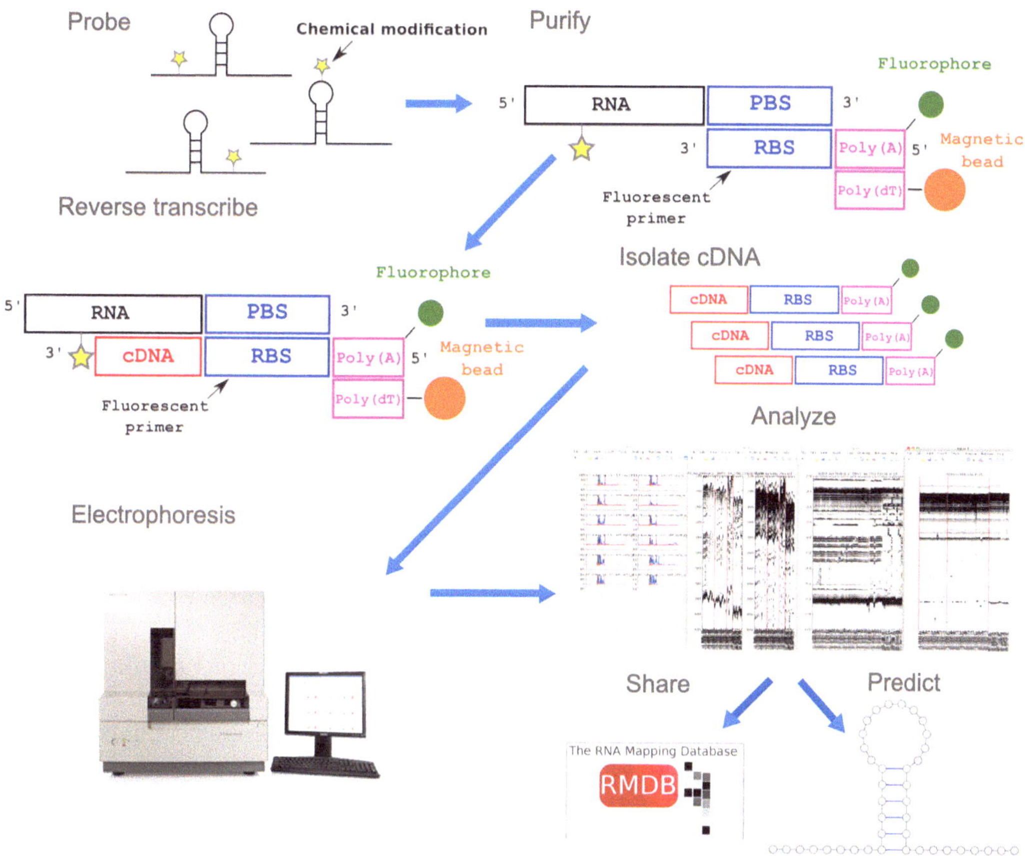

Fig. 3 Workflow of the chemical modification for capillary electrophoresis using bead purification. The RNA is chemically modified and pulled down. Oligo dT beads that bind the poly(A) tail of a fluorescent primer that contains an RNA binding site (RBS) complementary to the primer binding site (PBS) of the RNA. The RNA is then reverse transcribed. The reverse transcriptase stops at modified sites and produces a modification pattern that can be mapped to the sequence of interest using capillary electrophoresis. The readout can be simultaneously obtained for all mutants thanks to the 96-well format of the protocol. The resulting data are then analyzed using freely available software packages. Finally, the RNA Mapping Database can be used to share and visualize the data as well as to predict experimentally informed computational secondary structure models

2. 120 μL of 1 M DTT (stored at −20 °C) per sample plate.
3. 300 μL of 10 mM NTPs (stored at −20 °C) per sample plate.
4. 300 μL of 40 % PEG 8000 (stored at −20 °C) per sample plate.
5. 30 μL of T7 RNA polymerase (50 U/μL; stored at −20 °C) per sample plate.

2.5 RNA Folding

1. 240 μL of 0.5 M Na-HEPES, pH 8.0 (stored at room temperature), per sample plate. Stock should be prefiltered with a 0.2 μm filter. Other folding buffers can be used as well (*see* **Note 4**).

2. 240 μL of 100 mM $MgCl_2$ (stored at room temperature) per sample plate, if needed for proper folding of the nucleic acid system.
3. The reagents above may be different depending upon the conditions desired for proper folding/assembly of the RNA or DNA/RNA hybrid. For example, you can use a different buffer, different ions, and/or small molecules that bind the RNA.

2.6 Chemical Mapping Solutions

Different chemical modifiers can be used to interrogate different aspects of a nucleic acid of interest. Each one affects different parts of the nucleic acid in different ways. The solutions needed to quench each reaction vary from modifier to modifier and are added after chemical modification has occurred. The following chemical modifiers are the ones most commonly used for nucleic acid structure probing; volumes prepared using these recipes can be used for up to two 96-well sample plates. Ideally, the molecules should be modified with single-hit frequency; the necessary reaction conditions will vary depending on temperature, time of the chemical modification step, and length of the RNA (*see* **Note 5**). The conditions should be optimized ahead of time on, e.g., the wild-type sequence of the RNA. To avoid confusion between modification mixes or quenches, we recommend choosing only one of the following modifiers to carry out the M^2 experiment and focusing on no more than two 96-well plates (192 constructs) during a given experimental session. Following are recipes for 4× concentrated modifier mixes, and concentrated quenches (*see* below for final amounts used in each reaction).

1. DMS mix for dimethyl sulfate mapping (stored at room temperature), which primarily gives a signal for exposed Watson–Crick edges of adenines and cytosines. Mix 10 μL fresh dimethyl sulfate into 90 μL 100 % ethanol, then add 900 μL of sterile water. The quench for this reaction is 2-mercaptoethanol (stored at 4 °C).
2. CMCT mix for carbodiimide mapping using 1-cyclohexyl-(2-morpholinoethyl)carbodiimide metho-p-toluene sulfonate (CMCT—stored at −20 °C), which primarily gives a signal for exposed Watson–Crick edges of guanines and uracils in RNA and DNA. Mix 42 mg/mL of CMCT in H_2O. This reagent is usually kept at −20 °C. Before mixing, let the solid stock come to room temperature for 15 min before use. The quench for this reaction is 0.5 M Na-MES, pH 6.0 (stored at room temperature).
3. NMIA mix for 2′ OH acylation mapping (SHAPE) using N-methylisatoic anhydride (NMIA—stored at room temperature in a desiccator), which primarily gives a signal at dynamic RNA nucleotides. Mix 24 mg/mL in anhydrous dimethyl sulfoxide (DMSO—stored at room temperature in a desiccator).

The quench for this reaction is 0.5 M Na-MES, pH 6.0. SHAPE reactions (achieved with NMIA or 1M7) can also be quenched with 2-mercaptoethanol [21].

4. 1M7 mix for 2′ OH mapping using 1-methyl-7-nitroisatoic anhydride (1M7—stored at room temperature in a desiccator), a fast-acting reagent which primarily gives a signal at dynamic RNA nucleotides [27]: Mix 8.5 mg/mL of 1M7 in anhydrous DMSO. The quench for this reaction is 0.5 M Na-MES, pH 6.0, or 2-mercaptoethanol.
5. It is important to make sure that enough quench mix to stop the chemical modification reactions is available: 1 mL of quench lasts for up to two 96-well sample plates.

2.7 Chemical Quench and Bead Solution

1. Clean oligo-dT beads, here "Poly(A) purist" magnetic beads (Ambion/Applied Biosystems—stored at 4 °C). To remove any preservatives in the supplied stock solution, clean the beads by taking out 200 μL of bead stock solution, adding 40 μL of 5 M NaCl, and separating them on a magnetic stand. Let the beads collect into a pellet and remove supernatant. Resuspend the beads in 200 μL sterile water, separate again, and remove the supernatant. Resuspend in 200 μL sterile water. This clean bead stock lasts for at least 2 weeks if kept at 4 °C.
2. Fluorescent primer: Prepare a 0.25 μM solution of a fluorescent primer (labeled at its 5′ end with fluorescein, available as a modification from synthesis companies—usually stored at −20 °C) that is complementary to the 3′ end of the RNA of interest. In the following protocol, this primer should have a poly(A) stretch that will be used to bind to the oligo-dT beads in the purification steps immediately before and after reverse transcription (Fig. 3).
3. 8 mL of 5 M NaCl per sample plate.

2.8 Reverse Transcription Components

1. 120 μL of 5× First Strand buffer per sample plate (Invitrogen—stored at −20 °C).
2. 12 μL of 0.1 M DTT per sample plate (stored at −20 °C).
3. 12 μL of Superscript III reverse transcriptase (200 U/μL, Invitrogen—stored at −20 °C) per sample plate.

2.9 RNA Hydrolysis and Acid Quench Solutions

1. 840 μL of 0.4 M NaOH (stored at room temperature) per sample plate. This 0.4 M NaOH can be prepared in 50 mL volumes and kept at room temperature for at least 2 months.
2. Acid quench mix, prepared by mixing 1 volume of 5 M NaCl, 1 volume of 2 M HCl, and 1.5 volumes of 3 M sodium acetate, pH 5.2. This stock can be kept at room temperature for at least 2 months (*see* **Note 6**).

2.10 ROX–Formamide Elution

1. ROX–formamide mix: Mix 2.75 μL of ROX 350 ladder in 1,200 μL of Hi-Di formamide (both available from Applied Biosystems). The ROX ladder will be used for aligning and normalizing several samples in the downstream data analysis. This volume is enough for one 96-well sample plate; larger stocks can be prepared and stored frozen at −20 °C.

3 Methods

All procedures are performed at room temperature unless specified. For pipetting mixes into each well of a 96-well plate, use an extra plate and a multichannel pipettor to reduce pipetting efforts: pre-aliquot the mix to 8 wells in a separate, auxiliary 96-well plate (*see* **Notes 7** and **8**). Then use an 8-channel pipettor to pipette the desired amount per well to the 12 columns in the main reaction plate.

3.1 Mutant DNA Assembly

Primer assembly by PCR is used to produce the DNA templates used to transcribe RNA for all mutants. This strategy allows for rapid generation of single nucleotide mutants—to produce a mutant for a specific position, only a few of the primers in the assembly need to be changed—and is amenable to the 96-well plate format. To facilitate mixing the correct primers together, it is recommended to organize the primers for all mutants into 96-well plates: each plate should contain all variants of a given primer in the assembly for all mutants with wild-type primers filling the other well (Figs. 1 and 2).

1. Dilute the primer stocks to the appropriate concentration. Usually, the first and last primers in the assembly are set to 100 μM concentration, and any intermediate primers are set to a lower initial concentration (e.g., 1 μM). The final concentration of these primers in the PCR reactions will be 25-fold less than the stocks.
2. Pipette 2 μL of the first primer plate into a 96-well PCR plate using a multichannel pipette. With the addition of the rest of the PCR mix to a volume of 50 μL, this will give a final concentration of 4 μM. Repeat this procedure for the rest of the primer plates so that the first and last primers are at a concentration of 4 μM and the intermediate primers are at a final concentration of 40 nM. Here and throughout, dispose tips after each pipetting step to avoid cross-contamination.
3. Prepare PCR mix. Each 50 μL reaction will involve 10 μL of 5× HF buffer, 1 μL 10 mM dNTPs, 1 μL Phusion polymerase (2,000 U/mL), and 38 − 2 Y μL of sterile water, where Y is the number of primers in the assembly, per reaction (e.g., for an assembly of 2 primers, 34 μL of water would be added for each

Table 2
PCR program used to assemble the DNA sequence that will be transcribed into the desired RNA. Annealing temperatures may have to be adjusted depending on the melting temperatures of each primer in the assembly

Steps	Time/cycles	Temperature
Denaturation	30 s	98 °C
Denaturation	10 s	98 °C
Annealing	30 s	64 °C
Extension	30 s	72 °C
Repeat steps 2–4	29 cycles more	
Polymerization	10 min	72 °C

reaction). Scale the volumes by 1.2 times the number of samples to allow for volume lost while pipetting.

4. Add 50 − 2Y μL of the PCR mix into each well of the PCR plate, where Y is the number of primers in the assembly, to give 50 μL reactions. It is easiest to use the "auxiliary" plate with a column filled with the PCR mix and pipette to the whole plate from there.
5. Cover the plate with a PCR plate seal and put it on a thermocycler with the program described in Table 2 and also *see* **Note 9**.
6. Assembly of the DNA templates with the correct length may be confirmed by agarose gel electrophoresis of 10 μL aliquots and use of a 20 bp length standard. If a 96-sample gel casting system is not available, a subset of mutants can be chosen for agarose gel electrophoresis.

3.2 DNA Purification

The purification method described here yields approximately 20–50 pmol of linear DNA template per sample well.

1. After the PCR is completed, add 72 μL (1.8 volumes) Agencourt AMPure XP beads to 40 μL of the PCR mix in each well, mix by pipetting up and down and leave at room temperature for 10 min. For purifying constructs smaller than 100 nucleotides, *see* **Note 10**.
2. Separate the beads by setting the plate on a 96 post-magnetic stand for 7 min.

   Discard supernatant and rinse each sample with 200 μL of 70 % ethanol (*see* **Note 1**); incubate at room temperature for 1 min and discard.

3. Repeat the previous 70 % ethanol wash, discard the supernatant and leave the beads to dry for at least 15 min on the magnetic stand. Residual ethanol will impair transcription.
4. Take the plate off the magnetic stand and elute each sample in 35 μL of sterile water. After 5 min, re-separate the beads using the magnetic stand.
5. Pipette out the solution to the new plate, leaving the magnetic beads behind.
6. The concentration of DNA can be checked via UV absorbance at 260 nm on, e.g., a NanoDrop system.

3.3 RNA Transcription and Purification

The following transcription and purification method results in approximately 20–40 pmol of RNA per sample well.

1. Prepare DNA dilutions by adjusting the stock concentration to be 0.2 μM. Aliquot 2.5 μL of the DNA dilutions into a new 96-well plate.
2. Prepare the transcription mix. Each 25 μL transcription will involve 2.5 μL of 10× transcription buffer, 1 μL 1 M DTT, 2.5 μL 10 mM NTPs, 2.5 μL 40 % PEG 8000, 13.75 μL sterile water, and 0.25 μL T7 RNA polymerase (50 U/μL) per sample. Scale the volumes of these shared components by 1.2 times the number of samples to account for pipetting errors (e.g., around 2,880 μL for 96 samples).
3. Add 22.5 μL of transcription mix into each well and mix well by pipetting up and down.
4. Incubate at 37 °C for 3 h—use a water bath or a thermocycler.
5. Successful RNA transcription can be confirmed by denaturing agarose gel electrophoresis (e.g., formaldehyde and agarose gel), although RNA bands may not be clear due to incomplete denaturing. Alternatively, reverse transcription provides a check on the RNA quality and sequence, but occurs downstream (*see* Subheading 3.7).
6. Purify the RNA following the same steps as in DNA purification (Subheading 3.2; *see* **Note 10**). Residual low amounts of DNA template do not interfere in later steps. However, if desired, DNAse I and calcium-containing DNAse I buffer can be added to the transcription mix before purification to degrade the DNA template.
7. Measure the concentration of each RNA mutant via UV absorption on, for example, a NanoDrop.

3.4 Folding or Complex Assembly

1. Prepare folding buffer: Mix 2 μL of 0.5 M Na-HEPES, pH 8.0, 2 μL 100 mM $MgCl_2$, and any additional reagents needed for proper folding of the RNA in question, per sample.

2. Transfer $4 + X$ μL of folding buffer to each well in a new 96-well plate per sample plate, where X is the volume of additional reagents used in the folding buffer; add Y μL of nuclease-free water such that $4 + X + Y = 14$ μL. Use the 96-well "auxiliary" plate to transfer the corresponding volume to the sample plate using a multichannel pipettor.
3. Prepare a plate of diluted RNA stocks at a concentration of at least 1.2 μM. Transfer 1 μL of each RNA mutant to the folding buffer plate to achieve at least 1.2 pmol of RNA molecules. The total volume in each well will be 15 μL.
4. Equilibrate at room temperature for 20 min or incubate the plate at the temperature and conditions needed to fold the RNA.

3.5 Chemical Modification

1. Add 5 μL of the freshly prepared chemical modification solution to each sample well and mix well by pipetting up and down. It is again easiest to pre-aliquot this buffer to the separate "auxiliary" plate (e.g., 65 μL in 8 wells for pipetting 5 μL into 12 columns) to use the multichannel pipettor to mix into the reaction plate. For modifier reagents that are prepared in DMSO, reagent volumes smaller than 5 μL can be used to reduce final DMSO content in reactions, but will give less modification.
2. Cover the sample plate with sealer tape and incubate at room temperature. The incubation time with the modification reagent should be varied according to the length of the RNA and the chemical modifier used (e.g., for a 200 nt RNA at 24 °C, DMS and CMCT requires an incubation time of 15 min; 1M7 modification is complete in 5 min, while NMIA requires 30 min). Modifier concentration or modification time should be decreased for longer RNAs to maintain single-hit kinetics on a longer construct (*see* **Note 5**).

3.6 Quench and Purification

1. Prepare bead-quench mix: Mix 3 μL of 5 M NaCl, 1.5 μL of clean oligo-dT beads, 5 μL of quench reagent and 0.25 μL of the 0.25 μM fluorescent primer stock (9.75 μL final volume) per sample. Scale the volumes according to 1.2 times the number of samples (e.g., around 1,120 μL for 96 samples). *See* **Note 11** for reactions using a 2-mercaptoethanol quench; and *see* **Note 12** for NMIA reactions.
2. Remove and discard the seal from the sample plates. Add 9.75 μL of the bead-quench mix to all sample wells (Fig. 4a). As before, pre-aliquot the quench mix to a set of 8 wells in the separate "auxiliary" plate and use an 8-channel pipettor to transfer the mix. Wait for 7 min to allow the primer to bind to the RNA.

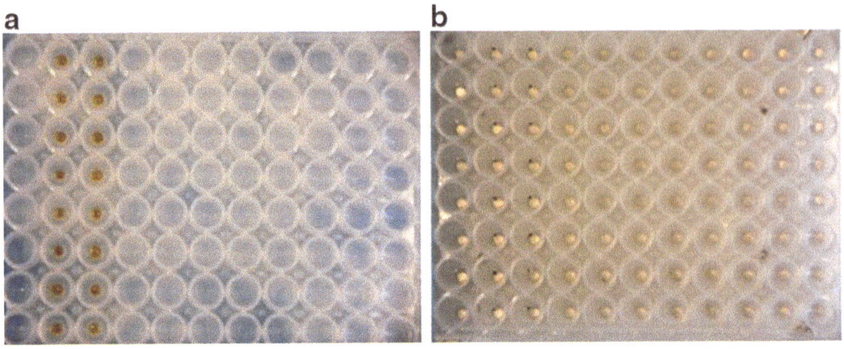

Fig. 4 Mutate-and-map experiments can include samples in all 96 wells of a plate; snapshots from a trial experiment with 16 wells are shown here. (**a**) Example wells after adding the bead-quench mix into modified RNA. (**b**) The sample beads collect into pellets after putting the sample plate on the magnetic stand for 7 min

3. Set the sample plate on a 96-post magnetic stand and wait for 7–10 min for the beads to collect (Fig. 4b).
4. Remove and discard supernatant while the sample plate is still on the separator. After discarding supernatant, rinse the beads by applying 100 μL of 70 % ethanol onto each well (*see* **Note 1**). The beads will stay collected at the bottom of the plate. Wait for 1 min and remove the ethanol from samples while the sample plate is on magnetic separator.
5. Repeat the 70 % ethanol wash above: add 100 μL, wait for 1 min, and remove while keeping plate on the magnetic separator (*see* **Note 13**). Let the sample plate dry for 10 min while still on the magnetic separator. After the samples are dry, make sure to reinspect the wells for any residual ethanol. If any drops are visible, pipette them out and leave to dry for 5 min longer. During this time, prepare reverse transcription mix (*see* Subheading 3.7).
6. Resuspend each sample in 2.5 μL of RNase-free water and take the plate off the magnetic stand. If reverse transcription mix is not ready, adhere a fresh plastic seal to plate to prevent evaporation.

3.7 Reverse Transcription and Final Purification

1. Prepare the Superscript III reverse transcription mix: Mix 1 μL of 5× First Strand buffer, 0.25 μL of 0.1 M DTT, 0.4 μL of 10 mM dNTPs, 0.75 μL sterile water, and 0.1 μL Superscript III reverse transcriptase (200 U/μL) per sample (the final volume of mix per sample will be 2.5 μL). Scale the volumes accordingly for approximately 1.2 of the number of samples (e.g., around 300 μL for 96 samples). *See* **Note 14** for reaction mixture to create sequencing ladders.
2. Take the sample plate off the magnetic separator, and add 2.5 μL of the Superscript III mix to the sample wells. The total

volume will be 5 μL. Again, pre-aliquot larger amounts to eight wells in the separate "auxiliary plate," and use the 8-channel pipettor to transfer the mix. Mix very well at least 10 times, or until the beads go fully back into solution. Make sure the reaction plate is not near the magnets for this step.

3. Seal the sample plate, and put it in an incubator set to 42 °C for 30 min. For longer RNAs, leaving the samples for longer times may lead to slightly better reverse transcriptase extension.
4. To remove the RNA template after reverse transcription, add 5 μL of 0.4 M NaOH to each well (again use the "auxiliary" plate to reduce pipetting effort). Incubate the samples at 90 °C for 3 min to hydrolyze RNA, while leaving fluorescent cDNA behind.
5. Take plate out of the incubator and cool on ice for 3 min.
6. Add 5 μL of acid quench mix to each well, using the "auxiliary" plate to reduce pipetting effort.
7. After 1 min, put the sample plate on a magnetic stand and wait for 7 min until the beads aggregate. Remove and discard supernatant with a multichannel pipettor.
8. Residual salt in the wells may interfere when injecting the samples into the capillary sequencer. To remove these impurities, rinse the beads by applying 100 μL of 70 % ethanol onto each well. Wait for 1 min and remove the ethanol.
9. Repeat the 70 % ethanol wash above. Remove the residual ethanol thoroughly and let the plate dry for 10 min.
10. Resuspend the beads in 11 μL of ROX–formamide mix while the plate is still in the magnetic stand. Wait for 15 min to complete the elution.
11. Transfer the supernatant to a capillary sequencing optical plate and remove any bubbles by centrifuging the plate with a benchtop plate centrifuge or using a pipettor.
12. The samples are now ready to run in an ABI 3100 capillary sequencer or similar with a standard sequencing or fragment analysis protocol.

3.8 Data Analysis, Curation, and Publication

The resulting data after the capillary electrophoresis run, saved in one .ab1 file (or one .fsa file) per sample well, can be analyzed using HiTRACE [28], ShapeFinder [29], CAFA [30], or other similar software packages. Hereafter we assume that the HiTRACE MATLAB toolkit is used to analyze the data, as it has been optimized for quantitation of data sets involving hundreds of traces, which benefit from alignment and global sequence annotation.

1. HiTRACE is a toolkit that helps align, normalize, map to sequence, and quantify the data obtained from chemical

mapping experiments of RNA, including M^2 assays. HiTRACE analysis is now freely available via an online, user-friendly interface at http://hitrace.org. Alternatively, HiTRACE is available as a set of flexible MATLAB scripts, and the full procedure of analyzing chemical mapping data with these scripts is given in the HiTRACE documentation and tutorials—*see*, e.g., https://sites.google.com/site/rmdbwiki/hitrace. The manual covers everything from installing the toolkit to analyzing the raw capillary data, to saving the quantified chemical reactivities in the RDAT annotated file format. Once an RDAT file with the experimental data is obtained, tools from the RNA Mapping Database (RMDB) [31] can be used to share, visualize, and obtain computationally predicted secondary structure models of the RNA guided by the M^2 data.

2. Researchers are encouraged to share their analyzed data sets by submitting them to the RMDB or SNRNASM [32] repositories. To submit to the RMDB, go to http://rmdb.stanford.edu/repository/register/ to register to the site and to http://rmdb.stanford.edu/repository/submit/ to submit your data. After submission and approval by the RMDB curators, the entry can be visualized in its details page (Fig. 5a, b).

3. The M^2 data can be used to guide secondary structure algorithms, such as RNAstructure [26]. Calculating the Z-scores across each nucleotide positions reveals the most significant perturbations due to each mutation. This information can be used as bonuses in the secondary structure prediction algorithms to compute improved secondary structure models [25]. Further, repeating this procedure by resampling the data with replacement using bootstrapping gives helix-wise confidence estimates [22, 25].

4. This pipeline is available at the RMDB structure server; to obtain mutate-and-map guided secondary structure models: Visit http://rmdb.stanford.edu/structureserver/. In "Input options" click "Upload RDAT" to expand the RDAT uploading options. In the expanded pane, click choose file and choose the RDAT file containing the mutate-and-map data. When the file is uploaded, the fields in the structure server forms will be auto-populated with the annotations and data of the uploaded RDAT file. In "Bonus options," click "2D bonuses" to expand the 2D bonus option pane. You will see the data as a matrix of bonuses in "Input bonuses" (some entries in this matrix will be zeros if some nucleotides in the sequence are missing data). Check "Filter by Z-scores" and input the number of bootstrap iterations for obtaining helix-wise support values. The larger the iterations, the longer it will take but the more accurate the support estimates will be. For most RNAs, 100–400 iterations appear sufficient to get converged support values. Optional

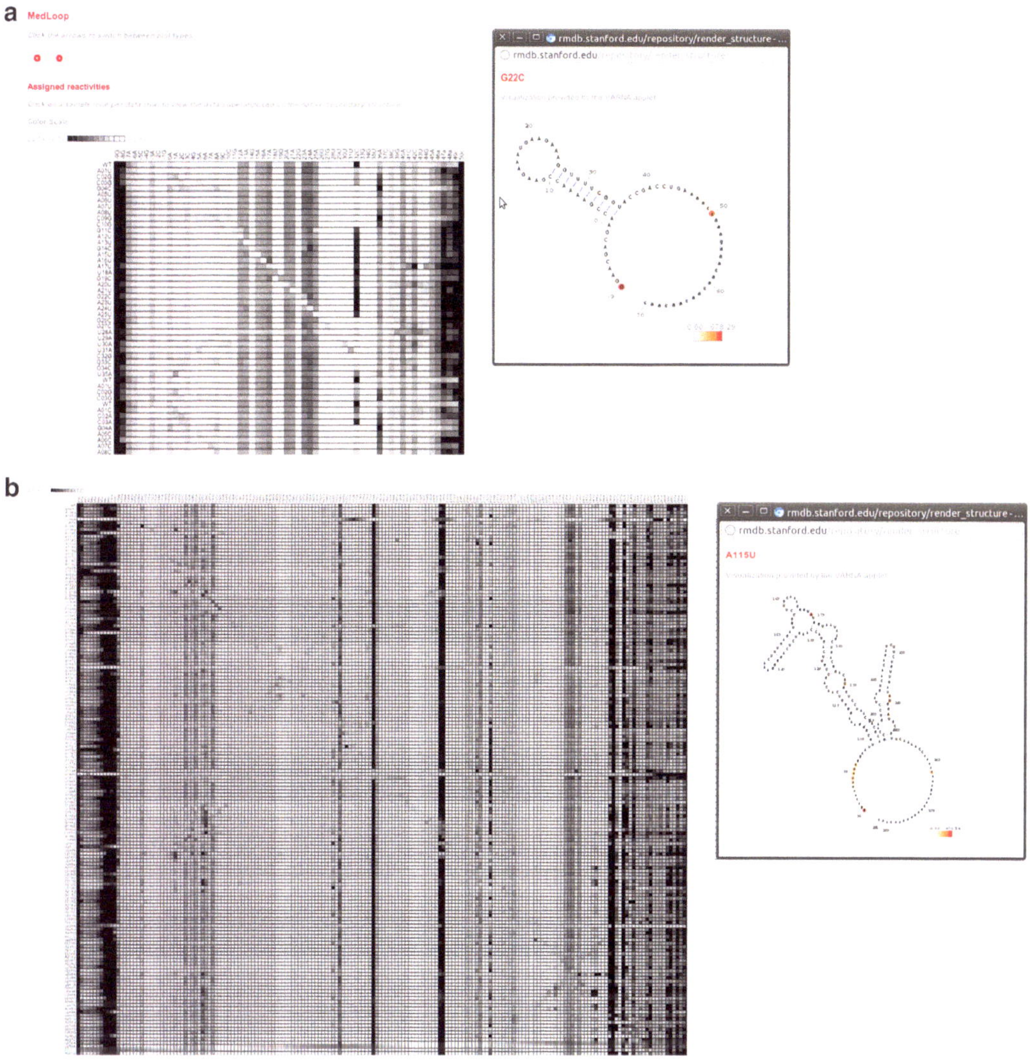

Fig. 5 After data collection, reactivities of each position for each mutant can be quantified using HiTRACE. These data can be made publicly available through the RNA Mapping DataBase (RMDB). The RMDB provides tools to visualize and share structure models of the RNA. Here, we show the RMDB details page with supporting secondary structure models of (**a**) the MedLoop motif (entry MDLOOP_DMS_0001) and the (**b**) P4–P6 domain of the *Tetrahymena* ribozyme (entry TRP4P6_SHP_0003). The secondary structure models are colored by chemical reactivity value at each nucleotide, as determined by the chemical mapping experiment (from *white* [low reactivity] to *red* [high reactivity]; *gray* nucleotides indicate places where the reactivity value is not available or considered to be an outlier as determined from "box-plot" statistics)

settings can be changed in "Other options," such as temperature and reference secondary structure (given in dot-parentheses notation). Click on "Submit"; you will be directed to a page containing the predicted model, including helix-by-helix support values (Fig. 6a, b).

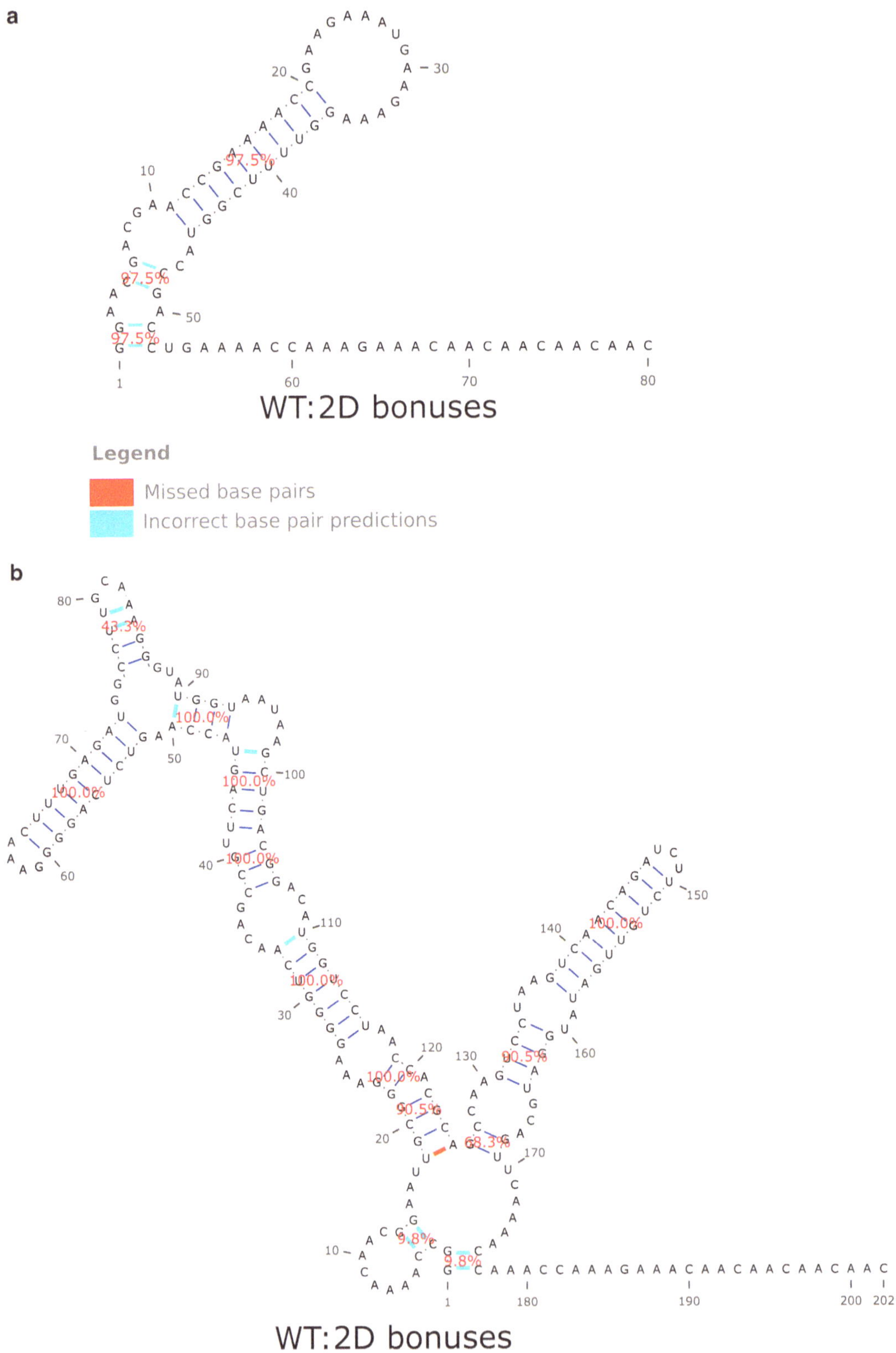

Fig. 6 The RMDB also provides a server that can use M² data to guide secondary structure modeling algorithms. Predicted secondary structure models are shown for the (**a**) MedLoop motif and the (**b**) P4–P6 domain of the *Tetrahymena* ribozyme using mutate-and-map data from entries MDLOOP_DMS_0001 and TRP4P6_SHP_0003 respectively. Base pairs are colored *orange* if they are false negatives or *blue* if they are false positives respect to a reference structure. Helix-wise bootstrap support values are shown in *red*

4 Notes

1. It is easiest to have a large amount of 70 % ethanol stored in a sterile 96-well rack in which each well holds at least 2 mL of volume. Care should be taken to avoid ethanol evaporation, as lower ethanol content washes will remove bound cDNA sample. Seal the reservoir when not in use during the protocol, discarding and freshly filling its contents before the experiment starts and before each major wash step in the protocol.
2. In the example experiments above we use AAAAAAAAAA AAAAAAAAAGTTGTTGTTGTTGTTTCTTT as our fluorescent primer sequence. Furthermore, we add a T7 promoter (TTCTAATACGACTCACTATA) and a buffer of "unstructured" nucleotides beginning with two guanines (e.g., GGCC AAAACAACGGAA) at the 5' end of our sequence of interest to make the RNA amenable to in vitro transcription. Similarly, we add a buffer, unstructured region of approximately ten nucleotides and a "tail" (AAAACAAAACAAAGAAACAACAAC AACAAC) that serves as the binding region for the fluorescent primer. *See* example sequences for these buffers in Figs. 1 and 2 as well as Table 1.
3. The NA_Thermo MATLAB toolkit (Das et al. unpublished software; freely available at https://simtk.org/home/na_thermo), or similar software [33, 34] can be used to design the primers to assemble the sequence designed in the above step as well as the mutants of interest. The primers for the wild-type sequence should be prepared or ordered first; all the steps below from PCR assembly to RNA transcription to sequencing ladders should be carried out to confirm that the primers produce the desired RNA. To obtain primer sequences using NA_Thermo, download the toolkit from https://simtk.org/home/na_thermo and unpack in any directory. As with any MATLAB toolkit, the unpacked directory should be added to the MATLAB path. In MATLAB, type:

   ```
   primers = design_primers(sequence)
   ```

   where *sequence* is your designed sequence from **step 2**. This will output (and save in the *primers* variable) the primers needed to build the designed sequence using PCR assembly. The design_primers function also accepts parameters such as desired melting temperature between primers and primer size. To easily obtain plate layouts of the mutant library, first type:

   ```
   sequences_to_order = single_mutant_library
   (primers, sequence, offset, region, librar-
   ies, name);
   ```

Where *offset* is an integer that is added to the sequence index to arrive at the conventional numbering for the RNA (30 in the MedLoop example of Fig. 1), *region* is the region of the sequence to be mutated (normally excluding the T7 promoter, buffer, and primer binding regions; 1:35 in the MedLoop example of Fig. 1), *libraries* is a vector with integers that code for the type of libraries to generate, that is, to which base each nucleotide should be mutated (we usually mutate to the complement of each base, which corresponds to library 1), and *name* is an assigned name of the sequence. Finally, type:

```
output_sequences_to_order_96well_diagram
(sequences_to_order, primers, name);
```

This will generate tab-delimited files with the sequences for each primer for each mutant, as well as images depicting the plate layouts (*see* Fig. 1 for a plate layout example of a simple RNA and Fig. 2 for a larger RNA with a more complex assembly). You should peruse the files in, e.g., Microsoft Excel, to confirm that the mutants are in the desired region.

The primers obtained above can be prepared in 96-well format on standard synthesizers or ordered from any oligo synthesis company, such as IDT.

4. Besides Na-HEPES, pH 8.0, other RNA folding buffers can be used, but note that the acylation reaction rate decreases by tenfold with each decreasing unit of pH.
5. To maintain single-hit kinetics on large RNA constructs, concentration of the chemical modifier should be reduced. We suggest that 1/2 the concentration of the chemical modifier be used for RNAs with lengths between 100 and 150 nucleotides, and 1/4 for RNAs with lengths between 150 and 250.
6. It is critical that the acid quench neutralizes the 0.4 M NaOH before the last wash. It is recommended to confirm that the final pH of a mix of 5 μL of 0.4 M NaOH and 5 μL of acid quench mix is between 5 and 7.
7. It is best practice not to reuse pipette tips when using the multichannel pipettors in any step of the protocol other than the ethanol washes to avoid sample cross-contamination.
8. The 96-well format of the protocol is best exploited when using an "auxiliary" plate to pre-aliquot premixes. This allows the use of multichannel pipettors to transfer the mixes to the sample plates with less pipetting effort.
9. The annealing temperature may need to be adjusted based on the primer melting temperatures under the PCR conditions; the above table uses Finnzyme/Thermo instructions for Phusion (+4 °C over the melting temperature of primers, which are set to be greater than or equal to 60 °C in the NA_thermo primer design scripts).

10. Ampure XP purification of small nucleic acids can be optimized by adding polyethylene glycol (PEG) 8000 to the Ampure beads. We have found that a mix of 3 volumes 40 % PEG 8000 with 7 volumes of Ampure beads is optimal for recovering DNA and RNA greater than 50 nucleotides.
11. The 2-mercaptoethanol quenching reagent can interfere with magnetic bead collection if left for too long before performing the first 70 % ethanol wash after chemical modification. It is therefore recommended to mix the quench premix right before applying it. Ruined samples will appear viscous and green. Samples washed timely and properly will appear light brown.
12. For NMIA probing, after performing chemical modification, the beads may smear against the well walls and will refuse to collect at the bottom of the well when the plate is transferred to the 96-post magnet. It is best to not scrape the beads off the walls. Instead, pipette up and down against the walls to wash them into the bottom of the well.
13. In the second ethanol wash before reverse transcription it is important to remove the 70 % ethanol supernatant thoroughly to avoid affecting the reverse transcription step. This can be accomplished by removing most of the ethanol with a P100 multichannel pipettor and subsequently removing any remaining droplets with a multichannel P10 pipettor placed close to the beads collected at the bottom of the plate.
14. As a quality check and also to prepare references for assigning bands in the capillary electrophoresis traces, acquisition of Sanger sequencing ladders for at least the wild-type construct are recommended. Prepare 1.2 pmol RNA, 1.5 μL oligo(dT) beads, 0.25 μL 0.25 μM fluorophore-labeled primer, and water to 2.5 μL volume for each sequencing reaction. Then add the following 2.5 μL mix: 1.0 μL of 5× First Strand buffer, 0.25 μL of 0.1 M DTT, 0.4 μL of 1.0 mM dNTPs, 0.35 μL sterile water, 0.4 μL of 1 mM ddATP (or one of the other three 2′,3′ dideoxynucleotide triphosphates) and 0.1 μL of Superscript III reverse transcriptase (200 U/μL) per sample (the final volume will be 5 μL). Carry out remaining reverse transcription and purification steps as described in Subheading 3.7.

Acknowledgements

We thank B. Alford, F. Ortega, A. Becka, T. Mann, and S. Tian for comments on the manuscript. This work is supported by the Burroughs-Wellcome Foundation (CASI to R.D.), a CONACyT fellowship (to P.C.), and the National Institutes of Health (T32 HG000044 to C.C.V. and R01 GM102519 to R.D.).

References

1. Mandal M, Lee M, Barrick JE et al (2004) A glycine-dependent riboswitch that uses cooperative binding to control gene expression. Science 306:275–279
2. Kulshina N, Baird NJ, Ferre-D'Amare AR (2009) Recognition of the bacterial second messenger cyclic diguanylate by its cognate riboswitch. Nat Struct Mol Biol 16:1212–1217
3. Gesteland RF, Cech TR, Atkins JF (2006) The RNA World, 3rd edn. CSHL Press, Cold Spring Harbor
4. Leontis NB, Westhof E (1998) The 5S rRNA loop E: chemical probing and phylogenetic data versus crystal structure. RNA 4:1134–1153
5. Staley JP, Guthrie C (1998) Mechanical devices of the spliceosome: motors, clocks, springs, and things. Cell 92:315–326
6. Lukavsky PJ, Kim I, Otto GA et al (2003) Structure of HCV IRES domain II determined by NMR. Nat Struct Biol 10:1033–1038
7. Brostoff SW, Ingram VM (1967) Chemical modification of yeast alanine-tRNA with a radioactive carbodiimide. Science 158:666–669
8. Moazed D, Stern S, Noller HF (1986) Rapid chemical probing of conformation in 16 S ribosomal RNA and 30 S ribosomal subunits using primer extension. J Mol Biol 187: 399–416
9. Cordero P, Kladwang W, VanLang CC et al (2012) Quantitative dimethyl sulfate mapping for automated RNA secondary structure inference. Biochemistry 51:7037–7039
10. Wilkinson KA, Merino EJ, Weeks KM (2006) Selective 2′-hydroxyl acylation analyzed by primer extension (SHAPE): quantitative RNA structure analysis at single nucleotide resolution. Nat Protoc 1:1610–1616
11. Regulski EE, Breaker RR (2008) In-line probing analysis of riboswitches. Methods Mol Biol 419:53–67
12. Lucks JB, Mortimer SA, Trapnell C et al (2011) Multiplexed RNA structure characterization with selective 2′-hydroxyl acylation analyzed by primer extension sequencing (SHAPE-Seq). Proc Natl Acad Sci USA 108:11063–11068
13. Sood VD, Beattie TL, Collins RA (1998) Identification of phosphate groups involved in metal binding and tertiary interactions in the core of the *Neurospora* VS ribozyme. J Mol Biol 282:741–750
14. Rhee Y, Valentine MR, Termini J (1995) Oxidative base damage in RNA detected by reverse transcriptase. Nucleic Acids Res 23:3275–3282
15. Chow CS, Cunningham PR, Lee K et al (2002) Photoinduced cleavage by a rhodium complex at G.U mismatches and exposed guanines in large and small RNAs. Biochimie 84:859–868
16. Kertesz M, Wan Y, Mazor E et al (2010) Genome-wide measurement of RNA secondary structure in yeast. Nature 467:103–107
17. Underwood JG, Uzilov AV, Katzman S et al (2010) FragSeq: transcriptome-wide RNA structure probing using high-throughput sequencing. Nat Methods 7:995–1001
18. Kladwang W, Das R (2010) A mutate-and-map strategy for inferring base pairs in structured nucleic acids: proof of concept on a DNA/RNA helix. Biochemistry 49:7414–7416
19. Wells SE, Hughes JM, Igel AH et al (2000) Use of dimethyl sulfate to probe RNA structure *in vivo*. Methods Enzymol 318:479–493
20. Tijerina P, Mohr S, Russell R (2007) DMS footprinting of structured RNAs and RNA-protein complexes. Nat Protoc 2:2608–2623
21. Spitale RC, Crisalli P, Flynn RA et al (2013) RNA SHAPE analysis in living cells. Nat Chem Biol 9:18–20
22. Kladwang W, VanLang CC, Cordero P et al (2011) Understanding the errors of SHAPE-directed RNA structure modeling. Biochemistry 50:8049–8056
23. Washietl S, Hofacker IL, Stadler PF et al (2012) RNA folding with soft constraints: reconciliation of probing data and thermodynamic secondary structure prediction. Nucleic Acids Res 40:4261–4272
24. Kladwang W, Cordero P, Das R (2011) A mutate-and-map strategy accurately infers the base pairs of a 35-nucleotide model RNA. RNA 17:522–534
25. Kladwang W, VanLang CC, Cordero P et al (2011) A two-dimensional mutate-and-map strategy for non-coding RNA structure. Nat Chem 3:954–962
26. Mathews DH, Disney MD, Childs JL et al (2004) Incorporating chemical modification constraints into a dynamic programming algorithm for prediction of RNA secondary structure. Proc Natl Acad Sci USA 101:7287–7292
27. Mortimer SA, Weeks KM (2007) A fast-acting reagent for accurate analysis of RNA secondary

and tertiary structure by SHAPE chemistry. J Am Chem Soc 129:4144–4145
28. Yoon S, Kim J, Hum J et al (2011) HiTRACE: high-throughput robust analysis for capillary electrophoresis. Bioinformatics 27:1798–1805
29. Vasa SM, Guex N, Wilkinson KA et al (2008) ShapeFinder: a software system for high-throughput quantitative analysis of nucleic acid reactivity information resolved by capillary electrophoresis. RNA 14:1979–1990
30. Mitra S, Shcherbakova IV, Altman RB et al (2008) High-throughput single-nucleotide structural mapping by capillary automated footprinting analysis. Nucleic Acids Res 36:e63
31. Cordero P, Lucks JB, Das R (2012) An RNA Mapping DataBase for curating RNA structure mapping experiments. Bioinformatics 28:3006–3008
32. Rocca-Serra P, Bellaousov S, Birmingham A et al (2011) Sharing and archiving nucleic acid structure mapping data. RNA 17: 1204–1212
33. Rydzanicz R, Zhao XS, Johnson PE (2005) Assembly PCR oligo maker: a tool for designing oligodeoxynucleotides for constructing long DNA molecules for RNA production. Nucleic Acids Res 33:W521–W525
34. Thachuk C, Condon A (2007) On the design of oligos for gene synthesis. Paper presented at the IEEE 7th international symposium on bioinformatics and bioengineering, Cambridge, Boston, USA, pp 123–130. IEEE

Chapter 5

Mapping RNA Structure In Vitro Using Nucleobase-Specific Probes

Nora Sachsenmaier, Stefan Handl, Franka Debeljak, and Christina Waldsich

Abstract

RNAs have to adopt specific three-dimensional structures to fulfill their biological functions. Therefore exploring RNA structure is of interest to understand RNA-dependent processes. Chemical probing in vitro is a very powerful tool to investigate RNA molecules under a variety of conditions. Among the most frequently used chemical reagents are the nucleobase-specific probes dimethyl sulfate (DMS), 1-cyclohexyl-3-(2-morpholinoethyl) carbodiimide metho-p-toluenesulfonate (CMCT) and β-ethoxy-α-ketobutyraldehyde (kethoxal). These chemical reagents modify nucleotides which are not involved in hydrogen bonding or protected by a ligand, such as proteins or metabolites. Upon performing modification reactions with all three chemicals the accessibility of all four nucleobases can be determined. With this fast and inexpensive method local changes in RNA secondary and tertiary structure, as well as the formation of contacts between RNA and its ligands can be detected independent of the RNA's length.

Key words In vitro chemical probing, DMS, CMCT, Kethoxal, RNA structure, RNA-protein complexes

1 Introduction

In the past, a vast number of noncoding regulatory RNA molecules were discovered, suggesting that RNA is the driving force in the most essential processes in the cell. To shed light on the basic mechanism of these RNA-dependent processes, it is important to explore the structure of an RNA and the assembly with its ligands, like proteins, metabolites or other RNA molecules. The process of folding is hierarchical and describes how RNA undergoes the transition from an unfolded, disordered state to the native, functional conformation [1–9]. Thus, to understand the dynamic process of folding it is crucial to determine the structure of RNA folding intermediates and the conformation of the native state.

A variety of methods have been used to map the structure of RNA in vitro [10–13]. A rapid, inexpensive and widely used

Christina Waldsich (ed.), *RNA Folding: Methods and Protocols*, Methods in Molecular Biology, vol. 1086, DOI 10.1007/978-1-62703-667-2_5, © Springer Science+Business Media, LLC 2014

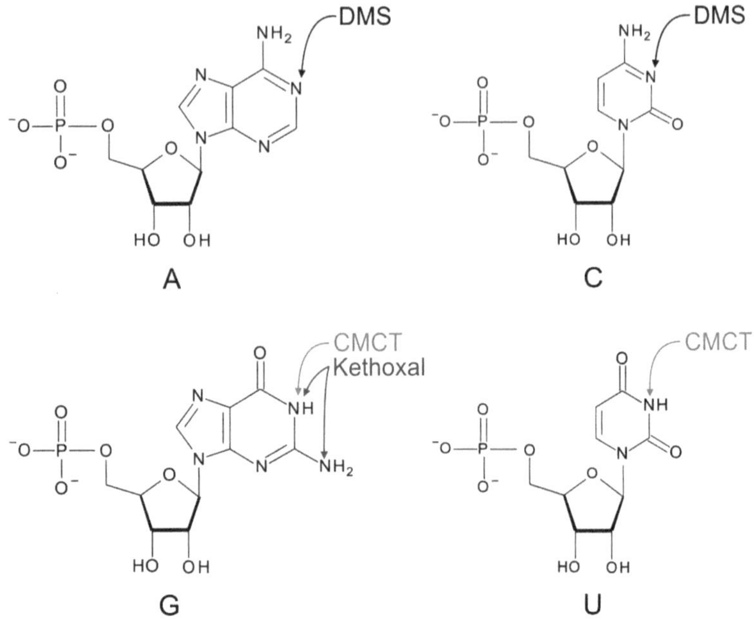

Fig. 1 The chemical reagents DMS, CMCT, and Kethoxal are nucleobase-specific in their modification reaction. DMS methylates N1 of adenine and N3 of cytosine. CMCT modifies mainly N3 of uracil and to a lesser extent the N1 of guanine, whereas Kethoxal modifies guanine at the N1 and N2 positions. *DMS* dimethyl sulfate, *CMCT* 1-cyclohexyl-3-(2-morpholinoethyl) carbodiimide metho-p-toluenesulfonate, *Kethoxal* β-ethoxy-α-ketobutyraldehyde

method to obtain structural information on RNAs is chemical probing in vitro. This technique employs the commercially available chemical reagents DMS, CMCT and Kethoxal that react with specific atoms of different nucleobases (Fig. 1), if they are not engaged in H-bonds [10, 14, 15]. While DMS methylates the N1 of adenine and N3 of cytosine, CMCT modifies the N1 of guanine and N3 of uracil and Kethoxal specifically attacks the N1 and N2 of guanine. Thus, it is possible to investigate the accessibility of all four nucleobases. The sites of modification can then be determined by reverse transcription (RT) using a radioactively or fluorescently labeled complementary DNA oligo [10, 14, 15], because the bulky modification groups at the Watson-Crick position of the nucleotides leads to termination of the reverse transcription (Fig. 2). After separating the cDNA pool on a standard denaturing polyacrylamide gel, residues, which are not involved in secondary or tertiary structure formation and/or are not protected from modification by binding to a ligand, become visible as strong DMS-dependent RT stops (or CMCT- and Kethoxal-dependent stops), which can be mapped onto the secondary structure of a given RNA. This provides a good picture of the specific structure adopted

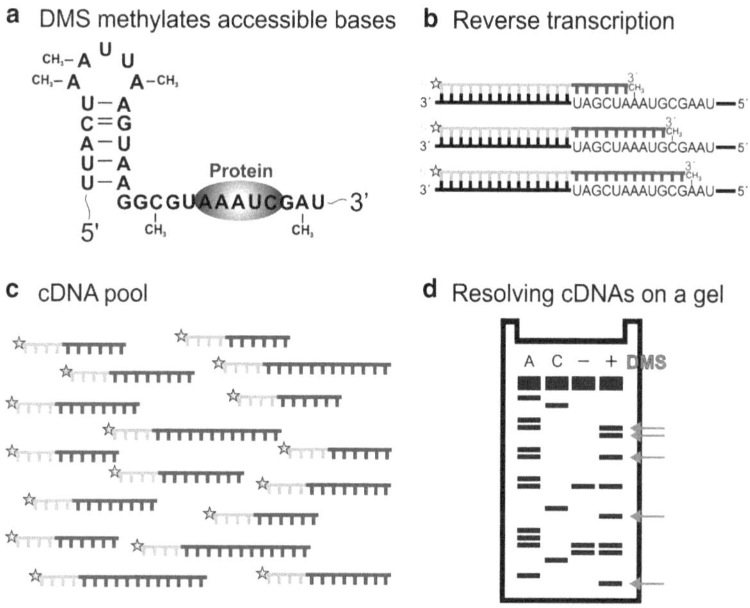

Fig. 2 Experimental framework of in vitro chemical probing with DMS. (**a**) Only A and C residues that are not engaged in H-bonding and are not interacting with proteins (or another ligand) are accessible to DMS and can thus be methylated. (**b**) Using a 5′ end-labeled primer, the sites of modification can be mapped by reverse transcription, as they cause a stop in the extension by the reverse transcriptase due to the bulky methyl group. RNA is shown in *black*, the annealed primer in *dark gray* and the extension by the RT in *light gray*; the *star* indicates the 5′end-label of the primer. (**c**) The generated cDNA pool is then resolved on a standard denaturing polyacrylamide gel (**d**), resulting in a distinct modification pattern for a given RNA. A and C denote sequencing lanes; *lane 3* is the RT stop control to detect natural stops of the extension (unmodified RNA is used in RT reaction); *lane 4* shows the DMS pattern of the RNA (*gray arrows*)

by the RNA. As such, the conformation of different stable intermediates along a folding pathway can be explored with this method. In addition, residues that interact with a protein (or another ligand) and the concomitant structural changes induced by the protein can also be characterized with this methodology (e.g., [10, 14–23]).

The chemical probing technique can be easily adapted for a given RNA and is largely insensitive to the folding conditions of the RNA and its size. In other words, it can be used to study the structure of RNAs which cannot be easily subjected to NMR and crystallization techniques. Therefore it has proven to be a very powerful tool, as it has been used to map the structure of various RNAs including group I and group II introns, ribosomal RNAs, and ligand-induced conformational changes (e.g., [10, 14–23]). We recently investigated the structure of the human telomerase RNA (hTR) in vitro using DMS chemical probing (Fig. 3; *Auer*

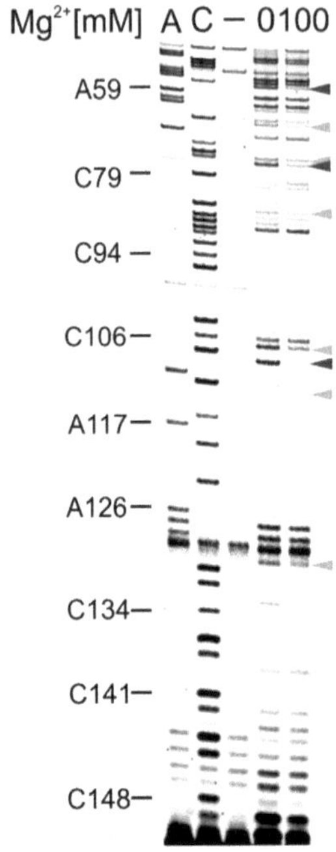

Fig. 3 Mg^{2+}-induced structural changes in human telomerase RNA. Using DMS, which methylates A-N1 and C-N3, if they are not involved in H-bonding, the structure of the human telomerase RNA has been mapped. A and C denote sequencing lanes; *The-lane* is an RT stop control to detect natural stops of the extension (independent of DMS modification; RNA was folded, but not incubated with DMS). Lanes labeled 0 and 100: The *number* refers to the Mg^{2+} concentration [mM] used in the folding reaction and both samples were treated with DMS. *Gray arrow heads* indicate residues that became protected from DMS modification upon high [Mg^{2+}]; *light gray arrow heads* indicate weak protections and *dark gray* represents strong protections

and Waldsich, unpublished). Comparing the folding states of hTR in the absence and presence of Mg^{2+} ions indicates that most of secondary structure of hTR is able to form in the absence of Mg^{2+} ions. In contrast, the pseudoknot appears to require high [Mg^{2+}] to fold (Fig. 3; *Auer and Waldsich, unpublished*). Also, these chemical probing data imply that the human telomerase RNA holds several interesting tertiary structure features to discover, as supposedly single-stranded residues are protected from modification, suggesting that these are engaged in H-bonding. In summary, chemical probing provides detailed information on the secondary and tertiary structure of a given RNA and can be complemented with other structural probing methods, such as SHAPE and hydroxyl-radical footprinting, or chemogenetic approaches [11–13].

2 Materials

Prepare all solutions with RNase-free ultrapure water (deionized water with a sensitivity of ≥18 MΩ cm at 25 °C) (*see* **Note 1**) and store the sterile solutions at room temperature if not stated otherwise. Use RNase-free plasticware and glassware for the experiments.

2.1 In Vitro Transcription

1. 10× transcription buffer: 0.4 M Tris–HCl, pH 7.5, 0.26 M $MgCl_2$, 30 mM spermidine. Store at −20 °C in the dark.
2. 1 M DTT (dithiothreitol). Store at −20 °C.
3. 100 mM ribonucleotides rATP, rCTP, rGTP, and rUTP. Store at −20 °C.
4. RNase inhibitor (40 U/μL). Store at −20 °C.
5. T7 RNA polymerase (homemade). Store at −20 °C.
6. RNase-free DNase I (2 U/μL). Store at −20 °C.
7. 10× TBE: 0.89 M Tris base, 0.89 M boric acid, 20 mM EDTA.
8. Glycogen (10 mg/mL). Store at −20 °C.
9. 0.5 M EDTA, pH 8.0.
10. Ethanol/0.3 M NaOAc, pH 5.0.
11. 70 % (v/v) ethanol.
12. Loading buffer I: 7 M urea, 25 % (w/v) sucrose, 0.025 % (w/v) bromophenol blue, 0.025 % (w/v) xylene cyanol in 1× TBE.

2.2 Purification of In Vitro Transcribed RNA

1. 5 % acrylamide solution: 5 % (v/v) acrylamide/bisacrylamide (19:1), 7 M urea in 1× TBE (*see* **Note 2**). Store at 4 °C in the dark.
2. 10 % (w/v) APS (ammonium persulfate). Store at 4 °C (*see* **Note 3**).
3. N,N,N′,N′-tetramethylethylenediamine (TEMED). Store at 4 °C.
4. 10× TBE: 0.89 M Tris base, 0.89 M boric acid, 20 mM EDTA, pH 8.0 (*see* **Note 2**).
5. Elution buffer: 100 mM Tris–HCl, pH 7.5, 250 mM NaOAc, pH 5.0, 2 mM EDTA, pH 8.0.
6. Fluor-coated TLC silica gel plate.
7. Short-wavelength UV light source (254 nm).
8. Saran wrap.
9. 0.5 M EDTA, pH 8.0.
10. Glycogen (10 mg/mL). Store at −20 °C.
11. Ethanol p.A.

12. 70 % (v/v) ethanol.
13. Glass plates (22 (L) × 20 (W) cm).
14. Comb and spacer (1.0–1.5 mm thickness).
15. Vertical gel electrophoresis apparatus (adjustable).
16. High voltage power supply.

2.3 Chemical Probing

1. 0.5 M potassium borate buffer, pH 8.0, stock. Store at 4 °C.
2. 0.5 M potassium borate buffer, pH 7.0, stock. Store at 4 °C.
3. 1 M cacodylate buffer, pH 7.5, stock. Store at 4 °C.
4. 2 M KCl.
5. 1 M $MgCl_2$.
6. 1-cyclohexyl-3-(2-morpholinoethyl) carbodiimide metho-p-toluenesulfonate (CMCT; 42 mg/mL). Store at −20 °C.
7. 10.7 M dimethyl sulfate (DMS). Store at 4 °C.
8. Kethoxal (37 mg/mL). Store at −20 °C.
9. 25 % (v/v) ethanol.
10. 14.3 M β-mercaptoethanol. Store at 4 °C.
11. 0.5 M EDTA, pH 8.0.
12. Glycogen (10 mg/mL). Store at −20 °C.
13. Ethanol/0.3 M NaOAc, pH 5.0.
14. 70 % (v/v) ethanol.

2.4 5′ End-Labeling of Primer

1. [γ-^{32}P]-ATP (10 μCi/μL, 6,000 Ci/mmol). Store at 4 °C.
2. 10 μM stocks of gene-specific DNA primers. Store at −20 °C.
3. T4 polynucleotide kinase (PNK; 10 U/μL). Store at −20 °C.
4. 10× T4 polynucleotide kinase (PNK) buffer. Store at −20 °C.
5. 0.5 M EDTA, pH 8.0.
6. Glycogen (10 mg/mL). Store at −20 °C.
7. EtOH/0.3 M NaOAc, pH 5.0.
8. 70 % (v/v) ethanol.

2.5 Reverse Transcription

1. 4.5× hybridization buffer: 225 mM K-Hepes, pH 7.0, 450 mM KCl. Store at −20 °C.
2. ^{32}P- or Cy5-labeled DNA oligos. Store at −20 °C.
3. 5× reaction buffer for the Transcriptor reverse transcriptase (Roche). Store at −20 °C.
4. 10 mM dNTPs mix (Li-Salt). Store at −20 °C.
5. 10 mM ddNTPs. Store at −20 °C.
6. 7.9× extension buffer: 1.3 M Tris–HCl, pH 8.0, 100 mM $MgCl_2$, 100 mM DTT. Store at −20 °C.

7. Transcriptor reverse transcriptase (10 U/μL; Roche), or comparable product. Store at −20 °C.
8. AMV reverse transcriptase (10 U/μL). Store at −20 °C.
9. 1 M NaOH.
10. 1 M HCl.
11. 0.5 M EDTA, pH 8.0.
12. Glycogen (10 mg/mL). Store at −20 °C.
13. Ethanol/0.3 M NaOAc, pH 5.0.
14. 70 % (v/v) ethanol.
15. Loading buffer I: 7 M urea, 25 % (w/v) sucrose, 0.025 % (w/v) bromophenol blue, 0.025 % (w/v) xylene cyanol in 1× TBE.
16. Loading buffer II: 7 M urea, 25 % (w/v) sucrose in 1× TBE.
17. 10× TBE: 0.89 M Tris base, 0.89 M boric acid, 20 mM EDTA.

2.6 Denaturing Polyacrylamide Gel Electrophoresis (PAGE)

1. 8 % acrylamide solution: 7 M urea, 8 % (v/v) acrylamide/bisacrylamide (19:1) in 1× TBE (*see* **Note 2**). Store at 4 °C in the dark.
2. 10× TBE: 0.89 M Tris base, 0.89 M boric acid, 20 mM EDTA (*see* **Note 2**).
3. 10 % (w/v) APS (ammonium persulfate). Store at 4 °C (*see* **Note 3**).
4. N,N,N′,N′-tetramethyl-ethylenediamine (TEMED). Store at 4 °C.
5. Glass plates (42 (L) × 20 (W) cm).
6. Metal plates which cover 2/3 of the glass plates.
7. Comb and spacer (0.4 mm thickness).
8. Vertical gel electrophoresis apparatus (adjustable).
9. High voltage power supply.
10. Whatman 3MM paper.
11. Saran wrap.
12. Phosphorimager exposure cassette and screen (GE Healthcare) or comparable product.
13. Phosphorimager Typhoon TRIO (GE Healthcare) or comparable equipment.

3 Methods

3.1 In Vitro Transcription Using the T7 RNA Polymerase

1. To set up the reaction mix 25 μg linear dsDNA template (containing a T7 promoter upstream of the gene of interest) with 100 μL 10× transcription buffer, 10 μL 1 M DTT, 20 μL of each 100 mM ribonucleotide stock (rATP, rCTP, rGTP and

rUTP, respectively), 5 μL RNase inhibitor (40 U/μL), 40 μL T7 RNA polymerase (homemade) and ddH_2O to a final volume of 1 mL (*see* **Notes 1**, **4** and **5**).

2. Incubate the sample at 37 °C for 4 h.
3. To degrade the template, add 5 μL RNase-free DNase I (2 U/μL) and incubate at 37 °C for 30 min.
4. To precipitate the sample, transfer it in a 15 mL tube and add 2 μL glycogen (10 mg/mL), 25 μL 0.5 M EDTA, pH 8.0, and 2.5× volumes of ethanol/0.3 M NaOAc, pH 5.0. Incubate the sample at −20 °C for at least 60 min.
5. Centrifuge the samples at 4 °C for 30 min (18,000 × *g*).
6. Remove the supernatant, wash the pellet with 70 % (v/v) ethanol, and dry it at room temperature for 10 min.
7. Resuspend the pellet in 30 μL ddH_2O and 80 μL loading buffer I.

3.2 Purification of In Vitro Transcribed RNA

1. To purify the transcript pour a denaturing polyacrylamide gel (*see* **Note 4**). The percentage of the gel depends on the length of the transcribed RNA (Table 1). In case of human telomerase RNA (451 nt) a 5 % acrylamide solution is used. Use 70 mL of the respective acrylamide solution and mix it with 70 μL TEMED and 700 μL 10 % (w/v) APS and let it polymerize for at least 60 min.
2. Load the entire sample on the gel and let it run at 20–30 W (the power depends on the percentage of the gel and the time of the gel run depends on the length of the transcript).
3. Disassemble the gel and cover it with a Saran wrap.

Table 1
Dye migration in denaturing polyacrylamide gels

AA solution (%)	Bromophenol blue	Xylene cyanol
5	35 nt	130 nt
6	26 nt	100 nt
8	19 nt	75 nt
10	12 nt	55 nt
20	8 nt	25 nt

Depending on the acrylamide concentration bromophenol blue and xylene cyanol co-migrate with nucleic acid fragments of different length. The appropriate percentage of acrylamide should therefore be chosen with respect to the size of the RNA or DNA molecules to be separated on the denaturing PAGE

4. Detect the transcript band by UV shadowing (254 nm). Place the gel on top of a fluor-coated TLC plate and hold it below a UV light source (254 nm). The TLC plate will appear in a yellow-green color, while the nucleic acid bands in the gel cast a shadow on the TLC plate, thereby becoming visible as a dark band.
5. Cut the marked band with a sterile blade and put the gel piece in a 15 mL reaction tube.
6. Add 3 mL elution buffer per 1 mL transcription reaction.
7. Freeze the sample at −80 °C for at least 30 min.
8. Thaw and vortex the sample.
9. Put the sample on a turning wheel at 4 °C for 2–3 h.
10. Centrifuge at 18,000 × *g* for 10 min.
11. Pour the supernatant in a syringe and press the liquid through a 0.2 μm sterile filter. If necessary, rinse the syringe with additional 200 μL elution buffer.
12. To precipitate the transcript, add 2 μL glycogen (10 mg/mL) and 2.5× volumes of ethanol p.A. and keep the sample at −20 °C for at least 60 min.
13. Centrifuge the samples at 4 °C for 30 min (18,000 × *g*).
14. Remove the supernatant and dry the pellet at room temperature for 10 min.
15. Resuspend the transcript in 25–50 μL ddH_2O depending on the size of the pellet. Store the sample at −20 °C.

3.3 Chemical Probing of RNA In Vitro

3.3.1 Folding Reaction

1. In a volume of 45 μL assemble 10 pmol RNA with 5 μl 1 M KCl and 25 μL 160 mM modification buffer. For DMS and Kethoxal reactions use cacodylate buffer, pH 7.5, and for CMCT samples potassium borate buffer, pH 8.0, as modification buffer.
2. Denature the RNA at 95 °C for 1 min.
3. Incubate the sample at room temperature for 2 min.
4. Add 5 μL 1 M $MgCl_2$ (f.c. 100 mM) or 5 μL ddH_2O to the folded and unfolded RNA samples, respectively, (Fig. 3) and incubate at 42 °C for 30 min (*see* **Note 6**).

3.3.2 Modification Step

1. Add the modification reagent to the RNA samples as indicated in Table 2 (*see* **Notes** 7 and **8**).
2. Incubate the samples at room temperature for 20 min. In case of DMS the samples can be kept at 37 °C as well.
3. To stop the reaction, add 1 μL β-mercaptoethanol to the DMS-treated samples and 1 μL 0.5 M potassium borate buffer, pH 7.0, to the Kethoxal-treated ones (*see* **Note 9**).

Table 2
Prepare fresh solutions of the modifying reagents for each experiment

CMCT	DMS	Kethoxal
Add 10 μL of a 42 mg/mL CMCT stock solution; always prepare a fresh stock solution	Add 1 μL of a 1.07 M DMS solution; always prepare a fresh 1:10 dilution in ethanol p.A. from the commercially available 10.7 M DMS stock	Add 10 μL 7.4 mg/mL Kethoxal solution; always prepare a fresh 1:5 dilution in 25 % (v/v) EtOH from a 37 mg/mL Kethoxal stock

4. Precipitate the samples; add 2 μL glycogen (10 mg/mL), 1 μL 0.5 M EDTA, pH 8.0, and 2.5× volumes of ethanol/0.3 M NaOAc, pH 5.0, and keep them at −20 °C for at least 60 min.
5. Centrifuge the samples at 4 °C for 30 min (18,000 × *g*).
6. Remove the supernatant, wash the pellet with 70 % (v/v) ethanol, and dry it at room temperature for 10 min.
7. Resuspend the CMCT-modified sample in 2.5 μL 25 mM potassium borate buffer, pH 8.0, per 1 pmol RNA. In case of DMS- and Kethoxal-treated RNA, resuspend the pellet in 2.5 μL ddH_2O per 1 pmol RNA, yielding a concentration of 0.4 μM.

3.4 5′ End-Labeling of DNA Primers

1. To set up the primer kinase reaction, mix 10 pmol DNA oligo, 10 pmol [γ-^{32}P]-ATP (10 μCi/μL, 6,000 Ci/mmol), 2 μL 10× PNK buffer and add ddH_2O to a volume of 19 μL (*see* **Notes 10** and **11**). Add 1 μL T4 polynucleotide kinase (10 U/μL).
2. Incubate the sample at 37 °C for 30 min.
3. Stop the reaction by adding 1 μL 0.5 M EDTA, pH 8.0.
4. To denature the enzyme place the sample at 95 °C for 1 min, then immediately put the tubes on ice for 2 min.
5. To precipitate the sample, add 2 μL glycogen (10 mg/mL) and 2.5× volumes of ethanol/0.3 M NaOAc, pH 5.0, and put the tube at −20 °C for at least 60 min.
6. Centrifuge the samples at 4 °C for 30 min (18,000 × *g*).
7. Remove the supernatant, wash the pellet with 70 % (v/v) ethanol, and dry it at room temperature for 10 min.
8. Resuspend the oligo in 40 μL ddH_2O and store it at −20 °C.

3.5 Detecting the Modification Sites by Reverse Transcription

1. To set up the annealing reaction mix 2.5 μL 0.4 μM RNA, 1 μL 0.25 μM ^{32}P-labeled primer or 0.5 μM Cy5-labeled primer, and 1 μL 4.5× hybridization buffer (*see* **Notes 12–14**).
2. Incubate the sample at 95 °C for 1 min.

3. Cool the sample either rapidly (snap cooling) by placing it on ice for 2 min or slowly by continuous incubation on a thermomixer until its temperature decreased to 55 °C.
4. For the extension of the ^{32}P-labeled primer, add 3.4 μL extension mix to the sample consisting of 1 μL 7.9× extension buffer, 0.7 μL 10 mM dNTPs, 1.55 μL ddH_2O, and 0.15 μL AMV reverse transcriptase (10 U/μL). In case of Cy5-labeled primers, add 14.5 μL extension mix consisting of 4 μL 5× reaction buffer, 2 μL 10 mM dNTPs, 1 μL 0.1 M DTT, 7.5 μL ddH_2O, 0.5 μL RNase inhibitor (40 U/μL), and 0.5 μL Transcriptor reverse transcriptase (10 U/μL). For the sequencing lanes add 1.5 μL of a 1 mM ddNTP solution to the sample in addition to the extension mix (*see* **Notes 13** and **14**).
5. Incubate the samples at 55 °C for 60 min to synthesize the cDNA.
6. To degrade the RNA template, add 3 μL 1 M NaOH.
7. Incubate for 30–60 min at 55 °C.
8. Add 3 μL 1 M HCl to neutralize the pH.
9. To precipitate the samples, add 2 μL glycogen (10 mg/mL), 1 μL 0.5 M EDTA, pH 8.0, and 2.5× volumes of ethanol/0.3 M NaOAc, pH 5.0, and incubate them at −20 °C for at least 60 min.
10. Centrifuge the samples at 4 °C for 30 min (18,000 × *g*).
11. Discard the supernatant, wash the pellet with 70 % (v/v) ethanol, and dry it at room temperature for 10 min.
12. Resuspend the cDNA in 10 μL loading buffer I (for ^{32}P-labeled cDNA) or loading buffer II (for Cy5-labeled cDNA).

3.6 Denaturing Polyacrylamide Gel Electrophoresis (PAGE) to Separate the cDNA Pool

1. Assemble the glass plates and stabilize them with clamps.
2. Mix 50 mL 8 % acrylamide solution with 500 μL 10 % (w/v) APS and 50 μL TEMED and pour the gel (*see* **Note 3**).
3. Insert the comb and let the gel polymerize in a horizontal position for at least 60 min.
4. After removing the clamps and the comb adjust the gel in the apparatus and fill the buffer reservoirs with 1× TBE (*see* **Note 2**).
5. Rinse the wells with 1× TBE.
6. Attach a metal plate to the front of the gel and let it pre-run at 40 W for 30 min (*see* **Note 15**).
7. Denature the samples at 95 °C for 1 min and then put them on ice.
8. Load half of the sample (5 μL) on the gel. If fluorescently labeled primers were used for reverse transcription, load 2 μL

of loading buffer I in one of the empty wells as it contains dyes (xylene cyanol and bromophenol blue). This allows following the gel run.

9. Use a darkroom or cover the apparatus with a dark plastic bag if you are working with light sensitive samples.
10. Run the gel at 40 W for 90–120 min depending on the size of the cDNA fragments to be separated (Table 1).
11. To disassemble the gel, remove the spacers and carefully detach the coated glass plate from the gel. In case of fluorescently labeled samples, proceed directly to **step 15**.
12. Transfer the gel on a Whatman 3MM paper and cover it with Saran wrap.
13. Dry the gel by applying heat and vacuum (~1 h at 80 °C).
14. Expose the gel to a phosphorimager screen overnight.
15. Scan the screen using a phosphorimager Typhoon Trio or equivalent equipment for radioactively labeled cDNA. In case of Cy5-labeled cDNA, scan the gel directly.
16. For the analysis of the gel (*see* **Notes 16–21**), use a software like ImageQuant or SAFA [24].

4 Notes

1. RNA degradation is a frequent problem when working with RNA, therefore some precautions should be considered: Always wear gloves while handling the samples; plasticware and glassware as well as solutions used should be prepared and maintained RNase-free, since autoclaving is not sufficient to inactivate all RNases; always keep RNA on ice while working (unless indicated otherwise) and store it at −20 °C; avoid freezing and thawing of the RNA stock too often; clean the bench and the equipment with "RNaseZap" (Ambion) or a similar product; aliquot the reagents, so if a contamination is encountered, one can simply take a new aliquot without having to remake the stock solutions.
2. To avoid differences in salt concentrations always use the same TBE buffer stock for preparing the acrylamide solution and running the gel.
3. The 10 % (w/v) APS solution can be stored at 4 °C for up to 2 weeks. Alternatively, freeze it in aliquots at −20 °C, which are stable up to 6 months.
4. For in vitro transcription the T7 MegaScript kit (Ambion) can be used alternatively and the transcript can be purified using the MegaClear kit (Ambion).

5. For optimal transcription efficiency it is important to optimize the incubation time as well as the concentration of nucleotides, Mg^{2+} ions and the DNA template, as these can significantly affect the transcription yield. Increasing the amount of DTT can enhance the yield as well.
6. Incubation temperature and time as well as concentrations of monovalent and divalent ions have to be optimized for a given RNA.
7. Kethoxal and DMS are highly toxic and carcinogenic chemicals which can easily penetrate the skin. Thus, always wear gloves and work under a fume hood to avoid inhaling. For more detailed information about Kethoxal, DMS and CMCT read the Material Safety Data Sheet (MSDS) provided by the manufacturer.
8. It is recommended to perform a concentration series with the chemical reagent to determine the single-hit conditions, i.e., only one modification per RNA molecule should occur. Single-hit conditions typically result in a pool of modified RNA harboring less than one modification and therefore the same amount of full-length cDNA is observed, for example, in the DMS-treated sample and in the unmodified control. If the lanes are fading out (and barely any full-length product is observed), a lower concentration of the modifying reagent should be used.
9. To assure that the DMS reaction is successfully quenched by β-mercaptoethanol, perform a stop control. Therefore add β-mercaptoethanol to the sample before adding the DMS. The stop control should give a pattern comparable to the RT stop control prepared from unmodified RNA.
10. The PNK enzyme and the PNK buffer are highly sensitive to changes in temperature. Therefore aliquot the buffer to avoid frequent freezing and thawing.
11. It is highly recommended to purify the non-labeled DNA oligos using a standard denaturing PAGE in order to remove any organic remnants from the oligo synthesis, since these can impair the end-labeling reaction. Thus, gel-purification of the primer increases the labeling efficiency and in turn the signal intensity. Gel-purification of Cy5-labeled oligos is recommended as well, as it improves the reverse transcription results.
12. Primers have to be designed in a way that they bind efficiently to their target and do not form intramolecular or intermolecular interactions. Their length is typically between 18 and 25 nt and they should preferably contain a G or C at their 3′ end. Varying primer length (and its T_M) can increase binding efficiency of the oligo to the RNA. To map RNAs consisting of

more than 100 nt, several primers are required to map the modification sites throughout the entire RNA. When designing this primer set, ensure that there in a certain overlap between the sequences mapped by subsequent primers.

13. Experimental conditions like incubation temperature and time, concentrations of dNTPs and ddNTPs (for the sequencing lanes), Mg^{2+} and RNA have to be adjusted for each construct individually to yield sufficient amounts of cDNA and in turn to obtain a good signal-to-noise ratio. If the ddNTP concentration is too high sequencing lanes tend to give a very faint signal in the upper part of the gel. If the concentration is too low, it may happen that some nucleotides cannot be detected, even though flanking residues are visible.
14. To distinguish between so-called natural stops of the reverse transcriptase enzyme and those induced by DMS (or CMCT, Kethoxal) modification, an RT stop control has to be included. This control is prepared by reverse transcribing RNA, which was refolded but not treated with DMS, CMCT or Kethoxal, thereby revealing the natural stops. These are of different origin: (1) Secondary structures that re-form after denaturing of the RNA as well as (2) G/C sequence stretches can block the reverse transcriptase enzyme and in turn stop the extension. (3) Due to a strain on the backbone RNAs often have "hot spots" for breakage. (4) Partial degradation by RNases results in RT stops as well.
15. If the probes on the gel run unevenly ("smiley effect"), ensure that the gel runs with sufficient temperature (45–55 °C) and attach a metal plate in front of the gel to achieve an equally distributed temperature all over the gel.
16. If the signal to noise ratio is weak, it is often advisable to switch from AMV reverse transcriptase to a more thermostable and sensitive enzyme, like Thermoscript (Invitrogen) or Transcriptor reverse transcriptase (Roche), since these show an improved performance in reverse transcribing highly structured and G/C- rich RNAs.
17. Signals with high intensity allow easier interpretation of data since background noise is reduced to a minimum. A low signal to noise ratio makes it difficult to analyze the gels properly. To avoid this problem, it is recommended to adjust the amount of starting material or loading more sample per lane. Another possibility is to extend the exposure time of the gels.
18. Fuzzy or diffuse bands mostly arise from excess salt in the samples or suboptimal buffer conditions (e.g., gel runs at too high voltage). The samples will not migrate uniformly through the gel and can even get stuck in the wells. To overcome this problem, wash the pellet with 70 % (v/v) ethanol and decrease the

precipitation time. Another solution is to increase the percentage of the gel and the running time, as the salt front will run out of the gel.

19. Dark smears on the gel can result from tiny air bubbles or fibers present in the gel or introduced when loading the samples.
20. Fuzzy bands can occur, if the RNA is not fully degraded after the reverse transcription. In this case it is important to control the pH after adding NaOH or prepare fresh NaOH and HCl stocks.
21. Uneven loading makes quantitation more difficult and less reliable, necessitating a rerun of the gel. It is often helpful to determine the counts of every sample and balance them before loading.

Acknowledgements

This work was supported by the Austrian Science Foundation FWF [grants Y401 and P23497 to C.W.]. We want to thank Katharina Auer for providing the data shown in Fig. 3.

References

1. Brion P, Westhof E (1997) Hierarchy and dynamics of RNA folding. Annu Rev Biophys Biomol Struct 26:113–137
2. Pyle AM, Fedorova O, Waldsich C (2007) Folding of group II introns: a model system for large, multidomain RNAs? Trends Biochem Sci 32:138–145
3. Schroeder R, Barta A, Semrad K (2004) Strategies for RNA folding and assembly. Nat Rev Mol Cell Biol 5:908–919
4. Sosnick TR, Pan T (2003) RNA folding: models and perspectives. Curr Opin Struct Biol 13:309–316
5. Treiber DK, Williamson JR (1999) Exposing the kinetic traps in RNA folding. Curr Opin Struct Biol 9:339–345
6. Treiber DK, Williamson JR (2001) Beyond kinetic traps in RNA folding. Curr Opin Struct Biol 11:309–314
7. Woodson SA (2000) Compact but disordered states of RNA. Nat Struct Biol 7:349–352
8. Woodson SA (2005) Structure and assembly of group I introns. Curr Opin Struct Biol 15:324–330
9. Woodson SA (2010) Compact intermediates in RNA folding. Annu Rev Biophys 39:61–77
10. Brunel C, Romby P (2000) Probing RNA structure and RNA-ligand complexes with chemical probes. Methods Enzymol 318:3–21
11. Shcherbakova I, Mitra S, Beer RH et al (2006) Fast Fenton footprinting: a laboratory-based method for the time-resolved analysis of DNA, RNA and proteins. Nucleic Acids Res 34:e48
12. Waldsich C (2008) Dissecting RNA folding by nucleotide analog interference mapping (NAIM). Nat Protoc 3:811–823
13. Wilkinson KA, Merino EJ, Weeks KM (2006) Selective 2′-hydroxyl acylation analyzed by primer extension (SHAPE): quantitative RNA structure analysis at single nucleotide resolution. Nat Protoc 1:1610–1616
14. Ehresmann C, Baudin F, Mougel M et al (1987) Probing the structure of RNAs in solution. Nucleic Acids Res 15:9109–9128
15. Moazed D, Stern S, Noller HF (1986) Rapid chemical probing of conformation in 16 S ribosomal RNA and 30 S ribosomal subunits using primer extension. J Mol Biol 187:399–416
16. Brunel C, Romby P, Westhof E et al (1991) Three-dimensional model of *Escherichia coli* ribosomal 5S RNA as deduced from structure

probing in solution and computer modeling. J Mol Biol 221:293–308

17. Moazed D, Noller HF (1986) Transfer RNA shields specific nucleotides in 16S ribosomal RNA from attack by chemical probes. Cell 47:985–994
18. Moazed D, Noller HF (1987) Interaction of antibiotics with functional sites in 16S ribosomal RNA. Nature 327:389–394
19. Moazed D, Noller HF (1989) Interaction of tRNA with 23S rRNA in the ribosomal A, P, and E sites. Cell 57:585–597
20. Moazed D, Robertson JM, Noller HF (1988) Interaction of elongation factors EF-G and EF-Tu with a conserved loop in 23S RNA. Nature 334:362–364
21. Moazed D, Samaha RR, Gualerzi C et al (1995) Specific protection of 16 S rRNA by translational initiation factors. J Mol Biol 248:207–210
22. Konforti BB, Liu Q, Pyle AM (1998) A map of the binding site for catalytic domain 5 in the core of a group II intron ribozyme. EMBO J 17:7105–7117
23. Waldsich C, Masquida B, Westhof E et al (2002) Monitoring intermediate folding states of the *td* group I intron *in vivo*. EMBO J 21:5281–5291
24. Das R, Laederach A, Pearlman SM et al (2005) SAFA: semi-automated footprinting analysis software for high-throughput quantification of nucleic acid footprinting experiments. RNA 11:344–354

Chapter 6

Massively Parallel RNA Chemical Mapping with a Reduced Bias MAP-Seq Protocol

Matthew G. Seetin, Wipapat Kladwang, John P. Bida, and Rhiju Das

Abstract

Chemical mapping methods probe RNA structure by revealing and leveraging correlations of a nucleotide's structural accessibility or flexibility with its reactivity to various chemical probes. Pioneering work by Lucks and colleagues has expanded this method to probe hundreds of molecules at once on an Illumina sequencing platform, obviating the use of slab gels or capillary electrophoresis on one molecule at a time. Here, we describe optimizations to this method from our lab, resulting in the MAP-seq protocol (Multiplexed Accessibility Probing read out through sequencing), version 1.0. The protocol permits the quantitative probing of thousands of RNAs at once, by several chemical modification reagents, on the time scale of a day using a tabletop Illumina machine. This method and a software package MAPseeker (http://simtk.org/home/map_seeker) address several potential sources of bias, by eliminating PCR steps, improving ligation efficiencies of ssDNA adapters, and avoiding problematic heuristics in prior algorithms. We hope that the step-by-step description of MAP-seq 1.0 will help other RNA mapping laboratories to transition from electrophoretic to next-generation sequencing methods and to further reduce the turnaround time and any remaining biases of the protocol.

Key words RNA structure, Chemical mapping, High-throughput sequencing, Footprinting, Multiplexing

1 Introduction

RNAs play an active role in all kingdoms of life [1, 2], with known functions including catalysis [2, 3], gene regulation [4], genome maintenance [5], and protein synthesis [1]. The functions of RNA molecules appear to be critically dependent on their structures or, in the case of gene translation, their lack of structure. Comparative sequence analysis, X-ray crystallography, and NMR [6, 7] remain the gold standards for inferring functional secondary structures and tertiary interactions in RNA molecules. However, all three of

Christina Waldsich (ed.), *RNA Folding: Methods and Protocols*, Methods in Molecular Biology, vol. 1086,
DOI 10.1007/978-1-62703-667-2_6, © Springer Science+Business Media, LLC 2014

these tools require significant investigator insight and effort and are not generally tractable for all RNAs. When these methods have not been applicable, chemical probing has been a valuable tool, particularly to falsify or support preexisting structural models [8–10]. Enhancing the information content of chemical probing to permit de novo structure determination, to leverage novel chemical probes, and to accelerate RNA design are ongoing areas of research [11–15]. Most chemical probes tend to react with nucleotides that are solvent exposed, flexible, or otherwise uninvolved in canonical or noncanonical base pairing [16–22]. These reactivities are often read out via a reverse transcriptase reaction that extends a DNA primer complementary to the modified RNA [16–22]. The reverse transcriptase stops at the site of a modification, leaving a complementary DNA (cDNA) whose length marks the modified nucleotide. To identify the sites of the reactive nucleotides, the pool of cDNAs is typically analyzed through length separation by gel or capillary electrophoresis (CE), using radiolabeled or fluorophore-labeled primers, respectively.

With magnetic bead purifications and 96-well plates, mapping several dozen RNAs with a set of different chemical reagents is now feasible on the time scale of a day [11]. However, some applications, such as characterizing RNAs discovered by high-throughput biological screens or created by computational design [13], require probing 1,000s of RNAs. Enzymatic mapping, which gives sparse reads, can be carried out on such large pools—or, indeed, at a genome-wide scale [23–25]—through next-generation-sequencing, but these data do not carry the information content of single nucleotide resolution chemical probes. An experimental protocol and bioinformatics pipeline developed by Lucks and colleagues has pioneered the use of 2′ OH acylating reagents and Illumina next-generation sequencing (NGS) to enable highly parallel chemical mapping of several RNAs [26, 27], but the method has not yet been applied to RNAs of many distinct sequences or with alternative chemical probes [26]. Furthermore, in our hands, this method leads to mapping profiles that do not agree within error with measurements from replicate mapping experiments based on capillary electrophoresis. The differences are at least partially due to the sequence-dependence of ligation and amplification steps that have been used to prepare libraries for Illumina sequencing (unpublished data).

To increase the rate at which a large number of RNAs may be probed and to reduce bias, we have further advanced this protocol into a MAP-seq (Multiplexed Accessibility Probing) method (Fig. 1). Multiple chemical reactions, such as DMS alkylation [12], carbodiimide modification of bases [12, 28], and acylation of the 2′ hydroxyl [21, 22] can be carried out in parallel (Fig. 2). Addition of an Illumina adapter sequence to reverse transcribed DNA is carried

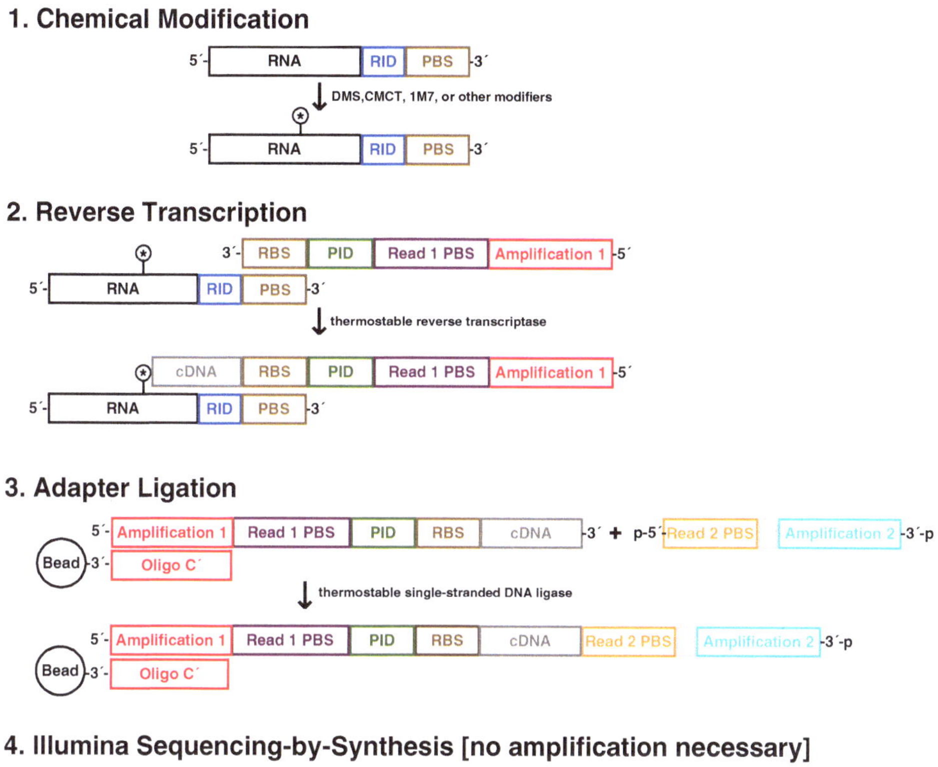

Fig. 1 Overview of the MAP-seq experimental protocol. Each RNA has a primer binding site at its 3′ end (PBS), preceded by an RNA ID (RID) unique to each sequence, which we usually sequester into a hairpin (Table 1). First, the RNA is subject to modification by a chosen chemical reagent. The site of modification is denoted with an asterisk. Second, the RNA is reverse transcribed using a primer that has at its 3′ end an RNA binding site (RBS) complementary to the primer binding site (PBS). The primer contains a sequence-balanced 12-nucleotide primer identification region (PID), and segments necessary for sequencing read 1 and bridge amplification in the Illumina flow cell. Third, the second Illumina adapter containing the read 2 PBS, and the other DNA necessary for bridge amplification is ligated to the 3′ end of each cDNA with a single-stranded DNA ligase. Additional sequence segments within the Illumina adapters (*gray*) can be used for further multiplexing of experimental conditions, but are not used in the current protocol

out under conditions that give near-quantitative single-stranded DNA ligation [29, 30]. Unique sequence identifiers at the 3′ ends of each RNA are sequestered into hairpins (Fig. 1, Table 1) to help prevent interference with the fold of the RNA segment of interest. Solution PCR steps, which can introduce significant length and sequence bias, are eliminated. Last, we noticed that current NGS sequence analysis packages [31–33] either became unusably slow and/or inaccurate in assigning reads at the 3′ ends of sequences and across highly similar sequences (e.g., single mutants in mutate-and-map experiments). We have therefore developed a new automated tool, called MAPseeker, which quantifies the reactivities of the nucleotides of each RNA under each condition.

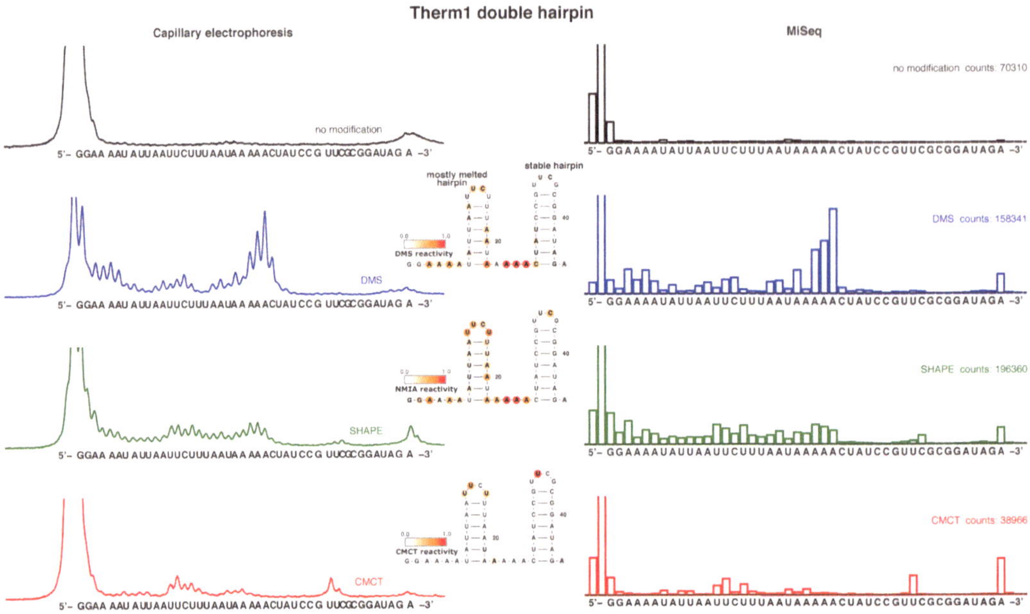

Fig. 2 Comparison of capillary electrophoresis data (CE) and MAP-seq data on a double hairpin RNA Therm1, probed under control conditions (no modifier), with dimethyl sulfate (DMS), N-methylisatoic anhydride (NMIA), and 1-cyclohexyl(2-morpholinoethyl) carbodiimide metho-p-toluene sulfonate (CMCT). Raw data are shown: experimental capillary electrophoretic traces (*left*, arbitrary units; longer products on left) and number of reads (*right*). When quantitated and displayed on the predicted secondary structure (*middle*; *red* indicates reactivity of most reactive nucleotide), the modifiers indicate that the first of two predicted hairpins in the Therm1 construct is significantly melted in the tested conditions (24 °C; 1 M NaCl; 50 mM Na-HEPES, pH 8.0), as compared to the second longer hairpin

Comparisons of this NGS approach and prior CE methods are shown for different chemical modifiers in Fig. 2. The data from the two techniques agree within the estimated error of each method, as can be seen in direct overlays for two RNAs (Fig. 3). These data suggest that MAP-seq biases have been reduced to the point that the NGS-based data are as trustworthy as CE data.

The MAP-seq protocol is rapidly evolving, with active developments to enhance multiplexing through "index" reads in Illumina machines, to probe oligonucleotide pools and sub-pools derived from massively parallel synthesis technologies, and to tackle RNAs with lengths of kilobases. Continuing improvements in single-stranded DNA ligation, through new enzymes and solution conditions [34, 35], may further reduce any residual bias at this step. This chapter gives a snapshot of our stable "version 1.0" lab protocol that runs on the tabletop MiSeq platform for RNA pools with lengths up to 200 nucleotides.

Table 1
Sequences used in this protocol, all shown from 5′ to 3′. *Color coding* corresponds to regions highlighted in Fig. 1. Abbreviations for modifications are those used by Integrated DNA Technologies: */5phos/* 5′-phosphate, */3phos/* 3′-phosphate, */iBioT/* internal biotinylated T, */3Bio/* 3′ biotin, */5-6FAM/* 5′-6-fluorescein

Example DNA Transcription Templates	TTCTAATACGACTCACTATAGGAAAATATTAATTCTTTAATAAAAACTATCCGTTCGCGGATAGAAAAGAAACAACAACAACAAC TTCTAATACGACTCACTATAGGAAACAAAAAAAAAAACGGCGATACGGATCGAGGCGAATTCGCCACATAGAAATATGGCGGAAA AAACAGACGGTTCGCCGTCTGAAAAGAAACAACAACAACAAC TTCTAATACGACTCACTATAGGAAACAAAAAAAAAAACGGCGATACGGATCGAGGCGAATTCGCCACATAGAAATATGGCGGAAA AAAATGCACGTTCGCGTGCATAAAAGAAACAACAACAACAAC TTCTAATACGACTCACTATAGGAAACAAAAAAAAAGGGCAGATCGAAAGATCAGTATAAGGGTACAGTACGAGGGTACAGCAAAA AAAACGCGACTTCGGTCGCGTAAAAGAAACAACAACAACAAC TTCTAATACGACTCACTATAGGAAACAAAAAAAAAGGGCAGATCGAAAGATCAGTATAAGGGTACAGTACGAGGGTACAGCAAAA AAAGGTTAGCTTCGGCTAACCAAAAGAAACAACAACAACAAC TTCTAATACGACTCACTATAGGAAACAAAAAAAAACTGCACGAAAAAATTCGAGCTGAAAACAGCAGGTTAAAAAGCCAGCCAAA AAACATAGCGTTCGCGCTATGAAAAGAAACAACAACAACAAC
Example RNAs	GGAAAAUAUUAAUUCUUUAAUAAAAACUAUCCGUUCGCGGAUAGAAAAGAAACAACAACAACAAC GGAAACAAAAAAAAAAACGGCGAUACGGAUCGAGGCGAAUUCGCCACAUAGAAAUAUGGCGGAAAAAACAGACGGUUCGCCGUCU GAAAAGAAACAACAACAACAAC GGAAACAAAAAAAAAAACGGCGAUACGGAUCGAGGCGAAUUCGCCACAUAGAAAUAUGGCGGAAAAAAAUGCACGUUCGCGUGCA UAAAAGAAACAACAACAACAAC GGAAACAAAAAAAAAGGGCAGAUCGAAAGAUCAGUAUAAGGGUACAGUACGAGGGUACAGCAAAAAAAACGCGACUUCGGUCGCG UAAAAGAAACAACAACAACAAC GGAAACAAAAAAAAAGGGCAGAUCGAAAGAUCAGUAUAAGGGUACAGUACGAGGGUACAGCAAAAAAAGGUUAGCUUCGGCUAAC CAAAAGAAACAACAACAACAAC GGAAACAAAAAAAAACUGCACGAAAAAAUUCGAGCUGAAAACAGCAGGUUAAAAAGCCAGCCAAAAAACAUAGCGUUCGCGCUAU GAAAAGAAACAACAACAACAAC
Reverse Transcription Primers A-D	AATGATACGGCGACCACCGAGATCTACACTCTTTCCCTACACGACGCTCTTCCGATCTACCAGGCGCTGGGTTGTTGTTGTTGTT TCTTT AATGATACGGCGACCACCGAGATCTACACTCTTTCCCTACACGACGCTCTTCCGATCTGAGGCCTTGGCCGTTGTTGTTGTTGTT TCTTT AATGATACGGCGACCACCGAGATCTACACTCTTTCCCTACACGACGCTCTTCCGATCTCTTTAAAATATAGTTGTTGTTGTTGTT TCTTT AATGATACGGCGACCACCGAGATCTACACTCTTTCCCTACACGACGCTCTTCCGATCTTGACTTGCACATGTTGTTGTTGTTGTT TCTTT
Illumina Adapter for cDNA ligation	/5phos/AGATCGGAAGAGCGGTTCAGCAGGAATGCCGAGACCGATCTCGTATGCCGTCTTCTGCTTG/3phos/
Oligo C′ [double biotinylated]	TGTGTAGATCTCGGTGGTCGCCGTATCATTTTTTTTTTTT/iBiodT//3Bio/
Loading Control 1	/5-6FAM/AAAAAAAAAAAAAAAAAAAAGTTGTTGTTGTTGTTTCTTT
Loading Control 2	/5-6FAM/GTTGTTGTTGTTGTTTCTTT

2 Materials

2.1 General Materials

We carry out some steps of this procedure within wells of a 96-well plate, which enables use of magnetic separators and multichannel pipettors and can reduce the experimentalist's effort. Alternatively, the entire protocol can also be carried out in RNAse-free nonstick Eppendorf tubes. We use a capillary electrophoresis (CE) sequencer and fluorescent primers to check the yield and length distribution at two steps prior to Illumina sequencing. Use of fluorescent DNA is particularly convenient for rapidly piloting improvements to the protocol and for directly comparing CE data to Illumina data. As an alternative, steps to monitor the lengths of DNA fragments can also be carried out without fluorescent primers on, e.g., Agilent Bioanalyzer systems [26].

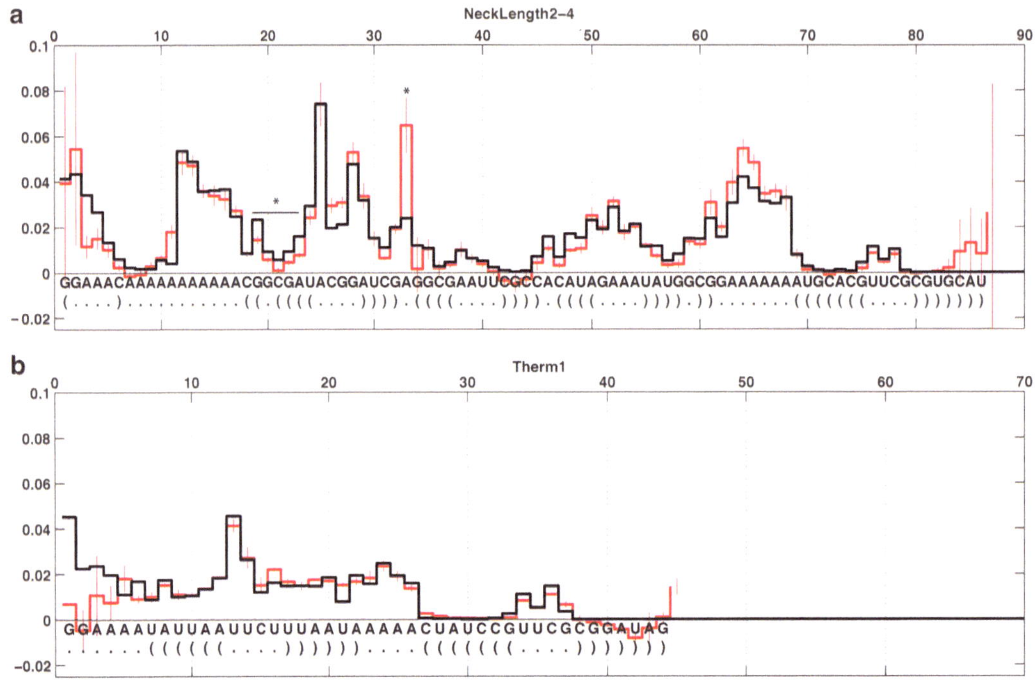

Fig. 3 Quantitative agreement of MAP-seq data (*black*) with "gold standard" capillary electrophoresis data (*red*) taken in replicate. (**a**) 2′ OH acylation data using the 1M7 (1-methyl-7-nitroisatoic anhydride) modifier for an RNA from the Eterna massively parallel design platform (from a player's project "Neck Length 2-4"). (**b**) 1M7 data for the Therm1 hairpin (*see* also Fig. 2). Error bars for the MAP-seq data were estimated from statistical error on the number of reads (smaller than the line width at most sequences), as carried out by MAPseeker analysis. Error bars for the CE data were estimated after HiTRACE quantitation [36] as the standard deviation across five measurements using 1M7 concentrations of 0.625, 1.25, 2.5, 5, and 10.0 mg/mL, and an additional two replicates in the 10 mg/mL condition. Two regions that show discrepancy between the CE and MAP-seq data beyond this error are highlighted with stars; both cases can be traced to band compression in the CE experiments (not shown), which introduces systematic uncertainty in band deconvolution near guanosine residues

1. One 96-well V-bottom plate.
2. Three or more plastic seal films.
3. 24-post and 96-post magnetic stand.
4. Plenty of RNase-free sterile water.
5. 10 mL 70 % ethanol, freshly prepared.
6. At least 120 μL 40 % polyethylene glycol 8000.
7. ROX–formamide mix: Mix 1 μL of ROX 350 ladder (Applied Biosystems) in 350 μL of Hi-Di formamide. This volume is enough for five 4-sample reactions and can be stored frozen at −20 °C.

8. Two 96-well format dry incubators, set to 55 °C and 90 °C before starting the chemical mapping part of this protocol. If these types of incubators are not available, a water bath or thermocycler may be substituted.
9. Access to an ABI 3100 sequencer or similar equipment for capillary electrophoresis.

2.2 Chemical Modification of RNA

It is assumed that the investigator has already obtained the RNA molecules of interest, either by chemical synthesis, in vitro transcription, or extraction from cells or viruses. The protocol below has been tested most thoroughly on pools of RNAs that have identical primer binding sites (PBS) at their 3′ ends. In addition, it is critical that the last segment of each sequence (RID in Fig. 1), immediately 5′ to the reverse transcription binding site, is unique in the pool [26, 27]. These nucleotides (8 in our example cases below; Table 1) are used later to identify the RNA with which each NGS read is associated. In our examples, these eight nucleotides are arbitrary octamers, but we also included additional complementary nucleotides to sequester the RID into a hairpin capped by a UUCG tetraloop (Table 1), which reduces the potential for this added sequence interfering with the RNA fold. Aside from these PBS and RID segments, there are few other constraints. We have carried out MAP-seq on RNA pools with between one and 4,000 distinct sequences in a single experiment.

Different chemical modification reagents probe different nucleotides and different aspects of the RNA structure. Each reagent may need its own quench. We list the recipes for common modification reagents and their quenches at concentrations and reaction times appropriate for RNAs under 100 nucleotides in length. Longer RNAs should use lower concentrations and/or shorter reaction times. When deviating from these conditions or using alternative reagents, the RNA molecules should be modified with single-hit frequency, which may depend on reagent concentration, reaction time, and temperature. The protocol given below assumes that four chemical probes will be tested on the pool of RNAs: no modification (addition of water or DMSO), DMS, CMCT, and 1M7. Every 4-sample reaction will require the following:

1. Each reaction should use 2.5–5.0 pmol of total RNA (5 μL of 0.5–1.0 μM RNA).
2. 10.0 μL of 0.5 M Na-HEPES, pH 8.0 (from a sterile stock filtered with a 0.2 μm filter). Other pH stocks can be used, but note that carbodiimide and 2′ hydroxyl modification reactions, as well as other modifiers, are sensitive to pH.
3. 20 μL of 50 mM $MgCl_2$ if the RNA requires magnesium for proper folding. If not, 20 μL 5 M NaCl can be used to give 1 M NaCl conditions; or other folding buffer.

4. 280 μL Agencourt AMPure XP beads (Beckman Coulter) for RNA purification (*see* **Notes 1** and **2**). Stored at 4 °C.
5. DMS mix for dimethyl sulfate mapping, which attacks exposed Watson–Crick faces of adenines and cytosines in RNA and DNA: Mix 1.0 μL fresh dimethyl sulfate into 9.0 μL 100 % ethanol, then add 90.0 μL of sterile water. The quench for this reaction is 2-mercaptoethanol. Each DMS reaction needs 5 μL DMS mix.
6. CMCT mix for carbodiimide mapping using 1-cyclohexyl-(2-morpholinoethyl)carbodiimide metho-p-toluene sulfonate (CMCT), which attacks exposed Watson–Crick faces of guanines and uracils in RNA: Mix 42 mg/mL of CMCT in H_2O. The solid stock for reagent is usually kept at −20 °C. Before mixing into water, incubate it at room temperature for 15 min, immediately prior to experiment. The quench for this reaction is 2-mercaptoethanol. Each CMCT reaction needs 5 μL CMCT mix.
7. 1M7 mix for 2′ hydroxyl acylation mapping using 1-methyl-7-nitroisatoic anhydride (1M7), a fast-acting reagent which attacks dynamic nucleotides by modifying the 2′ hydroxyl of an RNA: Mix 10 mg of 1M7 into 1 mL anhydrous DMSO. The quench for this reaction is 2-mercaptoethanol. Each 1M7 reaction needs 5 μL 1M7 mix. An alternative reagent for 2′ hydroxyl acylation is N-methylisatoic anhydride (NMIA).

2.3 Reverse Transcription

1. 4 μL 5× First Strand buffer (Invitrogen). Store at −20 °C.
2. 1 μL 0.1 M DTT. Store at −20 °C.
3. 1.6 μL dNTPs mixture (10 mM each nucleotide; store at −20 °C) per sample plate.
4. 1.0 μL 40 % PEG 8000.
5. 0.4 μL 200 U/μL SuperScript III reverse transcriptase (Invitrogen). Store at −20 °C.
6. 0.5 μL each of at least four 1 μM reverse transcription primers per sample. These primers contain the Illumina adapter for bridge amplification at the 5′ ends, followed by the Illumina sequencing primer binding sequence, followed by a primer identifier sequence (PID; Fig. 1), and finished with a region to bind the conserved 3′ ends of the RNAs. The PID region is 12 nucleotides long and must be sequence-balanced as part of a set of four primers (*see* **Note 3**; Table 1). The different PIDs can be used to distinguish between the same RNA modified by different reagents or subject to different modification reaction conditions. The primers should be purified by polyacrylamide gel electrophoresis (PAGE) or HPLC and stored at −20 °C.
7. 0.25 μL of 1 μM of at least one of the reverse transcription primers with a 5′ fluorescein (FAM) modification. Store at −20 °C.

8. Streptavidin-coated magnetic beads that have bound double-biotinylated oligos complementary to Illumina's Oligo C sequence (Oligo C′ beads). These beads are assembled using DynaBeads (Invitrogen), using the manufacturer's instructions for coupling to oligonucleotides under RNAse-free conditions. The oligonucleotide sequence is in Table 1.
9. 20 μL 0.4 M NaOH.
10. Acid quench mix, prepared by mixing 1 volume of 5 M NaCl, 1 volume of 2 M HCl, and 1.5 volumes of 3 M sodium acetate, pH 5.2. This stock can be kept at room temperature for at least 2 months. Of this mix, 20 μL is needed per 4-sample reaction.

2.4 Ligation of Second Illumina Adapter

Every 4-sample reaction requires:

1. 2.5 μL 10 μM Read 2 PBS/Amplification 2 (Fig. 1, Table 1). This contains standard sequences used for bridge PCR amplification and 2nd read sequencing on Illumina platforms.
2. 2 μL CircLigase ssDNA ligase (100 U/μL) (Epicentre, Illumina). Store at −20 °C.
3. 2 μL 10× CircLigase ssDNA ligase buffer (Epicentre, Illumina, supplied with CircLigase). Store at −20 °C.
4. 1 μL 1 mM ATP (supplied with CircLigase). Store at −20 °C.
5. 1 μL 50 mM $MnCl_2$ (supplied with CircLigase).
6. 2 μL 40 % PEG 1500.
7. 1 μL fluorescent oligo loading control mix: 1 nM fluorescein-labeled 20-nucleotide DNA oligomer and 1 nM fluorescein-labeled 40-nucleotide DNA oligomer (*see* **Note 4**).

2.5 Final Preparation for Sequencing

For every 4-sample reaction, you will need:

1. 0.5 μL 10 nM PhiX genomic DNA (20 nM of single strands), prepared for sequencing (Illumina, *see* **Note 5**).
2. 2 μL 2.0 N NaOH.
3. 10.0 μL EB buffer (Qiagen).

3 Methods

All steps are at room temperature unless specified.

3.1 Chemical Modification

1. Prepare modification buffer at 2× concentration. For each 20 μL modification reaction, mix 2 μL 0.5 M Na-HEPES, pH 8.0, 4 μL appropriate salt solution (e.g., either 50 mM $MgCl_2$ or 5 M NaCl, as above), and 4 μL RNase-free H_2O. For four reactions, prepare enough premix for five reactions (e.g., 10 μL 0.5 M Na-HEPES, pH 8.0, 20 μL of 50 mM $MgCl_2$, and 20 μL RNase-free H_2O).

2. Add 5 μL of each RNA solution (~2.5–5 pmol) to as many samples in the 96-well plates as you have reagents and reaction conditions. Add 10 μL of modification buffer to each well. Seal the plate with sealer tape. Let the RNAs fold at room temperature for 20 min, or, if necessary, place into a thermocycler, and take the plate through the necessary temperature steps to ensure proper folding of the RNA or RNAs in question.
3. While the RNA is folding, prepare fresh stocks of modification reagents as above. Always make sure that one of the modification conditions is just H_2O or just anhydrous DMSO, as a control reading for background reverse transcriptase stops. Array the various modification reagents in the auxiliary plate so that they may be easily added to the main reaction plate.
4. Add 5 μL of the modification reagent to each corresponding well of the main reaction plate. For four modification conditions [(A) no modification, (B) DMS, (C) CMCT, and (D) 1M7], use 5 μL of each stock reagents freshly prepared as above. The total volume in each of four reactions is now 20 μL.
5. Seal the plate and let incubate. For RNAs under 100 nucleotides with the modification reagents listed above, 15 min is appropriate. For longer RNAs, less time may be necessary so as to achieve single-hit kinetics. Alternative reagents may require their own modifications to this protocol (*see* **Note 6**).
6. While the RNAs are incubating with the modification reagents, array the quenches in corresponding locations in the auxiliary plate (or in a new auxiliary plate) so that they may be easily added to the main reaction plate. After the appropriate incubation time, add 50 μL of the corresponding quench to each well. The volume of each sample is now 70 μL.
7. Prepare the bead purification mix: for each reaction well mix 70 μL AMPure XP beads with 30 μL 40 % polyethylene glycol 8000. Pipette up and down at least 10 times to mix. Add 100 μL bead purification mix (2.5× volumes) to each reaction. Again pipette up and down at least 10 times to mix. Wait 10 min for the beads to bind the RNA (*see* **Note 2**).
8. Set the plate on a 24-post magnetic stand and wait 5 min for the beads to collect near magnets.
9. Pipette off the supernatant and discard.
10. Rinse the beads with 100 μL 70 % ethanol. Shake briefly with the plate on the stand. Wait 30 s. Remove the 70 % ethanol supernatant.
11. Repeat **step 10**. This time, take extra care to remove the last traces of the 70 % ethanol supernatant with a P10 pipette (*see* **Note 7**).

12. Let the samples dry for 10 min while remaining on the magnetic stand. Remove any final visible droplets of 70 % ethanol with a P10 pipette.
13. Resuspend the beads in each well in 2.5 μL RNase-free H_2O. Take the plate off the magnetic stand. Let stand for 5 min to ensure release of RNAs.
14. Place the plate back on the magnetic stand. Wait 1–2 min for the beads to collect. Transfer the supernatant in each well, which now contain the released, modified RNA, to a new plate.

3.2 Reverse Transcription

1. Prepare fresh primer stocks (*see* **Note 4**), 1 μM each. Primers should be used in sets of 4, which will be referred to as A–D within one set—primer A for the first reaction in a set of four, (here, no modification) primer B for the second, (here, DMS) etc. All primers in a set need to be used, due to sequence balancing requirements of the Illumina platform (*see* **Note 3**; Table 1 for example sequences). For at least one primer (primer A in this example), however, instead of using pure unlabeled primer, prepare a mixture from equal parts 1 μM unlabeled primer A and 1 μM fluorescently labeled primer A to use as the primer for the reaction in a group of four. This fluorescent tracer will allow for a rapid quality check by standard capillary sequencing prior to running on the Illumina sequencer. It is also fine to label all primers in use. The fluorophore-labeled cDNA does not interfere with the Illumina sequencing protocol (it is washed away during the bridge PCR on the Illumina flow cell).
2. Add 0.5 μL of 1.0 μM corresponding primer to appropriate well of the plate containing the modified RNAs.
3. Prepare the SuperScript III reverse transcription mix: for each reaction, mix 1 μL of 5× First Strand buffer, 0.25 μL of 0.1 M DTT, 0.4 μL of 10 mM dNTPs, 0.25 μL 40 % PEG 8000, and 0.1 μL Superscript III reverse transcriptase (200 U/μL).
4. Add 2.0 μL of the SuperScript III reverse transcription mix: for each reaction mix to each well. Mix well by pipetting up and down at least 10 times. The total volume of the reactions is 5 μL.
5. Seal the sample plate well. Put it in an incubator set to 55 °C for 30 min. For longer RNAs, longer incubation times may be used to help complete reverse transcription.
6. After 30 min, hydrolyze the leftover RNA by adding 5 μL of 0.4 M NaOH to each well. Incubate the plate at 90 °C for 3 min.
7. Add 5 μL of the acid quench mix to each well. Let the plate cool on ice for 5 min.

8. Add 4 μL of the Oligo C′ beads to each well. Mix thoroughly by pipetting up and down at least 10 times. Incubate at room temperature 10 min to ensure cDNA binding to the beads.
9. Place the plate on a magnetic stand and wait 1–5 min for the beads to collect.
10. Pipette off the supernatant and discard.
11. Rinse the beads with 100 μL 70 % ethanol. Shake briefly by hand with the plate on the stand. Wait 30 s. Remove the 70 % ethanol supernatant.
12. Repeat **step 11**. This time, take extra care to remove the last traces of the 70 % ethanol supernatant with a P10 pipette (*see* **Note 7**).
13. Let the samples dry for 5–10 min while remaining on the magnetic stand. Remove any final visible droplets of 70 % ethanol with a P10 pipette, and let dry further for 5 min.
14. Take the plate off the magnetic stand. Resuspend the beads in each well in 2.5 μL RNase-free H_2O.
15. Take 0.5 μL from every sample reverse transcribed with the primer A/fluorescent primer A mixture and transfer to a new plate. Add 10 μL of the ROX–formamide mixture to each of these representative samples and mix. Wait 5 min to ensure release of cDNA.
16. Place this new plate on a magnetic stand. Wait 1–5 min for the beads to aggregate. Transfer the supernatant to an optical plate used by an ABI capillary electrophoresis sequencer. These samples can be loaded on an ABI sequencer to ensure reverse transcription worked properly before proceeding to the ligation step (Subheading 3.3); alternatively they can be loaded at the same time as the ligated samples to save effort.
17. The remaining samples (still with beads) may be pooled for convenience before proceeding. Ensure the beads are still well-suspended in water by pipetting up and down. In the example case with one pool of RNAs modified under four conditions, the total volume of the combined samples will be 8 μL.

3.3 Ligation of the Second Illumina Adapter

1. Set one incubator or water bath at 68 °C. Set a second one at 80 °C.
2. Prepare the CircLigase ligation mix: for every 4-sample reaction, combine 2.5 μL of 10 μM DNA oligo of second Illumina adapter sequence, phosphorylated at both the 5′ and 3′ ends; 2.0 μL 10× CircLigase ssDNA ligase buffer; 1.0 μL 1 mM ATP; 1.0 μL 50 mM $MnCl_2$; 2.0 μL 40 % polyethylene glycol 1500; 1.5 μL of RNAse-free H_2O; and 2.0 μL CircLigase ssDNA Ligase (100 U/μL) (*see* **Note 8**). The adapter can also be blocked at its 3′ end with a spacer modification instead of a phosphate.

3. Add 12 μL of CircLigase ligation mix to the combined cDNA sample pool of each 4-sample reaction. Incubate at 68 °C for 2 h, then 80 °C for 10 min to inactivate the ligase. This incubation can be carried out in a thermocycler. The oligo C′ beads do not appear to affect the efficiency of the ligation (unpublished data).
4. Add 5 μL 5 M NaCl and mix thoroughly by pipetting up and down. This will ensure proper resuspension of the beads, and raise the salt concentration to ensure proper hybridization of cDNA to oligo C′ beads.
5. Pipette 10 μL to a well of a new 96-well plate. This sample fraction will be another quality check prior to sequencing. The remaining 15 μL of the pool will be used for sequencing. Pipette it to a second well on the same plate.
6. Wait 5 min to ensure rebinding of the cDNAs to the beads. Place the plate on magnetic stand. Wait 1–5 min for the beads to collect.
7. Pipette off the supernatant from each sample and discard.
8. Rinse the beads with 100 μL 70 % ethanol. Shake briefly with the plate on the stand. Wait 30 s. Remove the 70 % ethanol supernatant.
9. Repeat **step 8**. This time, take extra care to remove the last traces of the 70 % ethanol supernatant with a P10 pipette (*see* **Note 7**).
10. Let the samples dry for 5 min while remaining on the magnetic stand. Remove any final visible droplets of 70 % ethanol with a P10 pipette.
11. For the sequencing samples, take the plate off the magnetic stand. Resuspend the beads of a pool in 8 μL of RNase-free H_2O per four samples in the pool. Leave these pools separate until after quantifying the amount of cDNA in each pool.
12. For the quality check samples, take the plate off the magnetic stand. Resuspend each pool of beads in 10 μL of ROX–formamide mix. Wait 5–20 minutes for cDNA to release from the beads. Place the plate back on a magnetic stand. Wait 1–5 min for the beads to collect. Transfer the supernatant to an optical plate appropriate for an ABI capillary electrophoresis sequencer. Add 1 μL fluorescent oligo loading control mix (*see* **Note 4**) and mix well by pipetting up and down. Run these samples on the ABI capillary electrophoresis sequencer.

3.4 Quality Check of Ligation and Quantification of Final Sample for Sequencing

The following steps are based on using the suite of HiTRACE scripts for MATLAB (https://simtk.org/home/hitrace) [36]. The general strategy for this step is to integrate the area of the peaks corresponding to the two fluorescent oligos in the loading control mix, each of which should correspond to approximately 1 fmol of sample. Average them, and then compare their areas to the combined area under the peaks corresponding to the ligated samples. Below are commands to be executed within MATLAB after downloading the HiTRACE scripts and adding them to the MATLAB path. MATLAB and other computer syntax, functions, and variable names are written in an alternate font.

1. Verify that the ligation proceeded correctly. Execute the following command:

   ```
   d = quick_look({'foldername_with_ABI_data'},
   lowerbound, upperbound)
   ```

   foldername_with_ABI_data corresponds to the path of the folder containing the data from the quality check immediately after reverse transcription, and the quality check after ligation. The numbers lowerbound and upperbound depend on the length of the RNA(s) in question. In the resulting view of the capillary traces, ensure that bands are visible for the reverse transcribed product, typically with a strongest band for the fully extended product (Fig. 4). Another band may be visible for unextended primer, although in many experiments we do not see this peak (Fig. 4a), as our primer (Table 1) tends to be efficiently extended. In the second, ligated sample, check that this pattern of peaks has shifted to longer lengths (later elution times) after ligation (Fig. 4b). In principle this ligation should be quantitative but this is not always the case, and the fraction ligated should be estimated (*see* **Note 8**).
2. Subtract the baseline from the data contained in d and store it in a matrix called d_bsub:

   ```
   d_bsub = baseline_subtract_v2(d);
   ```
3. Plot the baseline-subtracted data:

   ```
   figure(5) % This creates a new figure window within
   MATLAB in which to plot the data.
   plot(d_bsub(:,2))
   ```

   Here, it is assumed that the second column of d_bsub corresponds to the ligated sample. Use this plot to identify the bounds on the time axis for the shorter oligo peak, the longer oligo peak, and the fully ligated cDNA region. Record these values as shorter_lower, shorter_upper, longer_lower, longer_upper, cdna_lower, and cdna_upper.
4. Calculate the area of the peaks in the cDNA region by summation:

   ```
   cdna_area = sum(d_sub(cdna_lower:cdna_upper,2))
   ```
5. Calculate the area of the peak for the shorter oligo:

   ```
   shorter = sum(d_bsub(shorter_lower:shorter_upper,2))
   ```
6. Calculate the area of the peak for the longer oligo:

   ```
   longer = sum(d_bsub(longer_lower:longer_upper,2))
   ```
7. Calculate the number of fmols in the 15 μL of sample that can be loaded on the Illumina platform (*see* **Notes 5** and **9**):

   ```
   fmols = 24 * cdna_area / (shorter + longer)
   ```

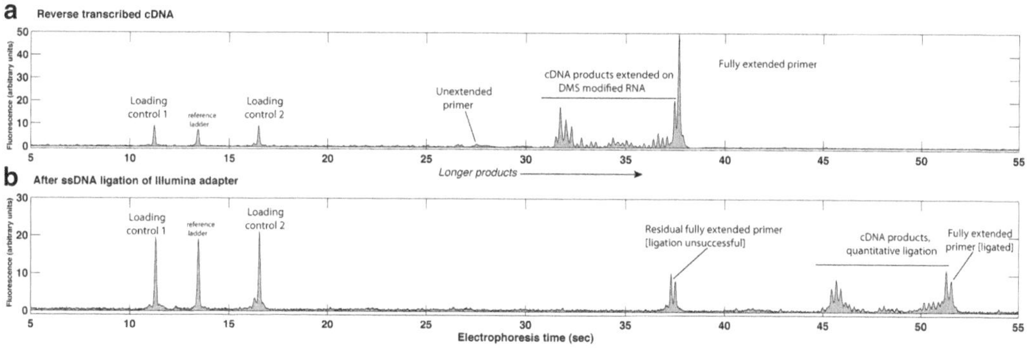

Fig. 4 Quality check and quantitation of NGS library concentration through the use of a fluorescent "tracer" primer and capillary electrophoresis. The sample shown corresponds to a four sample MAP-seq experiment (no modification, DMS, NMIA, CMCT) with a single RNA (Therm1; *see* Table 1), also shown in Fig. 2. Note that the left-to-right orientation of bands is the reverse of the orders in Figs. 2 and 3 and Fig. 5; signals corresponding to stops at more 3′ nucleotides are detected at earlier times in electrophoresis. (**a**) The data shown are for a fluorescein-labeled primer included in the reverse transcription reaction with the DMS sample; a fraction of this reaction was loaded on an ABI 3100 sequencer sensitive to fluorescein. (**b**) After the final Illumina adapter ligation step. The sample loaded here on the ABI 3100 sequencer was a fraction of the sample actually loaded on the Miseq Illumina platform. A clear shift to longer electrophoresis times (longer products; *right-hand peaks*) is apparent. Comparison of the integrated intensity of the ligated products to loading control peaks (1 fmol fluorescein-labeled controls; *left-hand peaks*) gives an estimate of the total ligated product

3.5 Final Preparation for Sequencing on a MiSeq

1. These instructions, especially the amounts of cDNA loaded, are specific for running on a MiSeq using Illumina's version 2 kits. If an alternative platform is used, this protocol will need to be appropriately adjusted. Place a MiSeq version 2 reagent kit in a pool of water as deep as the fill line on the kit to thaw. Place the HT1 buffer vial in the kit on the benchtop to thaw, and then keep on ice.
2. The target is to load 30 fmol of cDNA onto the MiSeq plus 5 fmol of the PhiX genome as a positive control. Thirty divided by the number of pools of samples is the number of femtomoles taken from each pool. Given the quantifications of the number of femtomoles in each pool calculated above, calculate the volume of sample needed to be taken from each pool and contributed to the final pool that will be used for sequencing.
3. If the yield of cDNA from a pool is high, such that the volume of sample to be taken from the pool is less than 0.5 μL, then first take a small aliquot of beads from the sequencing sample and add EB buffer to dilute it down such that the target number of femtomoles will be in a 0.5–4.5 μL volume.
4. If the yield of cDNA from a pool is low, such that the volume of the library needed to get the desired amount of cDNA is in excess of 4.5 μL, put the pool in a well on a 96-well plate, and place the plate on a magnetic stand. Wait 1–2 min for the

beads to collect. Discard the supernatant. Remove the plate from the stand. Resuspend the beads in the desired volume of EB buffer. Our experiments have shown that the beads do not release a significant amount of cDNA into the supernatant when resuspended in water at the end of the ligation stage. Instead, NaOH is necessary to denature the DNA and release it from the beads (*see* **steps** 7 and **8** below).

5. If the volume of the final combined pool of the cDNA library is less than 4.5 μL, add EB buffer such that the final volume equals 4.5 μL.
6. Take 1.0 μL of 10 nM PhiX control sample and put it in a well of a 96-well plate. Add 1.0 μL of EB buffer. Mix by pipetting up and down. Take 0.5 μL of this solution and add it to the final pool of cDNA (on beads).
7. Take 2.0 μL of 2.0 M NaOH and put it in a well of a 96-well plate. Add 18 μL of deionized water and mix by pipetting up and down. Add 5 μL of this solution to the final pool of cDNA beads plus PhiX. Mix thoroughly by pipetting up and down. The final volume will be 10 μL.
8. Let the solution stand for 5 min for the DNA to denature. Place the plate on a magnetic stand. Wait 1–5 min for the beads to collect. Transfer the supernatant to a 1.5 mL tube.
9. Add 990 μL of chilled HT1 buffer to the 1.5 mL tube. Mix thoroughly by pipetting up and down. Transfer 375 μL of this solution to a new 1.5 mL tube. Add 225 μL of additional HT1 buffer and again mix thoroughly by pipetting up and down (*see* **Notes 10** and **11**).
10. Proceed following Illumina's protocol for loading this final 600 μL solution in the reagent cartridge and loading the cartridge and flow-cell on the MiSeq. Use paired-end sequencing so as to be able to read out both the position of the reverse transcription stop and the identity of the RNA in the cluster. The RNA can be of arbitrary length as each sequence is identified by its 3′ RID (Fig. 1) during the first read on Illumina's 50-cycle kits. The first read must be long enough to read through the primer's PID, the 20-nucleotide RNA binding site (RBS), and then through enough nucleotides of the RID region to distinguish the sequence from any other in the pool. Our RNAs typically have RID regions of 8–15 nucleotides. Such an RNA pool would require a first read length of 47 nucleotides. The second read must be long enough to uniquely identify the position of the reverse transcription stop compared to any other position in the RNA. We typically use 20–30 nucleotides, although this may vary depending on the sequence of the RNA. If RNAs in the pool contain long repetitive sequences, the second read will need to be longer so as to

extend into the non-repetitive regions (The 50 cycle kit can, when used in the manner of this protocol, be used to read 76 nucleotides, summed over the first and second read.) Illumina's kits include read allotments for indexing above and beyond the stated read lengths that we do not use—we use the barcoding and identification schemes described here instead.

3.6 MAPseeker Analysis

1. The MAPseeker package is freely available for download at: https://simtk.org/home/map_seeker. Follow the instructions in the README to install on a Mac or other Unix-based machine.
2. Prepare two text files in FASTA format, with a label for each sequence on one line preceded by a ">" character, and the sequence following on the next line. The first FASTA file ("`RNA_sequences.fasta`") should contain the full-length sequences of every RNA in the run including the PBS. The second ("`primers.fasta`") should contain the full sequence of every reverse transcription primer used in the run. Label the primers with the corresponding name of the reagent, including the no modification control as "no mod." There are examples of these files included in the MAPseeker package.
3. The MAPseeker executable takes input in the following format:
```
$ MAPseeker -1 read1.fastq -2 read2.fastq -l RNA_
sequences.fasta -p primers.fasta -n 8
```
The files `read1.fastq` and `read2.fastq` are the FASTQ format sequence files from the MiSeq. The two .fasta files are those prepared in **step 1**. The flag "`-n`" specifies the length of the RID region of the RNA (8, in this example). If the `-n` flag is not specified, MAPseeker will automatically determine an RID length that will uniquely distinguish the library.
4. As output, MAPseeker gives summary statistics: the numbers of sequences found in the FASTQ files, the number in which the RNA binding site was found, the number that contained a primer ID that was specified in `primers.fasta`, and the number of sequences from read 1 and read 2 that matched an RNA sequence specified in `RNA_sequences.fasta`. It also gives text files named `stats_ID1.txt` through `stats_IDN.txt`, where *N* is the number of primers listed in `primers.fasta` (`stats_ID1.txt` corresponds to the first primer in primers.fasta, etc.). The text files contain tab-separated matrices of numbers. Each row corresponds to an RNA as they were listed in the `RNA_sequences.fasta` file. Each column corresponds to the site where reverse transcription stopped, beginning at 0 (full extension of the RT primer back to the RNA 5′ end) and ending at the 3′ end. Stops in the RID region as specified by "`-n`" above will all be zero, as they cannot be attributed conclusively to any one RNA. The values

correspond to the number of clusters from the sequencing run of the particular RNA that stopped at that position.

5. The data in these text files are tab-delimited, so they are amenable to analysis with most visualization or plotting programs, or to be further processed with custom analysis scripts. MAPseeker includes scripts in MATLAB (located in the src/matlab/ subdirectory) for quick data analysis and visualization. This protocol will describe the use of these MATLAB scripts. For ease of use, add the map-seeker/src/matlab/ directory to your MATLAB path, and set the present working directory to the directory containing the `stats_IDN.txt` files as well as the two input FASTA files. Run from within MATLAB:

```
[D, D_err, RNA_info, primer_info] = quick_look_
MAPseeker('RNA_sequences.fasta', 'primers.fasta',
'./', full_length_correction_factor, 0, 0);
```

This gives several windows. One is a histogram of counts per primer, and counts per RNA (figure window 1 within MATLAB; Fig. 5a). Further windows give visualization of the counts (with estimated errors) for the four most highly represented RNAs and "reactivities", corrected for reverse transcriptase attenuation as given by the following equation [26, 37]:

$$R(site\ i) = \frac{F(site\ i)}{\left[F(site\ 0) + F(site\ 1) + \cdots + F(site\ i)\right]} \quad (1)$$

If the ligation of the fully extended product [F (site 0)] was not quantitative, this factor can be taken into account in the full-length correction factor variable (*see* **Note 8**). The 1D profiles of the RNAs with the most reads are shown (figure windows 2 and 3 within MATLAB; Fig. 5b), as well as 2D representation of the entire data set (figure windows 4 and 5 within MATLAB; Fig. 5c). The 2D representations permit visual confirmation that the ID region of the RNA is folded into a hairpin; the 5′ strand of the stem should be protected to chemical modification, and the loop should be reactive. (The chemical reactivity of the 3′strand of the hairpin is not observable as it is used to identify the RNA; stops in this region result in reads that cannot be uniquely aligned to the library of RNA sequences.) All figure windows are also automatically saved to disk as encapsulated PostScript (EPS) files.

6. The matrix `D` contains the attenuation-corrected and background subtracted reactivities at each nucleotide position, and `D_err` contains its estimated error. They are cells with one matrix for each primer. `RNA_info` has information on the

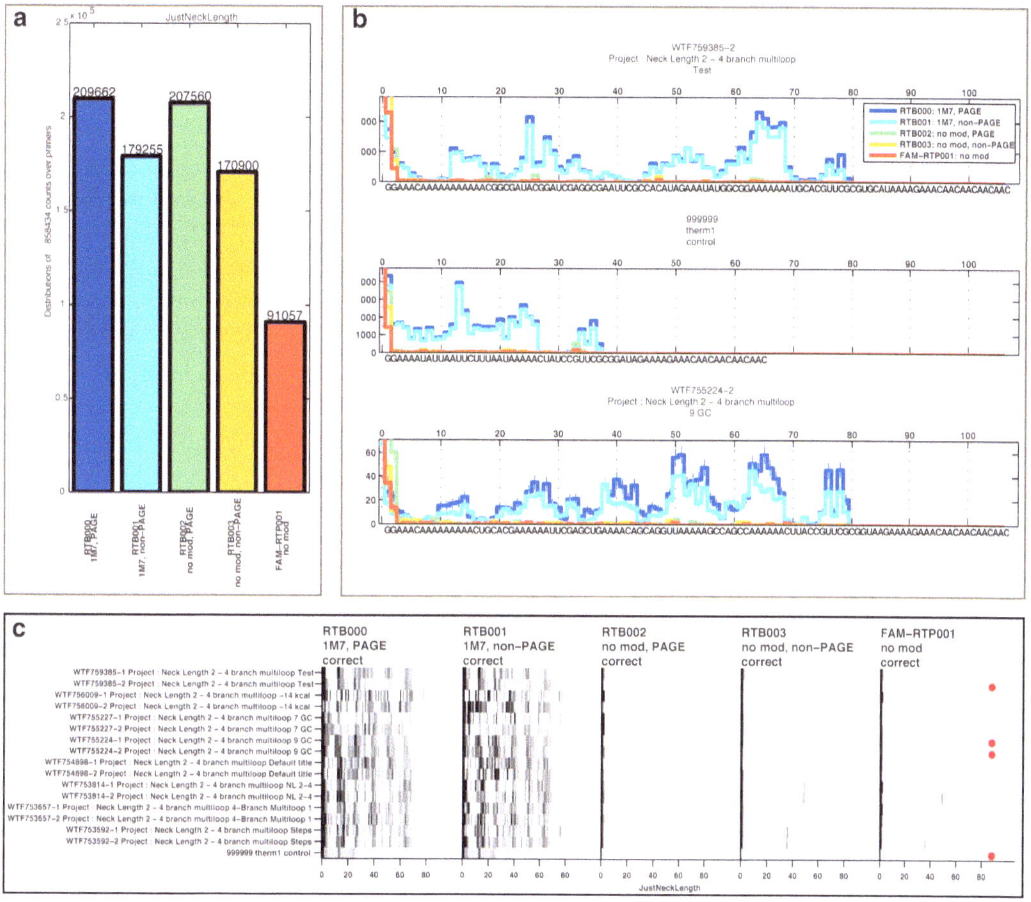

Fig. 5 Example MAPseeker output for a MAP-seq experimental data set involving 1M7 probing of Therm1 and the Neck Length 2-4 series of RNAs (*see* also Fig. 3). (**a**) Summary of number of counts across primers. In this experiment, four primers indexed two different library preparations (PAGE and non-PAGE) and modification type (1M7 or control). A fifth primer was used as a fluorescent tracer, doped into the control reactions. (**b**) Plot of raw counts and sequence for the RNAs with the most reads. (**c**) Summary display of all reactivities in heat-map format. Reactivities were automatically determined from raw counts by correcting for reverse transcription attenuation (*see* main text). *Red symbols* mark the four RNAs with the most reads

sequences and descriptions of the RNA library. The errors are based on Poisson statistics (note: sites with zero counts are assumed to have errors of ±1).

7. The output of MAPseeker is an RDAT file, which is formatted with the specifications available at http://rmdb.stanford.edu/repository/specs/. Structure analysis scripts, including helix-by-helix confidence estimation, are available on-line at: http://rmdb.stanford.edu/structureserver/. RDAT files can also be read and further analyzed via Python or MATLAB with the RDATkit scripts (https://simtk.org/home/rdatkit).

4 Notes

1. Alternative purification. If the number of wells used ends up being small (enough to fit in a microcentrifuge) or if the length of the RNA is less than approximately 50 nucleotides (under the cutoff for AMPure XP purification), ethanol precipitation (0.1 volumes of 3 M sodium-acetate, pH 5.2, then 3 volumes of cold ethanol; incubation in dry ice for 20 min; spin >15,000 × *g* in a microcentrifuge for 20 min) can be used instead of AMPure XP purification. However, AMPure XP beads are more convenient for purifying a large number of samples.
2. AMPure XP beads should not be mixed with the modification reagents for longer than the times prescribed in this protocol. The reagents and quenches may damage them and ruin the purification. If the beads turn from brown to green, then the purification has likely failed.
3. Illumina's MiSeq platform requires that the first 12 reads made by the sequencer are "sequence-balanced," namely, that there are approximately an equal number of A's, C's, G's, and T's read across all the clusters. This is not much of a problem for genomic DNA, but the cDNA libraries described here all begin with the same 20 nucleotide primer binding site, which would ruin the sequence balancing. So, immediately upstream of the primer region is a 12 nucleotide sequence-balancing region. We also use this region as a unique primer identifier (PID) for which chemical modification reaction a given cluster came from. Primers must be made and used in sets of four; within each set of four, one primer must have an A at the first position, another must have a C at the first position, etc. Our sequences (Table 1) were designed to minimize secondary structure with the primer region (as assessed with NUPACK) [38] while maintaining sequence balance and disallowing 4 or more Gs in a row in any primer.
4. The loading control sequences we use are in Table 1. We had these sequences on hand for other applications. Other sequences should work just as well, as long as their lengths are less than the reverse transcription primers used above, so that their signals appear at locations distinct from the library in the capillary electrophoretic trace. Note that at low 1 nM concentration, these oligos may stick to the sides of conventional 1.5 mL tubes, lowering the accuracy of the control over time; use of nonstick tubes and storage of stocks at concentrations greater than 100 nM are recommended.
5. Accurate quantification of the cDNA library is critical for acquiring high-quality sequencing data. Loading too much

sample will ruin the results due to too many overlapping clusters of DNA that cannot be distinguished from one another by the sequencer, and loading too little may not yield enough clusters for the Illumina platform to find the flow cell during initial auto-focus stages. Also note that double-stranded DNA, such as the PhiX genomic DNA, we recommend to be co-loaded with the single-stranded cDNA library, produces twice as many clusters per mole as single-stranded cDNA.

6. Only one site of modification can be identified by reverse transcription, so it is important to choose reaction conditions that modify each RNA once or not at all. Longer RNAs have more sites that can be modified, so it is important to lower reaction times and/or concentrations of reagents.
7. Ethanol is a strong inhibitor of reverse transcription and ligation. It needs to be completely removed by pipetting and evaporation, with no remaining liquid visible and with the beads appearing dried out.
8. Using CircLigase (i.e., TS2126 thermostable RNA ligase), we typically observe that the ligation of the adapter to reverse transcription products is near quantitative (>80 %) in a sequence-independent manner, within the detection limits of capillary electrophoresis. The one exception is that the fully extended product is ligated with poorer efficiency (~50 %; Fig. 4a, b); and may be due to a sequence bias (our transcripts begin with GG, leaving CC at the 3′ ends of these cDNA primer extension products). In any case, if the ligation efficiency of the full product is known, it can be specified at the time of data analysis and correction as a fourth argument in quick_look_MAPseeker. The default correction factor is 0.5; changing this number will affect the overall scale of the final reactivities and, less sensitively, the correction for attenuation which reweights the balance of short and long reverse transcription products (Eq. 1). Additionally, inclusion of an RNA sequence containing the same motif repeated at its 5′ and 3′ ends permits in situ estimation of attenuation correction and the appropriate correction factor for ligation of the full length product. We typically include the P4P6_double_hairpin construct as such an internal control. See the MAPseeker documentation for additional information.
9. The factor of 24 comes from the fact that the number of femtomoles of sample being read by the sequencer is 1/8 of the amount of primer used in the transcription, and the remaining sample is 1.5 times larger than the aliquot used in the quantification. The final factor of two comes from averaging the areas of the smaller and larger oligo peaks, by which the area under the library peaks is divided. If all primers are labeled, use a factor of 3 instead of 24.

10. While it is possible to design a protocol that could conceivably discard less material than this one, the volumes involved for the early steps may be so small, depending on the yield after ligation, that it would typically be necessary to dilute down some of the pools of beads after ligation, and then end up discarding the leftover anyway.
11. The 625 μL of solution left over after this step should be discarded, unless it is going to be used in a parallel MiSeq sequencing run on the same day. Our experience is that the solution is not useable for additional runs even if stored overnight at 4 °C.

Acknowledgements

We thank members of the Das laboratory for discussions and tests of the protocol. We thank Tom Mann and Frank Cochran for preparing the 1M7 acylating reagent, and S. Mortimer for an updated 1M7 synthesis protocol. Writing was supported by the NIH (IRACDA fellowship to MS; R01GM100953 to RD) and the Burroughs-Wellcome Foundation (Career Award to RD).

References

1. Ban N, Nissen P, Hansen J et al (2000) The complete atomic structure of the large ribosomal subunit at 2.4 Å resolution. Science 289:905–920
2. Guerrier-Takada C, Gardiner K, Marsh T et al (1983) The RNA moiety of ribonuclease P is the catalytic subunit of the enzyme. Cell 35:849–857
3. Cech TR, Zaug AJ, Grabowski PJ (1981) *In vitro* splicing of the ribosomal RNA precursor of *Tetrahymena*: involvement of a guanosine nucleotide in the excision of the intervening sequence. Cell 27:487–496
4. Fire A, Xu S, Montgomery MK et al (1998) Potent and specific genetic interference by double-stranded RNA in *Caenorhabditis elegans*. Nature 391:806–811
5. Greider CW, Blackburn EH (1985) Identification of a specific telomere terminal transferase activity in *Tetrahymena* extracts. Cell 43:405–413
6. Gutell RR, Lee JC, Cannone JJ (2002) The accuracy of ribosomal RNA comparative structure models. Curr Opin Struct Biol 12:301–310
7. Pace NR, Thomas BC, Woese CR (1999) Probing RNA structure, function, and history by comparative analysis. In: Gesteland RF, Cech TR, Atkins JF (eds) The RNA World, 2nd edn. Cold Spring Harbor Laboratory Press, Cold Spring Harbor, NY, pp 113–141
8. Sripakdeevong P, Kladwang W, Das R (2011) An enumerative stepwise ansatz enables atomic-accuracy RNA loop modeling. Proc Natl Acad Sci USA 108:20573–20578
9. Wang Z, Parisien M, Scheets K et al (2011) The cap-binding translation initiation factor, eIF4E, binds a pseudoknot in a viral cap-independent translation element. Structure 19:868–880
10. Flores SC, Altman RB (2010) Turning limited experimental information into 3D models of RNA. RNA 16:1769–1778
11. Kladwang W, Cordero P, Das R (2011) A mutate-and-map strategy accurately infers the base pairs of a 35-nucleotide model RNA. RNA 17:522–534
12. Cordero P, Kladwang W, VanLang CC et al (2012) Quantitative dimethyl sulfate mapping for automated RNA secondary structure inference. Biochemistry 51:7037–7039
13. Bida JP, Das R (2012) Squaring theory with practice in RNA design. Curr Opin Struct Biol 22:457–466

14. Mortimer SA, Johnson JS, Weeks KM (2009) Quantitative analysis of RNA solvent accessibility by N-silylation of guanosine. Biochemistry 48:2109–2114
15. Regulski EE, Breaker RR (2008) In-line probing analysis of riboswitches. Methods Mol Biol 419:53–67
16. Miura K, Tsuda S, Ueda T et al (1983) Chemical modification of guanine residues of mouse 5 S ribosomal RNA with kethoxal. (Nucleosides and nucleotides 46). Biochim Biophys Acta 739:281–285
17. Inoue T, Cech TR (1985) Secondary structure of the circular form of the *Tetrahymena* rRNA intervening sequence: a technique for RNA structure analysis using chemical probes and reverse transcriptase. Proc Natl Acad Sci USA 82:648–652
18. Ehresmann C, Baudin F, Mougel M et al (1987) Probing the structure of RNAs in solution. Nucleic Acids Res 15:9109–9128
19. Fritz JJ, Lewin A, Hauswirth W et al (2002) Development of hammerhead ribozymes to modulate endogenous gene expression for functional studies. Methods 28:276–285
20. Rocca-Serra P, Bellaousov S, Birmingham A et al (2011) Sharing and archiving nucleic acid structure mapping data. RNA 17:1204–1212
21. Merino EJ, Wilkinson KA, Coughlan JL et al (2005) RNA structure analysis at single nucleotide resolution by selective 2′-hydroxyl acylation and primer extension (SHAPE). J Am Chem Soc 127:4223–4231
22. Mortimer SA, Weeks KM (2007) A fast-acting reagent for accurate analysis of RNA secondary and tertiary structure by SHAPE chemistry. J Am Chem Soc 129:4144–4145
23. Kertesz M, Wan Y, Mazor E et al (2010) Genome-wide measurement of RNA secondary structure in yeast. Nature 467:103–107
24. Underwood JG, Uzilov AV, Katzman S et al (2010) FragSeq: transcriptome-wide RNA structure probing using high-throughput sequencing. Nat Methods 7:995–1001
25. Zheng Q, Ryvkin P, Li F et al (2010) Genome-wide double-stranded RNA sequencing reveals the functional significance of base-paired RNAs in *Arabidopsis*. PLoS Genet 6
26. Lucks JB, Mortimer SA, Trapnell C et al (2011) Multiplexed RNA structure characterization with selective 2′-hydroxyl acylation analyzed by primer extension sequencing (SHAPE-Seq). Proc Natl Acad Sci USA 108: 11063–11068
27. Aviran S, Trapnell C, Lucks JB et al (2011) Modeling and automation of sequencing-based characterization of RNA structure. Proc Natl Acad Sci USA 108:11069–11074
28. Ziehler WA, Engelke DR (2001) Probing RNA structure with chemical reagents and enzymes. Curr Protoc Nucleic Acid Chem Chapter 6, Unit 6 1
29. Blondal T, Thorisdottir A, Unnsteinsdottir U et al (2005) Isolation and characterization of a thermostable RNA ligase 1 from a *Thermus scotoductus* bacteriophage TS2126 with good single-stranded DNA ligation properties. Nucleic Acids Res 33:135–142
30. Li TW, Weeks KM (2006) Structure-independent and quantitative ligation of single-stranded DNA. Anal Biochem 349: 242–246
31. Langmead B, Trapnell C, Pop M et al (2009) Ultrafast and memory-efficient alignment of short DNA sequences to the human genome. Genome Biol 10:R25
32. Langmead B, Salzberg SL (2012) Fast gapped-read alignment with Bowtie 2. Nat Methods 9:357–359
33. Li H, Durbin R (2009) Fast and accurate short read alignment with Burrows-Wheeler transform. Bioinformatics 25:1754–1760
34. Torchia C, Takagi Y, Ho CK (2008) Archaeal RNA ligase is a homodimeric protein that catalyzes intramolecular ligation of single-stranded RNA and DNA. Nucleic Acids Res 36: 6218–6227
35. Zhelkovsky AM, McReynolds LA (2012) Structure-function analysis of Methanobacterium thermoautotrophicum RNA ligase—engineering a thermostable ATP independent enzyme. BMC Mol Biol 13:24
36. Yoon S, Kim J, Hum J et al (2011) HiTRACE: high-throughput robust analysis for capillary electrophoresis. Bioinformatics 27:1798–1805
37. Kladwang W, VanLang CC, Cordero P et al (2011) Understanding the errors of SHAPE-directed RNA structure modeling. Bio-chemistry 50:8049–8056
38. Zadeh JN, Steenberg CD, Bois JS et al (2011) NUPACK: analysis and design of nucleic acid systems. J Comput Chem 32:170–173

Chapter 7

Probing RNA Folding by Hydroxyl Radical Footprinting

Maria Costa and Dario Monachello

Abstract

In recent years RNA molecules have emerged as central players in the regulation of gene expression. Many of these noncoding RNAs possess well-defined, complex, three-dimensional structures which are essential for their biological function. In this context, much effort has been devoted to develop computational and experimental techniques for RNA structure determination. Among available experimental tools to investigate the higher-order folding of structured RNAs, hydroxyl radical probing stands as one of the most informative and reliable ones. Hydroxyl radicals are oxidative species that cleave the nucleic acid backbone solely according to the solvent accessibility of individual phosphodiester bonds, with no sequence or secondary structure specificity. Therefore, the cleavage pattern obtained directly reflects the degree of protection/exposure to the solvent of each section of the molecule under inspection, providing valuable information about how these different sections interact together to form the final three-dimensional architecture. In this chapter we describe a robust, accurate and very sensitive hydroxyl radical probing method that can be applied to any structured RNA molecule and is suitable to investigate RNA folding and RNA conformational changes induced by binding of a ligand.

Key words Hydroxyl radical footprinting, Fe(II)-EDTA, Structural probing of RNA, RNA folding, RNA tertiary structure, Group II intron, Intron–exon binding, Ribozyme

1 Introduction

RNAs with complex three-dimensional structures play fundamental roles in many cellular processes. These structured RNAs are indeed extremely versatile molecules in functional terms. For example, as ribozymes they can catalyze chemical reactions; as ribosomal RNAs and tRNAs they are essential to translate mRNA into proteins; as riboswitches, they regulate gene expression at the translational level [1]. In addition, new RNA sequences involved in other cellular pathways are constantly popping up from high-throughput sequencing studies and their structural characterization is crucial to understand their biological function.

Chemical reagents that modify atoms of bases such as dimethyl sulfate (DMS, which targets accessible N1 atoms of adenosines and N3 of cytosines) or kethoxal (which methylates the N1 and N2

Christina Waldsich (ed.), *RNA Folding: Methods and Protocols*, Methods in Molecular Biology, vol. 1086,
DOI 10.1007/978-1-62703-667-2_7, © Springer Science+Business Media, LLC 2014

positions of guanine) and RNases that cleave the substrate with specific sequence or structure preferences are frequently used as probes to assist in the structural characterization of RNAs in solution [2, 3]. More recently, two probing techniques were developed that allow to monitor the local backbone flexibility of every residue of an RNA molecule in a sequence-independent manner. Those are "in-line probing" [4–6], in which the rate of spontaneous cleavage is measured, and "Selective 2′ Hydroxyl Acylation analyzed by Primer Extension" (SHAPE) [7, 8], which monitors the nucleophilic reactivity of 2′ hydroxyl groups.

All these techniques provide valuable tools to define or confirm the secondary structure of an RNA molecule and, to a certain extent, these approaches can also be used to map either local conformational changes induced by a ligand or specific tertiary interactions involving previously identified secondary structure motifs [9–12]. Studies on the three-dimensional architecture of RNA molecules, however, cannot be carried out with these probes. By contrast, hydroxyl radical probing is a powerful technique specifically adapted to investigate the higher-order structure of nucleic acids and proteins [13–16]. Hydroxyl radicals (•OH) are highly reactive oxidative species [17] with short lifetimes. On DNA, they induce the oxidative degradation of the sugar ring by reacting with its various hydrogen atoms in the order 5′ H > 4′ H > 3′ H ~ 2′ H ~ 1′ H [18], resulting in breakage of the phosphodiester backbone. It is unclear whether this order stands also for RNA backbone cleavage since, to the best of our knowledge, experimental proof has never been reported. Sites of strand breaks can be easily identified either by prior end-labeling of the molecule under inspection or by subsequent primer extension reactions.

Applied to RNA molecules, one of the key features of hydroxyl radicals is that they induce backbone cleavage at *any* residue in a secondary-structure-independent manner, since single- and double-stranded regions are cleaved with the same efficiency [19]. Other key features are the small radius and water-like behavior of hydroxyl radicals, which ensure that the level of cleavage directly reflects the degree of exposition to the solvent of individual residues [20] and makes it possible to quantify the solvent accessibility of each section of a macromolecule with as fine as single nucleotide resolution. Hydroxyl radical probing has been frequently used to investigate the three-dimensional architecture of structured RNAs, at equilibrium, in a variety of ionic conditions [13] or to characterize structural intermediates along folding pathways. In fact, analysis of the cleavage pattern along the sequence allows one to define the "inside" and the "outside" of the RNA molecule under inspection [20]. Regions most protected from hydroxyl radical strand scission are those belonging to the solvent inaccessible, three-dimensional core formed during folding of the molecule, whereas sections exhibiting high levels of (•OH) cleavage are the ones directly exposed to the solvent. This probing information helps to reveal how helical segments lay out in space with respect to each other and can be very useful in three-dimensional modeling of RNA.

Another widely used application of hydroxyl radical probing on RNA molecules is in “footprinting” assays. This type of experiments consists of comparing the pattern of hydroxyl radical cleavage in the absence and in presence of a ligand (which can be a protein, a nucleic acid, a metabolite or any other molecule). As long as ligand binding is accompanied by changes in the solvent accessibility of the backbone of the RNA molecule under study, hydroxyl radical footprinting is a powerful method to investigate macromolecular interactions, ligand-induced conformational changes and ligand binding sites [21–24].

Hydroxyl radical species can be generated by a variety of methods such as synchrotron X-ray radiolysis [25], gamma-rays radiolysis [26] or by the use of peroxonitrite [27, 28]. However, the most frequent method to generate hydroxyl radicals uses the Fenton–Haber–Weiss chemistry [29, 30]. In this case, (•OH) species are produced from decomposition of hydrogen peroxide (H_2O_2) catalyzed by iron(II). In solution probing experiments, Fe(II) is chelated to EDTA to prevent its direct binding to nucleic acids backbone. During the chemical reaction Fe(II)-EDTA is oxidized to Fe(III)-EDTA; in order to recycle the latter back to Fe(II)-EDTA again and allow the reaction to proceed, a reducing reagent, usually ascorbic acid or dithiothreitol, is supplied to the reaction mix. The reaction is efficiently stopped by addition of a (•OH)-scavenger such as thiourea. The chemistry of the Fenton–Haber–Weiss reaction has been coupled to a variety of experimental setups. For example, Fe(II) can be tethered through a linker to a specific residue of a macromolecule, either a nucleic acid or protein [31, 32]. In this setup, and due to the transient nature of (•OH), backbone cleavages will happen in the vicinity of the tethered Fe(II), giving information about the three-dimensional molecular neighborhood around the residue to which the probe is attached. Another possibility is to combine the Fenton–Haber–Weiss chemistry to a quench-flow mixer apparatus in order to produce millisecond bursts of hydroxyl radicals [33]. This implementation allows to measure very fast changes in the solvent accessibility surface of macromolecules and, thereby, reveal transient folding intermediates along a folding pathway or a macromolecular assembly reaction.

This chapter describes a protocol for Fe(II)-EDTA footprinting of RNA molecules that is easy to implement, robust and very sensitive. This protocol is suitable to study different structural aspects such as RNA tertiary structure, ligand-induced conformational changes and even folding events as long as they take place within the minute (or longer) time scale. In the experimental approach outlined herein, Fe(II)-EDTA-catalyzed strand scissions are detected by primer extension reactions with ^{32}P-labeled primers; the resulting cDNA molecules are resolved by denaturing polyacrylamide gel electrophoresis and subsequently quantified. We successfully applied this protocol to study the three-dimensional folding of a group II self-splicing intron and to obtain very precise footprints of the 5′ exon on the ribozyme core (Fig. 1) [23]. The experimental data obtained not only confirmed known binding sites for the 5′ exon but also revealed, for the first time, conformational changes undergone by the intron's active site upon 5′ exon binding. Importantly, all these structural rearrangements were later confirmed by crystallographic data [34]. Although our protocol was optimized to probe large RNA molecules (> 600 nucleotides), it can be easily adapted to work on smaller RNAs, since the only prerequisite regarding the RNA sample to be analyzed is its ability to form homogeneous population of molecules with a well-defined, stable tertiary structure in solution.

2 Materials

Chemical reagents, especially the ones used for hydroxyl radical probing, should be of the highest purity, with minimal levels of heavy atoms that induce cleavage of the RNA backbone.

Prepare all solutions using ultrapure water. Carefully prepared solutions do not need to be sterilized but should always be stored at −20 °C.

We use silanized microcentrifuge plastic tubes throughout the entire protocol described herein. This treatment efficiently precludes sample loss by preventing RNA molecules to bind nonspecifically to the walls of the tubes (*see* **Note 1**).

2.1 Synthesis of RNA by In Vitro Transcription and Purification

1. DNA template. This can be a plasmid in which the DNA sequence corresponding to the RNA of interest is cloned immediately 3′ of the T7 promoter. Prior to transcription the plasmid needs to be linearized with a restriction enzyme that generates blunt or 5′ overhang ends. Alternatively, the template can be a PCR product generated with a forward primer that incorporates the T7 promoter sequence (*see* **Note 2**).
2. Native T7 RNA polymerase, either of commercial origin or home-purified from *E. coli* strain BL21 transformed with plasmid pAR1219. We determined experimentally that using ~1 μL of our T7 RNA polymerase per 100 μL of transcription reaction was already saturating and sufficient to generate high yields of RNA.
3. 10× T7 transcription buffer: 400 mM Tris–HCl, pH 7.9 (at 25 °C), 30 mM spermidine, pH 7.0, 60 mM $MgCl_2$, 0.1 % Triton X-100.

4. rNTP mix with 25 mM of each ribonucleotide (rATP, rCTP, rUTP, and rGTP) neutralized around pH 7.0–7.5.
5. 500 mM dithiothreitol (DTT).
6. 1 M $MgCl_2$.
7. RQ1 RNase-free DNase (1 U/μL; Promega).
8. 100 mM $CaCl_2$.
9. 500 mM Na_2EDTA, pH 8.0.
10. Formamide loading buffer: 98 % deionized formamide, 10 mM Na_2EDTA, pH 8.0, 10 mM Tris–HCl, pH 8.0, 0.025 % (w/v) bromophenol blue, 0.025 % (w/v) xylene cyanol.
11. For UV shadowing visualization of RNA: a silica gel, fluor-coated, thin layer chromatography (TLC) plate and a hand-held device with a shortwave UV light source (254 nm).
12. RNA elution buffer: 50 mM Na-MES, pH 6.2, 0.5 mM Na_2EDTA, pH 8.0, 0.01 % (w/v) SDS (*see* **Note 3**).
13. 1-butanol.
14. Spin-X centrifuge tube filters (Costar) with a 0.45 μm cellulose acetate filter.
15. 3 M sodium acetate.
16. Ethanol, absolute and diluted to 70 %.

2.2 Fe(II)-EDTA-Catalyzed Hydroxyl Radical Probing

1. Amount of RNA sample per probing reaction: 0.67 pmol of renatured RNA in a final volume of 46 μL.
2. 2× "+ $MgCl_2$" buffer for probing the native tertiary structure or for ligand-footprinting of the RNA sample: 80 mM Na-HEPES, pH 7.6 (at 37 °C), 200 mM NH_4Cl, 20 mM $MgCl_2$, 0.04 % (w/v) SDS (*see* **Note 4**).
3. 2× "– $MgCl_2$" buffer which only allows stabilization of the secondary structure: 80 mM Na-HEPES, pH 7.6 (at 37 °C), 0.04 % (w/v) SDS.
4. 50 mM ammonium iron (II) sulfate. Prepare a fresh 50 mM solution in H_2O for each experiment.
5. 100 mM Na_2EDTA, pH 8.0.
6. 500 mM DTT. This stock does not need to be prepared fresh each time if stored at –20 °C but prepare a fresh dilution to 250 mM DTT in H_2O for each experiment.
7. 250 mM l-ascorbic acid (sodium salt). If using ascorbic acid instead of DTT as a reducing agent, prepare a fresh 250 mM solution in H_2O for each experiment.
8. 30 % H_2O_2 (commercial stock). In the presence of DTT as a reducing agent, use a freshly prepared 5 % H_2O_2 dilution in H_2O. If using ascorbic acid instead of DTT, prepare a fresh 1 % H_2O_2 dilution in H_2O for each experiment.

9. 20 mM thiourea. Prepare a fresh 20 mM solution in H_2O for each experiment.
10. Bulk yeast tRNA (20 μg/μL).
11. 3 M sodium acetate.
12. Ethanol absolute and ethanol diluted to 70 %.

2.3 Radioactive 5′ End-Labeling of DNA Primers

1. 20 to 22mer DNA oligonucleotide primer, previously purified from shorter oligonucleotide fragments (*see* **Note 5**).
2. 10× kinase buffer: 500 mM Tris–HCl, pH 7.6, 100 mM $MgCl_2$, 50 mM DTT, 1 mM spermidine, pH 7.0, 1 mM Na_2EDTA, pH 8.0.
3. [γ-^{32}P]-ATP (3,000 Ci/mmol, 10 μCi/μL).
4. T4 polynucleotide kinase (10 U/μL).
5. Formamide loading buffer: 98 % deionized formamide, 10 mM Na_2EDTA, pH 8.0, 10 mM Tris–HCl, pH 8.0, 0.025 % (w/v) bromophenol blue, 0.025 % (w/v) xylene cyanol.
6. DNA elution buffer: 10 mM Tris–HCl, pH 7.5, 250 mM NaCl, 1 mM Na_2EDTA, pH 8.0, 0.02 % (w/v) SDS.
7. For visualization of 5′ ^{32}P-labeled primers: autoradiography film.
8. 1-butanol.
9. Spin-X centrifuge tube filters (Costar) with a 0.45 μm cellulose acetate filter.
10. 3 M sodium acetate.
11. Ethanol absolute.

2.4 Primer Extension of Hydroxyl Radical Probed Samples and Sequencing Reactions

1. For each elongation reaction use 0.08 pmol of hydroxyl radical- (or mock-) treated RNA sample and ~0.2 pmol of 5′ ^{32}P-labeled primer. For sequencing reactions use 0.16 pmol of non-treated RNA sample and ~0.4 pmol of 5′ ^{32}P-labeled primer.
2. 10× annealing buffer: 500 mM Tris–HCl, pH 8.3, 500 mM NaCl.
3. 10× elongation buffer: 250 mM Tris–HCl, pH 8.3, 350 mM NaCl, 30 mM $MgCl_2$, 100 mM DTT.
4. 10× sequencing buffer: 400 mM Tris–HCl, pH 8.3, 500 mM NaCl, 30 mM $MgCl_2$, 100 mM DTT.
5. dNTP mix at 5 mM each prepared from commercial stocks of 100 mM of dGTP, dATP, dTTP, dCTP.
6. 5 mM commercial stock of each dideoxynucleotide (ddGTP, ddATP, ddTTP, ddCTP).
7. SuperScript II RNase H$^-$ reverse transcriptase (200 U/μL; Invitrogen).

8. "SuperScript stop" mix: 60 mM Na_2EDTA, pH 8.0, 0.04 % (w/v) SDS.
9. 3 M NaOH: dilute fresh from a 10 M NaOH stock.
10. 3 M HCl.
11. 3 M sodium acetate.
12. Ethanol absolute and ethanol diluted to 75 %.
13. Formamide loading buffer: 98 % deionized formamide, 10 mM Na_2EDTA, pH 8.0, 10 mM Tris–HCl, pH 8.0, 0.025 % (w/v) bromophenol blue, 0.025 % (w/v) xylene cyanol.

2.5 Denaturing Polyacrylamide Gel Electrophoresis (PAGE)

1. 10× TBE: 500 mM Tris, 615 mM boric acid, 10 mM Na_2EDTA.
2. 40 % (w/v) acrylamide/bisacrylamide (19:1) stock solution. Store at 4 °C in the dark.
3. Urea.
4. 25 % (w/v) ammonium persulfate (APS) stock solution in H_2O.
5. N,N,N′,N′-tetramethylethylenediamine (TEMED).
6. Equipment for setting up the gels: For resolving primer extension reactions we use glass plates of 30 cm (W) × 40 cm (L) with 0.4 mm spacers and a ~40 wells comb. For RNA preparative purification, glass plates are 20 × 20 cm with 3 mm spacers and a ~8 wells comb. For 5′ ^{32}P-labeled DNA primers purification, glass plates are 20 × 20 cm with 1.5 mm spacers and a ~10 wells comb.
7. Electrophoresis running buffer: For resolving primer extension reactions use 0.5× TBE (diluted from 10× TBE stock) in the upper reservoir of the electrophoresis apparatus and a TBE-sodium acetate buffer in the lower one (*see* next step). For purification of RNA and 5′ radiolabeled DNA primers use 1× TBE buffer (diluted from 10× TBE stock) in the upper and lower reservoirs.
8. TBE-sodium acetate buffer: 1× TBE, 1 M sodium acetate, pH 5.1. It is convenient to previously prepare a stock solution of sodium acetate at acid pH. For example, to prepare 300 mL of 2.5 M sodium acetate, pH 5.1 weigh out 68.04 g of sodium acetate, add 14.3 mL of acetic acid (glacial) and bring up to the final volume with H_2O.
9. Gel fixing solution: 10 % (v/v) acetic acid (glacial), 10 % (v/v) ethanol absolute in H_2O.
10. Whatman 3MM chromatography paper.
11. Vacuum pump and gel dryer.
12. Phosphor screen and phosphorimaging instrument (e.g., Storm 820, Molecular Dynamics).

3 Methods

3.1 Synthesis of RNA by In Vitro Transcription and Purification

1. In a 1.5 mL silanized microcentrifuge tube combine, at room temperature, ~2.5–5 pmol of DNA template, 10 μL of 10× transcription buffer, 20 μL of 25 mM rNTPs mix, 2 μL of 500 mM DTT, 2 μL of 1 M $MgCl_2$, and 1 μL of home-purified T7 RNA polymerase in a final volume of 100 μL (this volume of transcription allows to recover >200 pmol of RNA after completion of the entire purification procedure). Incubate the transcription reaction at 37 °C for 2–3 h. To degrade the DNA template, add to the reaction, 4 μL of RQ1 DNase (1 U/μL) and 1.5 μL of 100 mM $CaCl_2$ (the enzyme requires at least 1 mM $CaCl_2$ for activity), then continue incubation at 37 °C for 30 min more. Stop the reaction by adding 10 μL of 500 mM Na_2EDTA, pH 8.0 and 120 μL of formamide loading buffer. Store at −20 °C until purifying on a denaturing polyacrylamide gel.
2. Purify T7 RNA transcripts by denaturing PAGE using the procedure detailed in Subheading 3.5. The percentage of polyacrylamide to use depends on the length of the RNA. Since our RNA samples are generally > 600 nucleotides we use 4 % denaturing polyacrylamide gels (this percentage is suitable to purify any RNA over 250 nucleotides).
3. Immediately prior to loading on the gel, denature the RNA sample by incubating at 42 °C for at least 20 min. Run the gel quite warm to keep RNA unfolded during migration (*see* **Note 6**).
4. Once the migration is finished, put the gel in a horizontal position and start by carefully removing the top glass plate from the gel. Immediately cover the exposed side of the gel with plastic wrap before flipping the gel to remove the remaining glass plate and cover the other side of the gel with plastic wrap as well.
5. To visualize the RNA sample by UV shadowing place the wrapped gel over the TLC plate and expose it to a shortwave UV light (254 nm). In these conditions, RNA (or DNA) molecules will absorb UV light and appear as dark bands against a light green fluorescent background. The top-most band on the gel should correspond to the full-length RNA of interest. Using a marker, outline this band (*see* **Note 7**).
6. Excise the outlined RNA band from the gel using a razor blade, remove the plastic wrap that covers the gel and place the gel slice into a 2 mL microcentrifuge tube.
7. With a 1 mL pipette tip completely crush the gel. In order to improve RNA recovery from the gel, place the tube in dry ice for 20 min to freeze the polyacrylamide. Let the tube to defrost at room temperature and add at least 3 volumes (~1.4 mL) of RNA elution buffer. Passive RNA elution is carried out at 4 °C overnight in a tube rotator.

8. To remove acrylamide from the supernatant, use Spin-X centrifuge tube filters according to manufacturer's instructions.
9. Recover the filtered supernatant in a 2 mL microcentrifuge tube and reduce its total volume to ~400 μL by 1-butanol extraction as follows: fill up the supernatant tube with 1-butanol, mix the two phases by vortexing or vigorous shaking, centrifuge at ~15,000 × *g* for 3 min to separate the two phases and discard the upper phase (which is the water saturated-butanol phase). Refill the tube containing the aqueous phase with 1-butanol and repeat the procedure above. Typically, four rounds of 1-butanol extraction are needed to reduce the supernatant volume from 1.4 mL down to 400 μL.
10. Precipitate the RNA by adding 1/10 volume of 3 M sodium acetate (~40 μL) and 3 volumes of 100 % ethanol and incubating at −20 °C for at least 2 h. Pellet the RNA by centrifuging at 20,000 × *g* for 20 min at 4 °C. Discard the supernatant and wash the pellet by adding 500 μL of 70 % ethanol and centrifuging at 20,000 × *g* for 5 min at 4 °C. Carefully remove all the supernatant and air-dry the pellet.
11. Resuspend the RNA pellet in ~100 μL H_2O and quantify the RNA yield by UV absorbance measurements. Store at −20 °C.

3.2 Fe(II)-EDTA-Catalyzed Hydroxyl Radical Probing

1. Prior to hydroxyl radical probing, submit the RNA sample to a renaturation protocol that allows the RNA molecules to stably fold into their biologically relevant conformation (*see* **Note 8**). To our particular RNA sample (the ~640 nucleotide lariat form of a group II self-splicing intron, Fig. 1 [23]) we apply the following protocol: 2 pmol of lariat intron in 75 μL of water is incubated at 50 °C for 3 min before addition of 75 μL of 2× "+ $MgCl_2$" probing buffer (at this temperature, and even in the presence of the "+ $MgCl_2$" probing buffer, the tertiary structure of the intron is disrupted but not its secondary structure); at this point, incubation at 50 °C is carried on for 3 more min before initiating slow cooling to 30 °C (or below) and a further 15 min incubation at the final probing temperature. For the sake of consistency, in studies intended to probe $MgCl_2$-dependent three-dimensional folding of the RNA sample we apply the same renaturation protocol in the presence of 2× "− $MgCl_2$" buffer even if in this case the RNA will not acquire any tertiary structure.
2. In our footprinting experiments, we renature the lariat intron sample immediately prior to the addition of a small volume (~10 μL) of 1× probing buffer containing the ligand molecule (in our case, a non-reactive RNA oligonucleotide encompassing the last 13 residues of the wild type 5′ exon) at a saturating binding concentration (*see* **Note 9**). After mixing, incubation at the final probing temperature should be carried on for at least 15 min to ensure probing under equilibrium binding

conditions. To keep experimental conditions as similar as possible the final volume for the RNA–ligand complex should be 150 μL, as described in the previous step.

3. Distribute each renatured sample (total volume of 150 μL) into three 46 μL aliquots. One aliquot will be used to generate the hydroxyl radical mock-treated sample while the other two will be used to generate duplicates of hydroxyl radical-modified samples.
4. Production of hydroxyl radicals is initiated in an empty microcentrifuge tube (which does not need to be silanized) by combining in the *following order*: 4 μL of 50 mM Fe(II), 4 μL of 100 mM Na_2EDTA, pH 8.0, 4 μL of 250 mM DTT, and 4 μL of 5 % H_2O_2. Mix well, *immediately* withdraw 4 μL and add them to the 46 μL RNA sample to be (•OH)-modified (*see* **Notes 10–12**). Incubate the reaction at 30 °C (room temperature also works) for 5 min. Repeat this procedure to produce "fresh" hydroxyl radical species for each one of the RNA samples to be probed. The exact same protocol is used to generate the mock-treated samples with the difference that in this case 4 μL of 62.5 mM DTT, only, is added to the RNA samples.
5. Stop the reactions by adding 50 μL of 20 mM thiourea. To precipitate the RNA continue by adding 0.5 μL yeast tRNA (20 μg/μL) as a carrier, 10 μL of 3 M sodium acetate, 330 μL of 100 % ethanol and incubate at −20 °C for at least 2 h. Pellet the RNA by centrifuging at 20,000 × *g* for 20 min at 4 °C. Discard the supernatant and wash the pellet by adding 500 μL of 70 % ethanol and centrifuging at 20,000 × *g* for 5 min at 4 °C. Carefully remove all the supernatant and air-dry the pellet.
6. Resuspend the RNA pellet in 8 μL H_2O. This volume should ensure an RNA concentration of ~0.08 μM if one assumes no sample loss during the precipitation step. Store at −20 °C.

3.3 Radioactive 5′ End-Labeling of DNA Primers

1. In a 1.5 mL silanized microcentrifuge tube mix well: 1 μL of 10 μM purified DNA primer, 1 μL of 10× kinase buffer, 5 μL 3.33 μM [γ-^{32}P]-ATP, 1 μL T4 polynucleotide kinase (10 U/μL), and 2 μL H_2O. Incubate at 37 °C for 50 min. Stop the reaction by adding 20 μL of formamide loading buffer. Store at −20 °C until purifying on a denaturing polyacrylamide gel (*see* **Note 13**).
2. Purify the 5′ ^{32}P-labeled primers on a 15 % denaturing polyacrylamide gel using the procedure detailed in Subheading 3.5. Immediately prior to loading on the gel, denature the labeled primer by incubating the reaction tube at 42 °C for at least 20 min. Run the gel at ~20 W until the bromophenol blue gets down to the bottom of the gel.

3. Once the migration is finished, put the gel in a horizontal position and carefully remove the top glass plate from the gel. Cover the exposed side of the gel with plastic wrap but do not remove the gel from the remaining glass plate. On the top of the plastic wrap firmly Scotch tape phosphorescent markers and expose them to light for a few seconds before performing the next step.
4. To visualize 5′ ^{32}P-labeled primers expose the gel to an autoradiography film for ~30 s. The primers will appear as dark bands. The phosphorescent markers will also form dark spots on the autoradiography film. These spots are very useful to precisely position the film on the top of the gel and correctly identify the gel area containing the labeled primer to be recovered.
5. Excise the labeled primer band from the gel using a razor blade, remove the plastic wrap that covers the gel and cut each gel slice into small pieces.
6. Place the gel pieces into a 2 mL microcentrifuge tube and incubate the tube in dry ice for 20 min to freeze the polyacrylamide. Let the tube defrost at room temperature and add at least 3 volumes of DNA elution buffer (~1.2 mL). Passive elution is carried out at 4 °C overnight in a tube rotator.
7. Remove acrylamide from the elution supernatant and reduce its volume down to ~300 μL using the procedure detailed in **steps 8** and **9** in Subheading 3.1.
8. Precipitate the labeled primer by adding 1/10 volume of 3 M sodium acetate (~30 μL) and 3 volumes of 100 % ethanol and incubating at −20 °C for at least 2 h. Pellet the primer by centrifuging at 20,000 × *g* for 25 min at 4 °C. Discard the supernatant, add 300 μL of 100 % ethanol (do not use 70 % ethanol here to prevent sample loss as much as possible) and centrifuge at 20,000 × *g* for 10 min at 4 °C. Carefully remove all the supernatant (check it for radioactivity and discard as appropriate, since there is always some sample loss at this point). Air-dry the pellet.
9. Resuspend the 5′ ^{32}P-labeled primer pellet into 20 μL of H_2O. This should ensure a primer concentration between 0.25 and 0.5 μM depending on the yield of the total purification procedure. In practice, and based on many labeling experiments, we estimate the primer concentration obtained from this procedure to be around 0.3 μM.

3.4 Primer Extension of Hydroxyl Radical Probed Samples and Sequencing Reactions

1. Set up the annealing reactions for subsequent primer extension of modified (or mock-treated) samples (*see* **Notes 14** and **15**) by combining, in a 1.5 mL silanized microcentrifuge tube, 1 μL of RNA sample (~0.08 pmol), 0.65 μL (~0.2 pmol) of 5′ ^{32}P-labeled primer, 0.5 μL of 10× annealing buffer, and 2.85 μL of H_2O.

2. Annealing reactions intended for subsequent RNA sequencing are performed on an RNA sample with the same sequence, but that was not submitted to the probing procedure (set up the latter annealing reactions by using double the volume of each reaction component detailed in **step 1**).
3. Place all the annealing reactions (from **steps 1** and **2**) in a dry bath, cover the tubes with a plastic or metal lid in order to reduce evaporation, incubate at 80 °C for 2 min and cool slowly down to room temperature by removing the dry bath block and placing it on the bench. Once at room temperature, briefly centrifuge the tubes to collect condensation drops at the bottom of the tubes and immediately proceed to the next step.
4. Elongation (primer extension) reactions are prepared at room temperature by adding to each 5 μL annealing reaction 1 μL of 10× elongation buffer, 1 μL of 5 mM dNTP mix, and 2 μL H_2O (it is more convenient to prepare a mix with these three reaction components instead of adding each one separately, especially if there are many elongation reactions to be performed). For sequencing with each one of the four dideoxynucleotides, the appropriate annealing reaction should be split first into four new 1.5 mL silanized tubes, each one containing 2 μL. To each one of these tubes add (at room temperature) 1 μL of 10× sequencing buffer, 0.5 μL of 5 mM dNTP mix, 5 μL of one of the four ddNTP stocks, and 0.5 μL H_2O. At this point initiate all the elongation and sequencing reactions successively by mixing well 1 μL of SuperScript II reverse transcriptase (200 U/μL) in each tube and immediately incubating it at 46 °C for 1 h.
5. Stop the reactions successively by adding to each tube 10 μL of "SuperScript stop" mix.
6. To degrade the RNA template, add to each tube 2 μL of 3 M NaOH and incubate at 90 °C for 4 min followed by 45 min at 46 °C. Subsequently, bring pH close to neutrality by adding to each reaction 2 μL of 3 M HCl.
7. Precipitate the samples by adding 76 μL H_2O, 10 μL of 3 M sodium acetate, and 330 μL of 100 % ethanol and incubating on ice for ~1 h. Pellet the cDNA by centrifuging at ~20,000 × *g* for 20 min at ~15 °C. Before discarding the supernatant check it for radioactivity (if there is radioactivity remaining in it, continue the precipitation reaction by incubating the tubes at −20 °C for 1 h). Wash the pellets by adding 300 μL of 75 % ethanol and centrifuge at ~20,000 × *g* for 10 min at 4 °C. Carefully remove all the supernatant before air-drying the pellets (*see* **Note 16**).
8. Resuspend the pellets corresponding to the elongation reactions in 12 μL of formamide loading buffer. Sequencing reac-

tions are usually resuspended in 10 μL of formamide loading buffer. However, if the total amount of radioactivity differs greatly between the pellets, the volume of loading buffer added in each tube should be varied in order to equilibrate the number of counts per μL between the four sequencing reactions.

9. Use 3 μL aliquots of each elongation and sequencing reaction and separate products on a 5, 6, or 8 % denaturing polyacrylamide gel (*see* below) using the procedure detailed in Subheading 3.5. Immediately prior to loading on the gel, denature the 3 μL samples by incubating them at 80 °C for 10–15 min and then keep them on ice while loading the gel. Run the gel at ~60 W to achieve a gel temperature of 45–50 °C. To analyze an RNA stretch close to the primer use a 6 or 8 % polyacrylamide gel and run it until the bromophenol blue dye gets stacked on the bottom of the gel. Conversely, if one wants to inspect an RNA region far from the primer, a 5 % polyacrylamide gel should be used and the xylene cyanol dye should migrate >3/4 of the gel length (*see* **Note 17**).

3.5 Denaturing Polyacrylamide Gel Electrophoresis (PAGE)

1. Determine the total volume needed to pour the denaturing polyacrylamide gel according to the size of glass plates and the thickness of the comb. The polyacrylamide percentage used to resolve RNA or DNA samples in each major procedure of the protocol is indicated in the section subheading. To resolve primer extension reactions, the polyacrylamide percentage to be used depends on how far from the primer lies the section of the RNA molecule one wants to analyze (*see* **Note 18**). To prepare the gel mix, combine in a glass flask the appropriate volumes of acrylamide/bisacrylamide in order to achieve the final desired percentage of polyacrylamide and either 0.5× final TBE (for primer extension gels) or 1× final TBE (for all other gels). To these, add 50 % (w/total final volume of gel mix) of urea. The urea's weight should be divided by 1.32 (urea density) to determine its volume and thus calculate the H_2O amount needed to achieve the desired final volume of the gel mix. The final urea concentration in the gel is ~8 M. Dissolve the urea by heating the gel mix moderately. Although not always necessary it is optimal to filter the gel mix through Whatman paper or a disposable filter unit. This filtration step should be added to the procedure if smears start to form during electrophoresis.
2. Start gel polymerization by adding to the gel mix 1/1,000 volume of 25 % APS and 1/1,000 volume of TEMED (this volume can be slightly reduced to slow down the polymerization reaction of high polyacrylamide percentage gels). Mix well and pour immediately between perfectly clean and clipped glass plates in order to avoid any air bubbles and gel leakages. Allow the gel to polymerize for at least 2 h.

3. After polymerization remove the comb gently and install the gel plates in the electrophoresis apparatus. Fill the reservoirs with the appropriate running buffer as follows: for resolving primer extension reactions use 0.5× TBE in the upper reservoir and TBE-sodium acetate buffer in the lower one (*see* **Note 19**); for all other gels use 1× TBE buffer in the upper and lower reservoirs. Rinse the gel wells thoroughly with the upper reservoir buffer and pre-run the gel for ~20 min at >10 W.
4. While the gel is pre-running, denature the samples as described for each major procedure of the protocol in the appropriate subheading. Immediately prior to loading the samples rinse the wells very thoroughly again with the upper reservoir buffer, since urea diffuses into the wells during the pre-run.
5. Conduct electrophoresis as described for each major procedure of the protocol in the appropriate subheading. At the end of short electrophoretic runs of 6 or 8 % polyacrylamide gels used to resolve primer extension reactions, the 5′ ^{32}P-labeled primer should be stacked at the very bottom of the gel (by action of the salt-pH gradient), at the level of the bromophenol blue dye. This non-extended primer should be cut out from the gel with a razor blade before proceeding to the next step. Using a radioactivity counter to localize the exact primer position, cut only the minimal amount of gel necessary to remove the primer. After cutting, press a piece of Whatman 3MM paper on the top of the gel strip in order to peel it away from the glass plate and proceed to the next step.
6. In order to prevent the diffusion of cDNA bands on the gel and thus increase their sharpness, all primer extension resolving gels are fixed in an acetic acid–ethanol solution prior to drying as follows: put the glass plate with the gel attached to it in a plastic tray. Gently pour the fixing solution on the top of the gel (use only the amount of solution needed to completely cover the gel surface) and let it stand for 5–10 min. Remove the fixing solution from the gel by gently tilting the glass plate. Repeat the fixing bath two more times and immediately move to the next step.
7. Carefully take the glass plate out from the tray taking care not to create any folds on the gel surface, drain briefly and immediately press a sheet of Whatman 3MM paper on top of the gel. The wet gel will stick to the Whatman 3MM paper and can then be removed from the glass plate. Cover the gel with plastic wrap and put it in a gel dryer under vacuum for ~1 h at 80 °C.
8. Expose the dried gel to a phosphor screen overnight (or longer) and subsequently scan the screen using a phosphorimaging instrument such as the Storm 820.

3.6 Analysis of Hydroxyl Radical Probing Data from Primer Extension Resolving Gels

1. In gels generated following the protocol described herein, variations in the intensity of cDNA bands directly reflect variations in the extent of hydroxyl radical cleavage at each residue along the RNA molecule (*see* Fig. 1c through g for a representative primer extension gel and data analyses output). Thus, lighter regions in a gel lane (lane stemming from hydroxyl radical probing of a sample with a define tertiary fold) correspond to "protections," that is, positions of the RNA molecule that were not (or only slightly) cleaved by hydroxyl radicals because they are less exposed to the solvent (these positions most probably belong to the solvent inaccessible core of the RNA molecule). Conversely, in the same gel lane, darker regions correspond to nucleotides that were frequently cleaved by hydroxyl radicals, reflecting their higher solvent accessibility. Typically, visual inspection of the gel to compare band patterns generated in the different experimental conditions readily identifies nucleotides with conspicuous intensity variations. These intensity changes, and more subtle ones that are not detectable by eye, can then be quantified manually (**step 3**) or by using the SAFA software (**step 4**). Lanes corresponding to mock-treated samples generated in the same experimental conditions as hydroxyl radical-treated samples are essential controls used (1) to identify bands corresponding to reverse transcriptase pauses (generally structure-induced) happening during elongation and (2) to rigorously assign and quantify "protections" or "enhanced reactivity" bands specifically due to hydroxyl radical cleavages and not resulting from the action of other experimental factors, such as RNA degradation, for example.
2. Assignment of hydroxyl radical protections on the sequence of the RNA molecule is achieved by comparing to the sequencing lanes generated from the same RNA sequence, taking into account that dideoxy sequencing-cDNAs are one nucleotide longer than the corresponding cDNAs derived from hydroxyl radical-probed samples.
3. This step requires the use of a gel imaging quantification software such as ImageQuant (*see* **Note 20**). In order to quantify the level of "protection" or "enhanced reactivity" along a section of the molecule the best way is to draw lines (using a rectangle object can also work) across that section of the molecule. Perform this step for each one of the gel lanes to be compared. Line traces data can then be exported, processed, and graphically presented with a data analysis software such as Excel or KaleidaGraph. In order to directly compare lane profiles and quantify the extent of protection of specific nucleotides in different experimental conditions, band intensities between lanes need first to be normalized to correct for nonuniform loading. This normalization step requires to spot a "reference" band on the gel, that is, a band present in all gel lanes and whose intensity

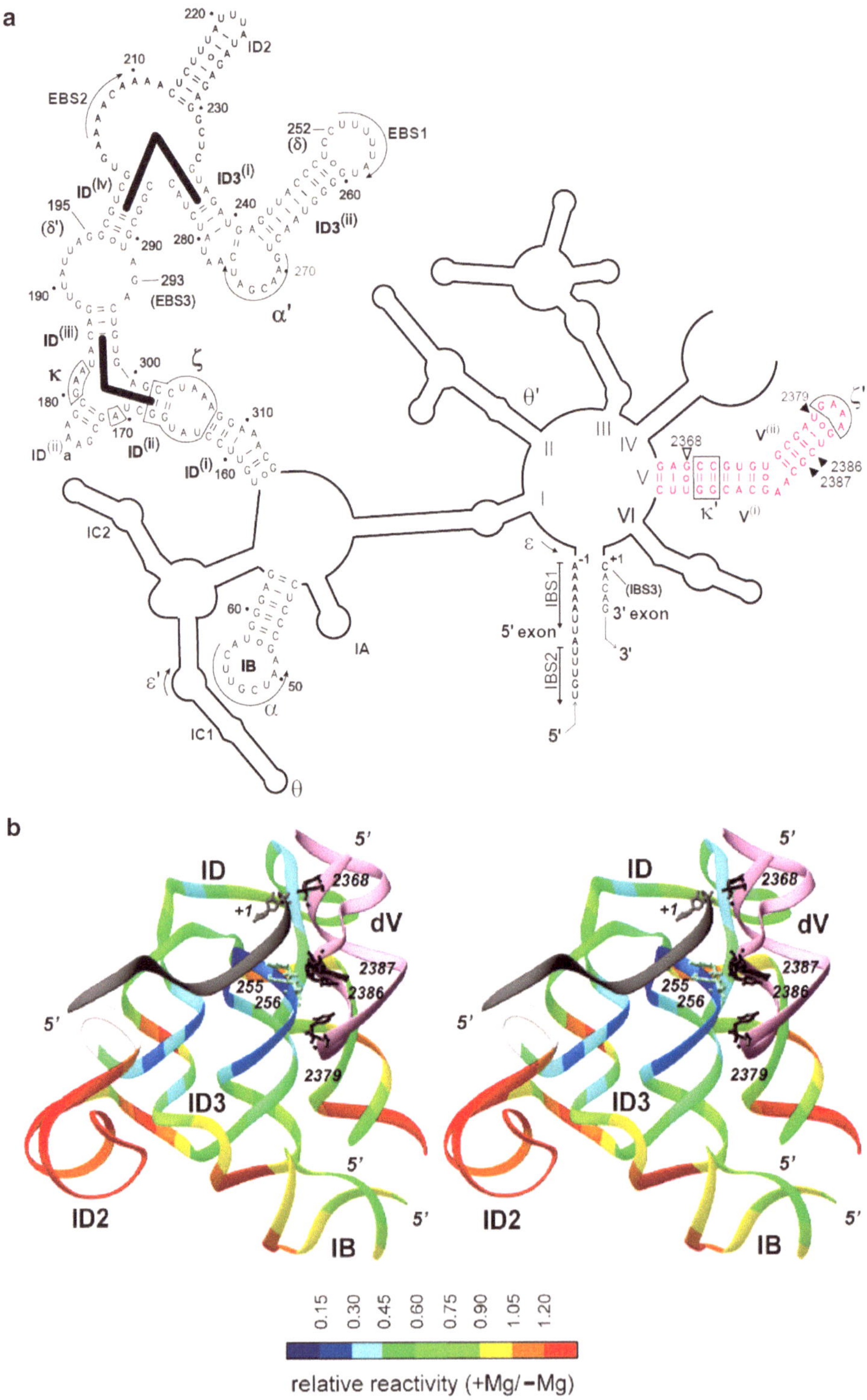

Fig. 1 Hydroxyl radical footprinting of the lariat form of mitochondrial intron *Pl*.LSU/2 from the alga *Pylaiella littoralis*. (**a**) Secondary structure model of the *Pl*.LSU/2 intron [23]. Individual nucleotides are shown only in those sections of the molecule that are modeled in *panel b*. *Greek letters* designate tertiary interactions within the intron. Recognition between the intron and its exons involves the EBS-IBS pairings. Pairs of helices that stack

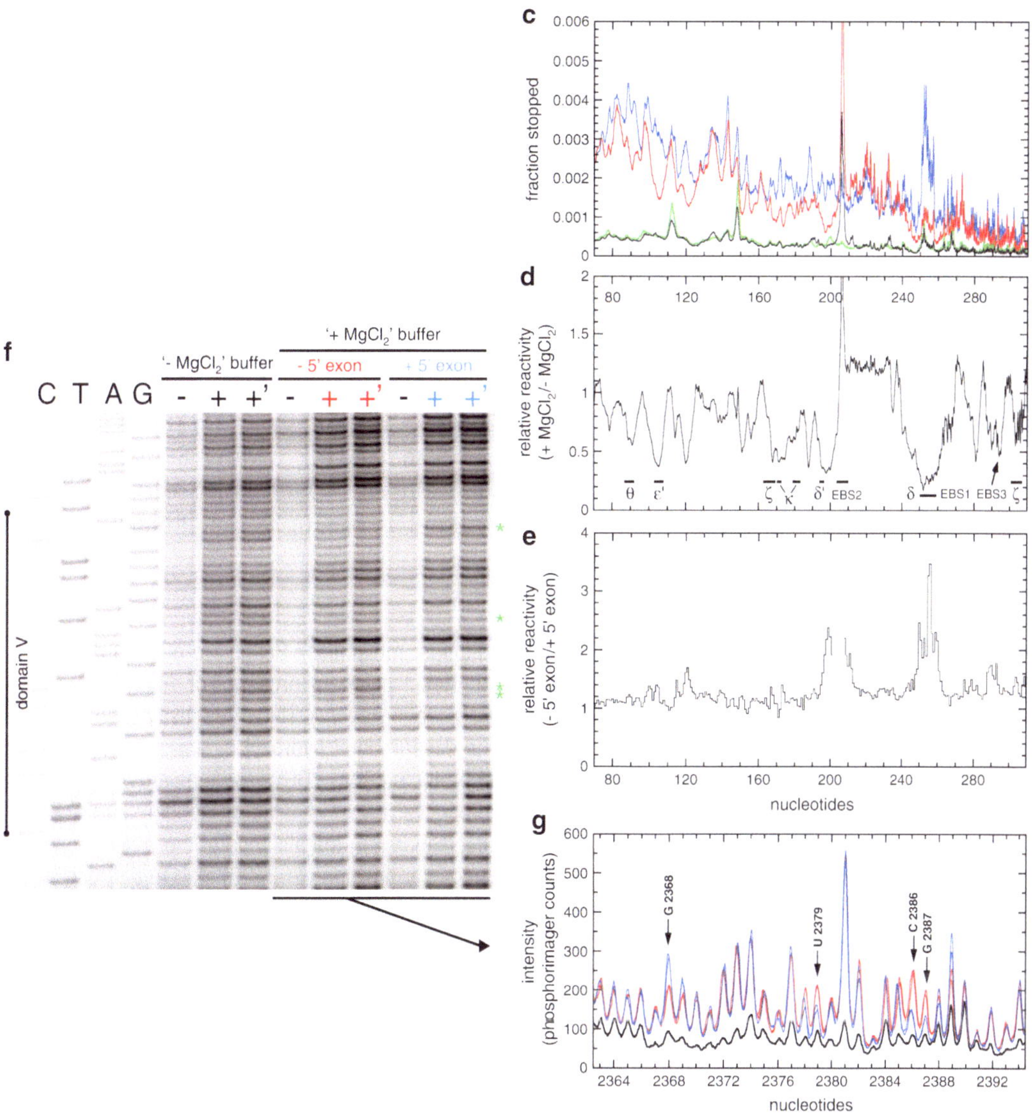

Fig. 1 (continued) end-to-end are indicated by *black bars*. *Empty and solid triangles* indicate domain V (*in pink*) residues whose reactivity to hydroxyl radicals is increased or decreased, respectively, upon binding of the 5′ exon (experimental data in *panels f* and *g*). Domain V numbering takes into account the ORF in domain IV, which was removed in the constructs used for in vitro experiments. (**b**) The three-dimensional model (generated from mutational data and phylogenetic analyses; [23]) of the complex between subdomain ID, domain V (*pink*) and the ligated exons (*gray*) was colored (except for domain V and the exons) according to the extent of protection from hydroxyl radical cleavage in the presence of magnesium (values from *panel d*; values could not be obtained for positions 202–207, shown in off-white, *see panel e*). Nucleotides shown in domain V and positions 255 and 256 in the EBS1 loop are the ones footprinted by the 5′ exon (*panels e–g*). (**c**) Comparison of graph profiles corresponding to primer extension reactions used to detect hydroxyl radical cleavages on the *Pl*.LSU/2 lariat intron under different experimental conditions. The intron was exposed to hydroxyl radicals in the presence of a saturating concentration of an unreactive 13mer RNA oligonucleotide analog of the 5′ exon at either 10 mM $MgCl_2$ (*red curve*, f_{+Mg}) or 0 mM $MgCl_2$ (*blue curve*, f_{-Mg}). *Black* and *green curves*, mock-treated samples with (b_{+Mg}) and without (b_{-Mg}) magnesium, respectively. Ordinates, fraction of primer extensions

seems insensitive to experimental conditions tested (such as a reverse transcription pause, for example). For every lane, divide the intensities along the lane profile by the intensity of the "reference" band lying in the same lane (if possible, or necessary, several "reference" bands can be used to confirm results). Normalized profile graphs for modified and mock-treated samples under different experimental conditions can then be superimposed and quantitative levels of solvent accessibility can be estimated for each nucleotide. These experimental data can in turn be interpreted in terms of tertiary RNA folding or ligand-induced protections or conformational changes if footprinting experiments are being analyzed.

4. As an alternative to the manual quantification procedure described above, one can use the SAFA (Semi-Automated Footprinting Analysis) software which assists the user in data normalization and allows to quantify the intensity of each cDNA band [35].

Fig. 1 (continued) ending 3′ of nucleotide *x* (relative to all longer molecules). Scans are over subdomains IC and ID and nucleotide numbering is according to *panel a*. (**d**) Reactivity in the presence of magnesium relative to that in its absence $[(f_{+Mg} - b_{+Mg})/(f_{-Mg} - b_{-Mg})]$ (from data shown in *panel c*). These reactivity estimates were used to color the three-dimensional model in *panel b*. The experimental hydroxyl radical probing data fully support the model, since the sections of the molecule most protected from (•OH) cleavages are the ones forming the internal core of the intron. (**e**) Footprints of the unreactive 5′ exon analog on subdomains IC and ID. Ordinates, ratio of the fraction of molecules ending 3′ of nucleotide *x* in the absence of the 5′ exon over that in the presence of the 5′ exon. Figures could not be obtained over positions 202–207 due to a strong stop in magnesium-containing lanes. (**f**) Representative denaturing polyacrylamide gel (8 % polyacrylamide—8 M urea) used to resolve primer extension reactions and quantitate hydroxyl radical footprinting data. The section of the gel shown is around domain V (the sequence of which is delimited by a vertical line) and allows one to visualize the footprints of the 13mer 5′ exon analog on domain V (footprinted nucleotides are indicated by *green asterisks*). Mapping of footprinted nucleotides is done relative to the sequencing reactions (G, A, T, C) and by taking into account that cDNAs from sequencing reactions are one nucleotide longer than the corresponding cDNAs from hydroxyl radical-modified samples. Experimental conditions (buffers, temperature, etc.) used for hydroxyl radical footprinting of the 5′ exon are the ones provided in Subheading 3.2 (the 5′ exon analog is always used at 6 μM). Lanes marked (−) correspond to mock-treated samples and lanes (+) and (+′) correspond to duplicates of hydroxyl radical reactions under each experimental condition tested. The *horizontal line* at the bottom of the gel image indicates the lanes whose graph profiles were compared in *panel g*. (**g**) Quantification of the primer extension resolving gel shown in *panel f*. *Red* and *blue scans* correspond to lariat intron exposed to hydroxyl radicals in the absence or in the presence of 6 μM of inactive 5′ exon analog, respectively. For each condition, the two scans corresponding to the (+) and (+′) experiments were superimposed to show the reproducibility of 5′ exon footprints. *Thick black scan*, mock-treated sample. Footprinted nucleotides in domain V are indicated by *arrows*. Altogether, our hydroxyl radical footprinting data allowed us to propose for the first time that upon 5′ exon binding, and formation of EBS-IBS pairings, a local conformational change takes place close to the catalytic site that pushes the EBS1 segment against domain V (*panel b*), resulting in the (•OH) protections identified in *panels e–g* [23]. This structural rearrangement has since been confirmed by crystallographic data [34]. Moreover, the fact that G2368, which is part of the catalytic site, is more reactive to hydroxyl radicals in the presence of the 5′ exon than in its absence shows that the conformational change identified here extends to the proximal part of domain V

4 Notes

1. Microcentrifuge plastic tubes are silanized by the vapor phase deposition method as follows: spread open microcentrifuge tubes inside a glass desiccator whose lid is equipped with a two-way stopcock connectable to a vacuum pump. *Under the hood*, use three microcentrifuge tubes standing vertically and add a total of 4.5 mL of dichlorodimethylsilane. Apply the vacuum until the level of liquid inside the tubes starts to go down. At this point, immediately turn the stopcock to the "close" position and wait until all of the dichlorodimethylsilane has gone into vapor phase. The deposition of the vapor phase takes several hours; usually, an overnight exposure of the tubes is enough. At the end of the silanization, open the desiccator, let the HCl vapors clear away and autoclave the tubes (this is an essential step).
2. T7 RNA polymerase is known for incorporating extra nucleotides (not encoded in the DNA template) at the 3′ end of transcripts. If the 3′ terminus of the RNA molecule under study needs to be well defined, an efficient method to achieve this is to use a PCR product having C2′ O-methyl riboses on the penultimate nucleotide or the last two nucleotides of the DNA template strand [36].
3. SDS is very useful to prevent RNA molecules to bind nonspecifically to the wall of the tubes. However, if subsequent probing experiments involve protein molecules, as is the case for RNA–protein hydroxyl radical footprinting, for example, SDS should be excluded from all buffers.
4. Do not perform (•OH) probing in the presence of polyhydroxylated compounds such as glycerol, ethanol or sugars since they are potent hydroxyl radical scavengers. Therefore, for footprinting experiments involving a protein stored in presence of glycerol, make sure to exchange the storage buffer against a glycerol-free one prior to (•OH) probing. Tris buffer can also quench the reaction and should be replaced by cacodylate or MOPS buffers. Even HEPES, used in our probing buffers, can quench to some extent. When quenching is not dramatic however, it is possible to adapt probing conditions in order to produce a higher amount of hydroxyl radicals to achieve sufficient RNA cleavage (*see* **Note 10**). Concerning ionic conditions, magnesium concentrations up to ~25 mM and monovalent ions concentrations up to ~100 mM do not affect the probing reaction.
5. Prior to labeling it is recommended to gel-purify a large amount of primer stock solution that can be stored at −20 °C and used for multiple labeling reactions (a 20 % denaturing

polyacrylamide gel allows to separate full-length primer from all shorter fragments). This will ensure that the only molecule to be 5′ end-labeled will be the full-length primer and the final yield of the labeling reaction will be according to expectations.

6. For optimal separation from possible smaller fragments (this can be the case if T7 RNA polymerase pauses at certain sequences of the DNA template), gel migration should be carried on until the RNA of interest has reached about halfway the length of the plates. RNA migration is deduced by the migration of the dyes present in the loading buffer. In order to have good gel resolution of RNA species it is important not to overload the gel (this can be achieved by loading the transcription reaction into two or more gel wells instead of only one).
7. The gel should not be exposed to UV light for more than 30 s to prevent potential UV-induced damage of the RNA.
8. Before performing a probing experiment it is essential to make sure that the RNA sample of interest is able to form a homogeneous population of molecules having a defined, correctly folded tertiary structure; otherwise, the hydroxyl radical cleavage pattern will reflect the weighted sum of the multiple structural conformations adopted by the RNA in solution and any signal that emerges from the background will be essentially uninterpretable. There are a number of methods used to check the structural homogeneity of an RNA sample such as UV melting curves or native polyacrylamide gels [37]. RNA molecules purified under denaturing conditions must be renatured under near-equilibrium conditions to attain a well-defined, biologically meaningful conformation before being used for any structural probing. Consequently, a renaturation protocol appropriate to the particular RNA molecule being studied should be optimized and applied to the sample immediately prior to each probing experiment (freeze-thaw cycles tend to disrupt the tertiary folding). These important remarks do not apply, however, if one is characterizing structural intermediate states of an RNA molecule along a folding pathway, since in this case probing in done under non-equilibrium conditions, before the molecule reaches its final, well-folded tertiary structure.
9. Hydroxyl radical footprinting experiments are very useful to characterize a ligand binding site (that ligand can be another RNA molecule, a protein or a small metabolite) and potential conformational rearrangements induced by the ligand on the target RNA. These footprinting experiments can also be used to estimate the dissociation constant (K_d) of the RNA molecule and its ligand. In this case, probing reactions should be performed over a wide range of ligand concentrations.

10. Following this procedure, final probing conditions are: 1 mM Fe(II)-2 mM Na_2EDTA, pH 8.0, 5 mM DTT and 0.1 % H_2O_2. Under our experimental conditions (buffer composition, size of RNA molecule, etc.) these concentrations of reagents generate an appropriate level of strand scission on the input RNA, allowing us to quantify ($^{\bullet}OH$)-cleavage signal without disrupting the three-dimensional structure of the sample due to too many backbone cuts (*see* **Note 20** for a discussion about this point as well). Similarly, levels of ($^{\bullet}OH$) cleavage should be optimized for each RNA sample. We found that the most efficient way to achieve this optimization is to test tight ranges of H_2O_2 concentrations, keeping all other probing reagents unchanged.
11. We experimentally determined that if 5 mM ascorbate is used instead of 5 mM DTT, the concentration of H_2O_2 should be lowered from 0.1 % to 0.02–0.03 % to achieve the same level of RNA backbone cleavage. This is probably due to the fact that DTT is a moderate ($^{\bullet}OH$)-scavenger.
12. It is also possible to perform hydroxyl radical probing in the absence of H_2O_2. In this case hydroxyl radicals are generated by the action of Fe(II) on O_2 molecules present in solution. Comparatively to "+ H_2O_2" conditions, however, the time required to achieve the same level of hydroxyl radical cleavage of the RNA is much higher. Indeed, we found that the reaction time has to be extended from 5 min to ~2 h when probing in the absence of H_2O_2 (with all the other reagents unchanged).
13. Alternatively to radioactive labeling, DNA primers can be 5′ end-labeled with fluorescent tags before being used for primer extension reactions (described in Subheading 3.4). In this case, however, fluorescent primer extension products are resolved by capillary electrophoresis (CE), instead of conventional polyacrylamide gel electrophoresis, and subsequently quantified in an automated way [38]. Thus, the use of 5′ end fluorescent primers is probably a better approach if hydroxyl radical probing is to be performed on a high-throughput scale. However, the use of radiolabeled primers ensures greater sensitivity and makes it possible to obtain reliable primer extension data from very small amounts of modified RNA samples (*see* Subheading 3.4).
14. For an efficient annealing reaction primer binding sites should be chosen in RNA regions without any stable secondary structure.
15. Strand scissions induced by hydroxyl radicals on RNA samples can also be directly detected in denaturing polyacrylamide gels if the RNA is radiolabeled, at either, its 5′ or 3′ end, prior to the probing reaction. However, one can only resolve the first ~200 nucleotides from the labeled end, so that RNA molecules

over ~400 nucleotides cannot be entirely analyzed by this end-labeling approach.

16. This step greatly improves the quality of cDNA migration on subsequent denaturing polyacrylamide gels since cDNA bands become sharper and one can read a longer section of the molecule at nucleotide resolution. Precipitation should be efficient to allow recovery of all smaller cDNA fragments. At the same time, however, the pelleted material should not contain too much non-extended labeled primer since its presence can disturb cDNA migration and cause smears on the gel (another possible source of "smeary" gels is the incomplete degradation of the RNA template).

17. Using short and long PAGE runs one can read ~200 nucleotides with a single primer, so that four primers are normally sufficient to read an RNA ~600 nucleotides long. Generally, we use primers lying ~120 nucleotides apart from one another along the RNA sequence. This allows us to scan the same section of the RNA from differently located primers and verify the reproducibility of the probing data. With our primer extension protocol the only segments of the RNA molecule of interest that cannot be analyzed are the binding site for the primer that lies at the 3′ end of the molecule and the ~7 nucleotides immediately 5′ from it. Should it be important to analyze this section of the molecule one possible strategy is to covalently attach the 3′ end of the molecule to a "tag" sequence that provides a new primer binding site without interfering with the native folding of the RNA. Another possibility is to analyze different circular permutated constructs of the same RNA in which its natural 3′ end has become linked to another section of the molecule (again, it should be checked that the circular permutations engineered are compatible with the native fold of the molecule).

18. For example, to start reading ~7 nucleotides away from the 3′ end of the labeled DNA primer, an 8 % polyacrylamide gel should be used. Conversely, a 5 % polyacrylamide gel allows to analyze sections of the molecule lying >130 nucleotides away from the primer.

19. The use of this TBE-sodium acetate buffer in the lower reservoir generates a salt and H^+ ion gradient that efficiently slows down the migration of cDNA fragments reaching the lower part of the gel. In this way, cDNA fragments stack on the bottom of the gel, the gap between the bands is reduced and one can read a longer section of the molecule from a single PAGE run.

20. First, quantify the full-length cDNA band in all lanes to estimate the level of ($^\bullet$OH) cleavage in hydroxyl radical-modified samples. Ideally, the level of ($^\bullet$OH) modification per molecule should be around 1 (this is achieved when the full-length

cDNA in a modified sample represents ~85 % of that measured in the corresponding mock-treated sample) but for very long RNAs (>400 nucleotides) these conditions generate a very low signal-to-noise ratio that makes quantification difficult and imprecise. For this reason we use moderately higher levels of (•OH) modification per molecule (with the full-length cDNA in modified samples being ~60 % of that for mock-treated samples). It is important to mention that we checked experimentally that under these modification conditions hydroxyl radical hits affecting the same RNA molecule remain structurally unrelated and, by consequence, they do not cause unfolding of the molecule (this is an essential premise to extract meaningful structural information from these probing conditions). Truly over-modified samples are easily detected since there is a bias toward small cDNAs fragments on primer extension gels.

Acknowledgements

We are grateful to François Michel for support and critical reading of the manuscript. This work was supported by an ANR (Agence National de la Recherche) grant (2010 BLAN 1502 01).

References

1. Atkins JF, Gesteland RF, Cech TR (eds) (2011) RNA Worlds: from life's origins to diversity in gene regulation. Cold Spring Harbor Laboratory Press, Cold Spring Harbor
2. Ehresmann C, Baudin F, Mougel M et al (1987) Probing the structure of RNAs in solution. Nucleic Acids Res 15:9109–9128
3. Brunel C, Romby P (2000) Probing RNA structure and RNA-ligand complexes with chemical probes. Methods Enzymol 318:3–21
4. Soukup GA, Breaker RR (1999) Relationship between internucleotide linkage geometry and the stability of RNA. RNA 5:1308–1325
5. Nahvi A, Sudarsan N, Ebert MS et al (2002) Genetic control by a metabolite binding mRNA. Chem Biol 9:1043
6. Winkler W, Nahvi A, Breaker RR (2002) Thiamine derivatives bind messenger RNAs directly to regulate bacterial gene expression. Nature 419:952–956
7. Merino EJ, Wilkinson KA, Coughlan JL et al (2005) RNA structure analysis at single nucleotide resolution by selective 2′-hydroxyl acylation and primer extension (SHAPE). J Am Chem Soc 127:4223–4231
8. Wilkinson KA, Merino EJ, Weeks KM (2006) Selective 2′-hydroxyl acylation analyzed by primer extension (SHAPE): quantitative RNA structure analysis at single nucleotide resolution. Nat Protoc 1:1610–1616
9. Costa M, Christian EL, Michel F (1998) Differential chemical probing of a group II self-splicing intron identifies bases involved in tertiary interactions and supports an alternative secondary structure model of domain V. RNA 4:1055–1068
10. Fedorova O, Waldsich C, Pyle AM (2007) Group II intron folding under near-physiological conditions: collapsing to the near-native state. J Mol Biol 366:1099–1114
11. Weinberg Z, Regulski EE, Hammond MC et al (2008) The aptamer core of SAM-IV riboswitches mimics the ligand-binding site of SAM-I riboswitches. RNA 14:822–828
12. Low JT, Weeks KM (2010) SHAPE-directed RNA secondary structure prediction. Methods 52:150–158
13. Celander DW, Cech TR (1991) Visualizing the higher order folding of a catalytic RNA molecule. Science 251:401–407
14. Laggerbauer B, Murphy FL, Cech TR (1994) Two major tertiary folding transitions of the *Tetrahymena* catalytic RNA. EMBO J 13: 2669–2676

15. Brenowitz M, Chance MR, Dhavan G et al (2002) Probing the structural dynamics of nucleic acids by quantitative time-resolved and equilibrium hydroxyl radical "footprinting". Curr Opin Struct Biol 12:648–653
16. Loizos N (2004) Mapping protein-ligand interactions by hydroxyl-radical protein footprinting. Methods Mol Biol 261:199–210
17. Buxton GV, Greenstock CL, Helman WP et al (1988) Critical review of rate constants for reactions of hydrated electrons, hydrogen atoms and hydroxyl radicals in aqueous solution. J Phys Chem Ref Data 17:513
18. Balasubramanian B, Pogozelski WK, Tullius TD (1998) DNA strand breaking by the hydroxyl radical is governed by the accessible surface areas of the hydrogen atoms of the DNA backbone. Proc Natl Acad Sci USA 95:9738–9743
19. Celander DW, Cech TR (1990) Iron(II)-ethylenediaminetetraacetic acid catalyzed cleavage of RNA and DNA oligonucleotides: similar reactivity toward single- and double-stranded forms. Biochemistry 29:1355–1361
20. Latham JA, Cech TR (1989) Defining the inside and outside of a catalytic RNA molecule. Science 245:276–282
21. Powers T, Noller HF (1995) A temperature-dependent conformational rearrangement in the ribosomal protein S4.16 S rRNA complex. J Biol Chem 270:1238–1242
22. Powers T, Noller HF (1995) Hydroxyl radical footprinting of ribosomal proteins on 16S rRNA. RNA 1:194–209
23. Costa M, Michel F, Westhof E (2000) A three-dimensional perspective on exon binding by a group II self-splicing intron. EMBO J 19: 5007–5018
24. Dann CE 3rd, Wakeman CA, Sieling CL et al (2007) Structure and mechanism of a metal-sensing regulatory RNA. Cell 130:878–892
25. Sclavi B, Woodson S, Sullivan M et al (1997) Time-resolved synchrotron X-ray "footprinting", a new approach to the study of nucleic acid structure and function: application to protein-DNA interactions and RNA folding. J Mol Biol 266:144–159
26. Hayes JJ, Kam L, Tullius TD (1990) Footprinting protein-DNA complexes with gamma-rays. Methods Enzymol 186:545–549
27. King PA, Jamison E, Strahs D et al (1993) 'Footprinting' proteins on DNA with peroxonitrous acid. Nucleic Acids Res 21:2473–2478
28. Swisher JF, Su LJ, Brenowitz M et al (2002) Productive folding to the native state by a group II intron ribozyme. J Mol Biol 315: 297–310
29. Fenton HJH (1894) The oxidation of tartaric acid in presence of iron. J Chem Soc 65: 899–910
30. Haber F, Weiss J (1934) The catalytic decomposition of hydrogen peroxide by iron salts. Proc R Soc Lond A 147:332–351
31. Heilek GM, Noller HF (1996) Site-directed hydroxyl radical probing of the rRNA neighborhood of ribosomal protein S5. Science 272:1659–1662
32. Newcomb LF, Noller HF (1999) Directed hydroxyl radical probing of 16S ribosomal RNA in 70S ribosomes from internal positions of the RNA. Biochemistry 38:945–951
33. Shcherbakova I, Mitra S, Beer RH et al (2006) Fast Fenton footprinting: a laboratory-based method for the time-resolved analysis of DNA, RNA and proteins. Nucleic Acids Res 34:e48
34. Toor N, Keating KS, Fedorova O et al (2010) Tertiary architecture of the *Oceanobacillus iheyensis* group II intron. RNA 16:57–69
35. Das R, Laederach A, Pearlman SM et al (2005) SAFA: semi-automated footprinting analysis software for high-throughput quantification of nucleic acid footprinting experiments. RNA 11:344–354
36. Kao C, Zheng M, Rudisser S (1999) A simple and efficient method to reduce nontemplated nucleotide addition at the 3 terminus of RNAs transcribed by T7 RNA polymerase. RNA 5: 1268–1272
37. Costa M, Fontaine JM, Loiseaux-de GS et al (1997) A group II self-splicing intron from the brown alga *Pylaiella littoralis* is active at unusually low magnesium concentrations and forms populations of molecules with a uniform conformation. J Mol Biol 274:353–364
38. Vasa SM, Guex N, Wilkinson KA et al (2008) ShapeFinder: a software system for high-throughput quantitative analysis of nucleic acid reactivity information resolved by capillary electrophoresis. RNA 14:1979–1990

Chapter 8

Monitoring Global Structural Changes and Specific Metal-Ion-Binding Sites in RNA by In-line Probing and Tb(III) Cleavage

Pallavi K. Choudhary, Sofia Gallo, and Roland K.O. Sigel

Abstract

In this chapter we describe the use of two methods, in-line probing as well as terbium(III) cleavage. Both methods can be applied to RNAs of any size, structure, and function. Aside from revealing directly metal ion-binding sites these techniques also provide structural information for longer RNA sequences that are out of range to be analyzed with other techniques such as NMR. The cleavage pattern derived from in-line probing experiments reflects local and overall conformational changes in RNA upon the addition of metal ions, metal complexes, or other ligands. On the other side, terbium(III) cleavage experiments are applied to locate specific metal ion-binding sites in RNA molecules.

Key words RNA, Riboswitch, Folding, Ribozymes, Metal ions, In-line probing, Terbium cleavage

1 Introduction

To reach full activity for their versatile tasks, RNA molecules have to fold into highly specific and complex tertiary structures. However, a close packing of the RNA strand is only possible upon binding to positively charged ions, i.e., mostly metal ions that allow neutralization of the negative charge of the sugar–phosphate backbone. Generally, in the first step of folding, monovalent metal ions like K^+ and Na^+ allow a pre-organization of the RNA by binding nonspecifically to the backbone [1–3]. At this stage, the RNA is mainly arranged in thermodynamically stable helical parts separated by random-coil regions. Upon binding to divalent metal ions such as Mg^{2+}, the pre-folded domains are then oriented correctly in three-dimensional space to give the complex and functional tertiary structure [4, 5].

Christina Waldsich (ed.), *RNA Folding: Methods and Protocols*, Methods in Molecular Biology, vol. 1086,
DOI 10.1007/978-1-62703-667-2_8, © Springer Science+Business Media, LLC 2014

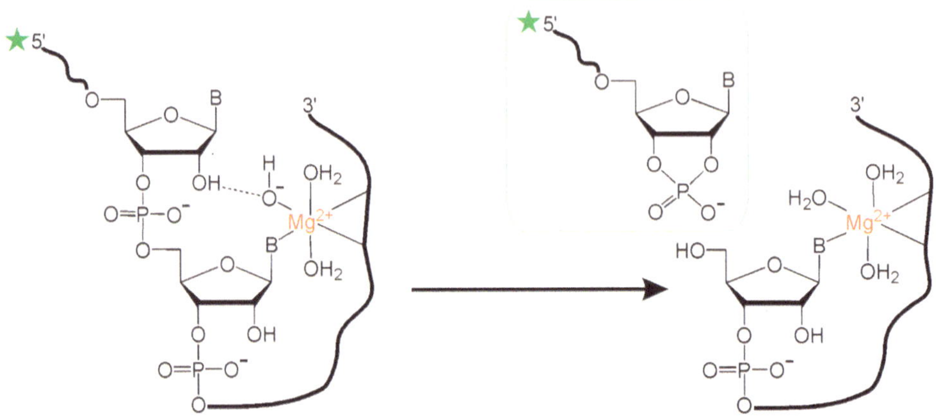

Fig. 1 Schematic view of the self-cleavage reaction in RNA. The 2′ OH of the ribose attacks the neighboring bridging phosphate in a Mg^{2+}-assisted reaction cleaving one phosphodiester linkage. The solvated metal ion hydroxo complex (*orange*) coordinates to the RNA and acts as a base-catalyst by activating the attacking 2′ OH. Cleavage can also occur in the absence of any metal ion if the geometry is ideal and the cleavage sites are therefore not necessarily identical with metal binding sites. By using 5′ labeled RNA (indicated as *green asterisk*), the 5′ cleavage products (*green*) can be detected on a denaturing polyacrylamide gel (the ^{32}P-label can also be attached to the 3′ end)

Given the large variety of structures and functions RNA can achieve, it is highly intriguing to investigate this class of molecules more closely. However, despite the continuous progress over the past decades to develop methods and tools to efficiently study, e.g., structural details, the investigation of the folding mechanism, conformational changes, or metal ion-binding sites of small and large RNA molecules still remains challenging.

The two methods described in this chapter, in-line probing and terbium(III) cleavage, are convenient techniques to study large RNA molecules. Both techniques exploit the inherent instability of RNA due to its 2′ OH group: In a spontaneous intramolecular nucleophilic reaction the 2′ OH group attacks the adjacent phosphodiester linkage leading to self-cleavage of the RNA backbone (Fig. 1). This reaction can be accelerated by the presence of divalent metal ions [6] or acid/base catalysis [7]. Interestingly, the cleavage rate is not only dependent on external factors but is also proportional to the "in-line character" of the phosphate linkage, a fact that is exploited in the in-line probing assay.

For an ideal "in-line geometry" the incoming 2′ OH group and the leaving 5′ OH are situated exactly opposite to each other of the central phosphorous. Such a geometry is however difficult to adopt in stable base paired regions such as double helices, and hence these regions are generally more protected than unstructured regions of the RNA [8]. In contrary, single-stranded regions are inherently more dynamic and consequently take up more commonly an "in-line geometry," thus being cleaved also more efficiently. The differences in cleavage intensities therefore provide direct information on the flexibility or rigidity of a specific RNA region.

Each RNA molecule subjected to in-line probing shows a distinct pattern of cleavage products. As this pattern is defined by the secondary and tertiary structure of the RNA, the pattern is altered by structural changes of the RNA molecule [8]. In-line probing is thus an ideal method to detect structural changes of RNA independent of their size. It is important to note that the cleavage sites detected by in-line probing do not necessarily coincide with specific metal-binding sites. Therefore this technique is mainly used to investigate structural changes in RNA upon interactions with metal ion complexes or other ligands, as for example in so-called riboswitches [9].

Apart from detecting conformational changes in RNA, the knowledge on the specific location of metal ion-binding sites is crucial to understand folding, structure and mechanism of any RNA [10–13]. Mg(II) is the key player in such metal ion interactions, but its spectroscopic detection is highly challenging and Mg(II) catalyzed hydrolytic cleavage of the RNA is very slow and often inefficient. As a way out, lanthanide(III) ions are often applied to map the metal ion-binding sites in RNA, their higher charge leading to higher affinities [14–16]. Importantly, the addition of small amounts of Ln(III) ions usually does not affect the activity of ribozymes. Terbium(III) has been used in the past to mimic earth alkaline ions, e.g., mostly Ca(II) [16], but also Mg(II) having similar properties [17, 18]. Cleavage reactions with Tb(III) can be carried out at physiological pH because the Tb(III)-aqua species $[Tb(H_2O)_n]^{3+}$ has a pK_a value of ~7.9 in contrast to hydrated Mg(II) with a pK_a of 11.4 [17]. Consequently, the acid–base assisted hydrolytic cleavage of the backbone is much faster with Tb(III) compared to Mg(II). It is assumed that Tb(III) replaces Mg(II) at its specific binding sites within the folded RNA tertiary structure and indeed it has been shown that the respective cleavage patterns are alike [17]. The $Tb(OH)(aq)^{2+}$ complex activates a 2′ OH close-by and subsequently the intramolecular nucleophilic attack of the neighboring 3′–5′-phosphodiester bond can occur as described in Fig. 1. While application of lower micromolar Tb(III) concentrations can specifically detect high affinity metal ion-binding sites, high millimolar concentrations of Tb(III) cleave the RNA backbone at single-stranded and non-Watson–Crick base paired regions, thereby revealing the secondary and tertiary structure of the RNA [15, 16, 19].

In this chapter we describe both, in-line probing and terbium(III) cleavage, and show their practical application using the example of the 202 nt long *btuB* riboswitch of *E. coli*, a natural aptamer sequence that recognizes coenzyme B_{12} (AdoCbl; Figs. 3 and 4) [20, 21].

2 Materials

All solutions must be prepared using freshly autoclaved (121 °C, 15 min, 210 kPa) bi-distilled water that is additionally filtered, e.g., through 0.2 μm syringe filters or 0.22 μm bottle top filter units. In the following we refer to this as "ultrapure water." If not stated otherwise, the buffers were prepared using this ultrapure water before being filtered again using the above mentioned filtering systems. All consumables such as reaction tubes and pipette tips must be purchased as DNase/RNase-free items or otherwise need to be autoclaved (121 °C, 20 min, 210 kPa) before usage, as has to be done with all glassware. All chemicals are at least puriss p.a. or biograde and are purchased from usual suppliers such as Sigma-Aldrich or Brunschwig Chemie. [γ-^{32}P]-ATP was purchased from Perkin Elmer as a batch of 5 mCi (6000 Ci/mmol).

Centrifuges used are a Sorvall RC6+ as well as standard tabletop centrifuges from Eppendorf (both cooling and non-refrigerated ones). All vertical gel electrophoresis apparatuses are homemade and high voltage power supplies are used (4,000 V, 300 W, 300 mA maximum). The gels are dried on Whatman chromatography paper (0.18 mm) on gel dryers. Screening is performed using storage phosphor screens (GE Healthcare), or comparable products. Phosphorimaging is performed on a Storm860 phosphorimager (GE Healthcare) or comparable equipment. The gels are analyzed with the ImageQuant TL software (GE Healthcare) and the intensity changes evaluated with Origin9 (OriginLab), or comparable software.

2.1 General Stock Solutions for Polyacrylamide Gel Electrophoresis (PAGE)

1. Denaturing electrophoresis buffer (1× TBE): The 1× TBE buffer is prepared by dilution from 10× TBE with bi-distilled water followed by autoclaving (121 °C, 15 min, 210 kPa). The autoclaved buffer can be stored at room temperature (r.t.) for several weeks. Approximately 1 L of 1× TBE buffer is needed to run 1 or 2 gels using one gel apparatus.
2. 10× TBE buffer: 0.89 M Tris, 0.89 M boric acid, pH 8.3, and 20 mM Na_2H_2EDTA.
3. Denaturing 10 % electrophoresis gel solutions: 7 M urea, 10 % Long Ranger gel solution in 1× TBE buffer. Use bi-distilled water and filter with 0.22 μm bottle top filter units (*see* **Note 1**, Fig. 2). The gel solutions are best prepared 1 day before use and stored in the dark at 4 °C. The gel solutions should be cool when used for casting the gels (*see* **Note 2**). For a gel with the dimensions of 30 cm × 45 cm × 0.5 mm roughly 80 mL of gel solution is needed, although we advise to prepare 100 mL. Gel solutions can be stored a couple of weeks in the dark at 4 °C.
4. Formamide loading buffer: 82 % (v/v) formamide, 0.16 % (w/v) xylene cyanol (XC), 0.16 % (w/v) bromophenol blue (BB), 10 mM EDTA, pH 8.0 (at 22 °C, i.e., r.t.) (*see* **Note 3**). The formamide loading buffer should be stored at 4 °C.

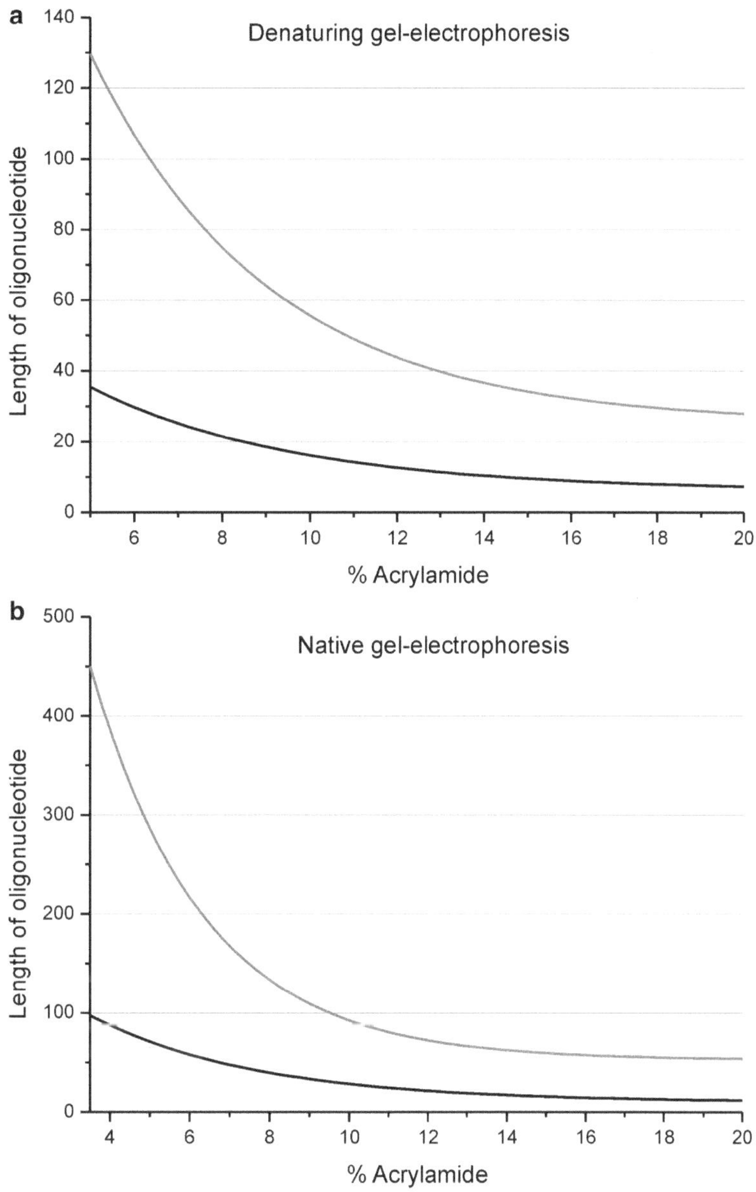

Fig. 2 Dye mobility for selecting the appropriate acrylamide percentage of denaturing and native gels. The mobility of bromophenol blue (*black, bottom curve*) and xylene cyanol (*gray, top curve*) is shown relative to the length of a single-stranded oligonucleotide on denaturing (*top*, 7 M urea) and native (*bottom*) polyacrylamide gels of different percentage [31]. Note that these mobilities can be used only as an approximation. The percentage of acrylamide in the gel should be chosen such that the oligonucleotide of interest migrates between the two dyes or above xylene cyanol. Aberration in migration speed of the oligonucleotide in both denaturing and native gels can however arise from its primary, secondary, and tertiary structure as well as the ratio between acrylamide and bisacrylamide

2.2 Buffers and Solutions Needed for In-Line Probing Assays

1. 1 M Tris–HCl buffer, pH 8.3 (at 22 °C): The buffer can be stored at r.t. and should be used within a few weeks.
2. 200 mM $MgCl_2$ (*see* **Note 4**). The solution can be stored at r.t. and should be used within a few weeks.
3. 1 M KCl (*see* **Note 4**). The solution can be stored at r.t. and should be used within a few weeks.

2.3 Buffers and Solutions Needed for Tb(III) Cleavage Assays

1. 5 mM sodium cacodylate buffer, pH 5.5 (at 22 °C): Prepare a 100 mM aqueous solution of sodium cacodylate trihydrate. Adjust the pH to 5.5 using a concentrated HCl solution (32 %) and dilute the 100 mM stock solution to 5 mM for further use. The buffer can be stored at r.t. and should be used within a few weeks.
2. 500 mM $TbCl_3$ solution: Dissolve anhydrous $TbCl_3$ in 5 mM sodium cacodylate buffer, pH 5.5 (at 22 °C); store at –20 °C.
3. 10× $TbCl_3$ solutions: 10× $TbCl_3$ solutions of the required concentrations are prepared in 5 mM sodium cacodylate buffer, pH 5.5 (at 22 °C) by diluting the original stock of 500 mM $TbCl_3$. All $TbCl_3$ solutions have to be stored at –20 °C.
4. 250 mM MOPS, pH 7.0 (at 22 °C): Adjust the pH to 7.0 by addition of a concentrated KOH solution (e.g., 3 M). The buffer should be stored in the dark at r.t. and used within a few weeks.
5. 2× quenching buffer: 82 % (v/v) formamide and 10 mM EDTA, pH 8.0 (at 22 °C). Store at 4 °C.
6. 3 M sodium acetate, pH 5.2 (at 22 °C).
7. 100 % (v/v) ethanol.
8. Formamide loading buffer as described in Subheading 2.1.

2.4 Solutions Needed for the Control Experiments

1. 10× sodium carbonate (Na_2CO_3) buffer: 0.5 M Na_2CO_3, pH 9.0 (at 22 °C), 10 mM EDTA, pH 8.0 (at 22 °C). The buffer can be stored at r.t. and used within a few weeks.
2. 10× sodium citrate buffer: 0.25 M trisodium citrate, pH 5.0 (at 22 °C). Adjust the pH by addition of concentrated HCl. The buffer can be stored at r.t. and used within a few weeks.
3. 2× quenching buffer as described in Subheading 2.3.
4. RNase T1 solution (1 U/μL): RNase T1 stock with a concentration of 1,000 U/μL was diluted to a final concentration of 1 U/μL in a buffer containing 50 mM Tris–HCl, pH 7.4 (at 22 °C) and 50 % (v/v) glycerol. The RNase T1 solution should be stored at –20 °C and used before its expiry date given by the company.

3 Methods

Both in-line probing and terbium(III) cleavage are applied on RNA which is singularly labeled either at its 5′ or 3′ end. Although other labeling strategies are feasible, labeling with ^{32}P is still the most applied technique, since it allows gel experiments with very high resolution. For the description of the in vitro transcription with T7 RNA polymerase [22, 23] and the 5′ end-labeling [24, 25] or 3′ end-labeling of RNA [26] we refer to the literature.

3.1 In-Line Probing Assay

In-line probing reactions are performed in a total volume of 10 μL. If ^{32}P-labeled RNA is used, the concentration of the labeled RNA should be around 2–3 nM or 20,000–40,000 cpm per reaction, while the total RNA concentration can vary depending on the experiment (*see* **Note 5**). Although some RNA sequences might be folded correctly after in vitro transcription and isolation, it is advisable to refold the RNA directly prior to the experiment. The refolding protocol has to be adapted to the specific sequence. The refolding protocol used in our laboratory for the *btuB* aptamer sequence is described, *see* **Note 6**.

1. To prepare negative and positive controls, mix ^{32}P-labeled RNA (2–3 nM) with the adequate amount of 1 M Tris–HCl buffer, pH 8.3 (at 22 °C), 200 mM $MgCl_2$, and 1 M KCl to reach a total volume of 10 μL (*see* **Note 7**). For a practical example of the negative and positive control, please *see* Table 1.
2. For the positive control add the desired amount of ligand (in our case this would be 1 μL of 0.1 mM AdoCbl; Table 1).

Table 1
Practical example of a positive and negative control experiment for in-line probing

Reagent	c_{stock}	c_{final}	$V_{negcontrol}$	$V_{poscontrol}$
^{32}P-RNA[a]	300 nM	30 nM	1 μL	1 μL
Tris–HCl, pH 8.3	1 M	50 mM	0.5 μL	0.5 μL
KCl	1 M	100 mM	1 μL	1 μL
$MgCl_2$	200 mM	20 mM	1 μL	1 μL
Coenzyme B_{12}	100 μM	10 μM	–	1 μL
Water			6.5 μL	5.5 μL

[a]This concentration indicates the total concentration of the RNA after labeling. The concentration of the actually labeled RNA molecules is however lower and varies depending on the RNA sequence (*see* **Note 5**)

3. If needed, adjust the final concentrations with ultrapure water to a total volume of 10 μL.
4. Incubate at 25 °C for 40 h.
5. Quench the reaction by adding 10 μL formamide loading buffer and analyze the reaction mixture by denaturing polyacrylamide gel electrophoresis.

3.2 Terbium(III) Cleavage Assay

The total volume of the reaction mixture is 10 μL.

1. Mix together ^{32}P-labeled RNA (about 2–3 nM or 20,000–40,000 cpm per lane) and unlabeled RNA (1–1.5 μM) (*see* **Note 8**) with 1 μL of 250 mM MOPS buffer, pH 7.0 (at 22 °C) (*see* **Note 9**) and KCl at its required concentration.
2. Keep the samples at 90 °C for 45 s for full denaturation of the RNA.
3. Cool the samples at room temperature for 1 min.
4. Add the required concentration of $MgCl_2$ from an appropriate stock for RNA folding.
5. Incubate the samples at the required temperature (varies with the system) (*see* **Note 10**) for 15 min for a complete folding of the RNA.
6. Add 1 μL of $TbCl_3$ from the desired 10× stock solution except for the negative control.
7. Incubate the reaction mixture on ice for an optimized time period (generally ranges from 30 min to 2 h) (*see* **Notes 11** and **12**).
8. Quench the reaction by adding 10 μL of 2× quench buffer.
9. Add 3 M sodium acetate, pH 5.2 (at 22 °C) to reach the final concentration of 0.3 M followed by the addition of 2.5 volumes of ice-cold ethanol (*see* **Note 13**).
10. Incubate the samples at −20 °C overnight.
11. Centrifuge the samples at 15,000 × *g* for 30 min at 4 °C.
12. Decant the supernatant carefully and dry the pellet in a SpeedVac evaporator (*see* **Note 14**).
13. Dissolve the pellet in 5 μL of formamide loading buffer.

3.3 Controls and Ladders

The same controls and ladders are used for both in-line probing and Tb(III) cleavage. For each control or ladder the equal amount and concentration of ^{32}P-labeled RNA is needed as applied in the experimental lanes. The total volume of the control and ladders is 10 μL. The control and ladders should be freshly made prior to loading on the gel.

3.3.1 Non-reacted RNA as a Control

This control is crucial to check the inherent extent of degradation of the RNA sample. In an ideal case, this lane should only show the

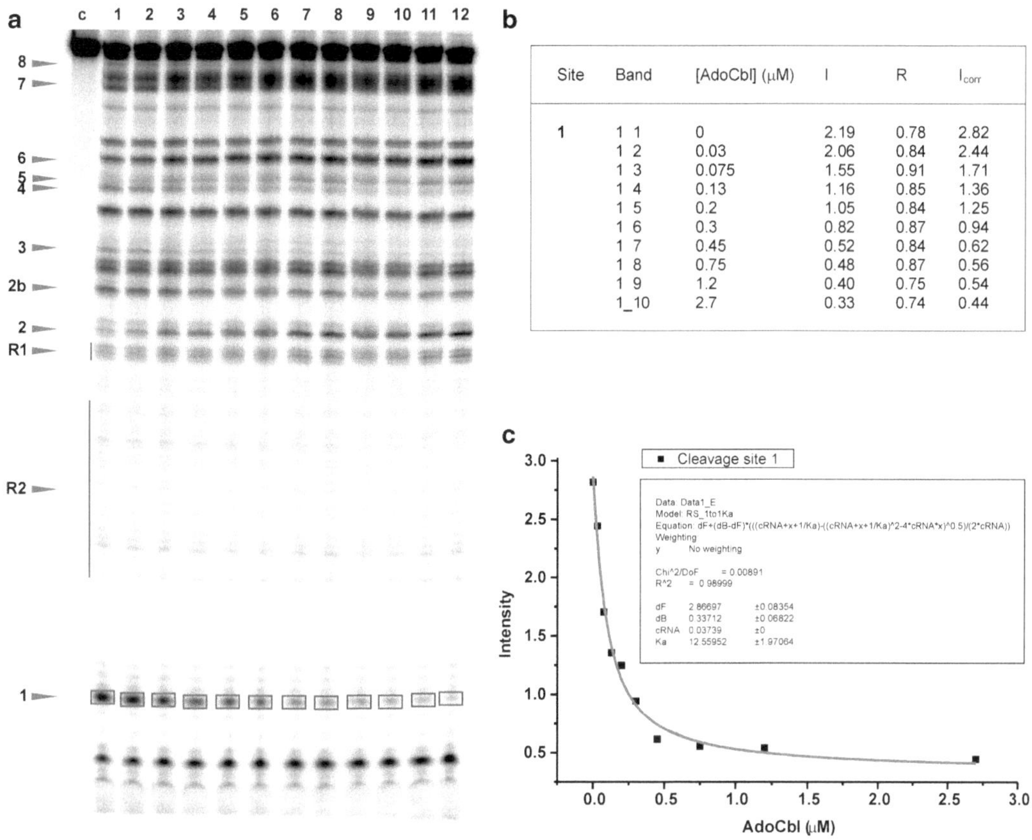

Site	Band	[AdoCbl] (μM)	I	R	I_{corr}
1	1_1	0	2.19	0.78	2.82
	1_2	0.03	2.06	0.84	2.44
	1_3	0.075	1.55	0.91	1.71
	1_4	0.13	1.16	0.85	1.36
	1_5	0.2	1.05	0.84	1.25
	1_6	0.3	0.82	0.87	0.94
	1_7	0.45	0.52	0.84	0.62
	1_8	0.75	0.48	0.87	0.56
	1_9	1.2	0.40	0.75	0.54
	1_10	2.7	0.33	0.74	0.44

Fig. 3 Example of the quantification of an in-line probing experiment, using a titration-experiment of the *btuB* aptamer sequence with coenzyme B_{12}. (**a**) Evaluation of the polyacrylamide gel. The *arrows* indicate the nine specific cleavage sites as well as the two constant regions R1 and R2 that are marked with a *line* and are used for background correction. The *gray boxes* frame the evaluated intensities at site 1. *Lane c* is the non-reacted RNA. The coenzyme B_{12} concentration is increasing from lane 1 to 12. (**b**) Evaluation of cleavage site 1 to obtain the corrected intensities I_{corr}. (**c**) Plot of the intensity changes at site 1 after background correction. The data is fit to a 1:1 binding isotherm [32], yielding the affinity constant between the RNA and its ligand

full-length RNA band. If this is the case, all cleavage products seen in the other lanes are "real" and can be evaluated unambiguously (Fig. 3).

1. Dilute ^{32}P-labeled RNA in ultrapure water.
2. Add 10 μL of formamide loading buffer and store the sample in liquid nitrogen until it is loaded on the gel.

3.3.2 RNase T1 Ladder

1. Dilute ^{32}P-labeled RNA in 2 x quenching buffer, e.g., dilute 1 μL of ^{32}P-labeled RNA in 7 μL 2× quenching buffer.
2. Add 1 μL of 0.25 M sodium citrate buffer.
3. Add 1 μL of RNase T1 (1 U/μL) and mix gently by pipetting up and down (*see* **Note 15**).
4. Incubate the mixture at 55 °C for 5 min (*see* **Note 16**).

5. Immediately add 3 μL formamide loading buffer and 7 μL of ultrapure water and load the sample directly on the gel. If the sample cannot be loaded immediately, it should be stored in liquid nitrogen until loaded on the gel.

3.3.3 Alkaline Hydrolysis Ladder

1. Dilute ^{32}P-labeled RNA in ultrapure water, e.g., dilute 1 μL of ^{32}P-labeled RNA in 8 μL ultrapure water.
2. Add 1 μL of 0.5 M Na_2CO_3 buffer.
3. Incubate the mixture at 90 °C for 5 min (*see* **Note 16**).
4. Immediately add 10 μL formamide loading buffer and load the sample directly on the gel. If the sample cannot be loaded immediately, it should be stored in liquid nitrogen until loaded on the gel.

3.4 Separation of Cleavage Products by Denaturing PAGE

1. Analyze the cleavage products of the respective experiment (from Subheadings 3.1 and 3.2) and the applicable controls (from Subheading 3.3) by denaturing PAGE (Figs. 3 and 4).
2. Run the gel at 15–20 W until the lower dye bromophenol blue reaches the lower end of the gel (*see* **Note 17**).
3. Preheat the gel dryers and set up the vacuum pump (*see* **Note 18**).
4. Remove first the spacers and then the glass plate on the top with the help of a spatula.
5. Put Whatman chromatography paper (0.18 mm) on the gel, tap it gently and lift it up quickly turning it with the gel being now on the top (*see* **Note 19**).
6. Cover the gel with plastic foil.
7. Put the gel on the gel dryer (paper pointing down) and dry under vacuum at 90 °C for 60–90 min (*see* **Note 20**).
8. Switch off the heating mode of the gel dryer and cool the gel for about 1 h under vacuum. To prevent breaking of the gel, the vacuum should not be released until the gel is cooled down to around 40 °C.
9. Carefully release the vacuum and put the gel for exposure overnight.
10. Scan the gel using a phosphorimager and analyze the gel with the respective software, e.g., ImageQuant TL.

3.5 Quantification

1. Draw a rectangle around the band of a cleavage site to be analyzed (Fig. 3a). The whole band has to be included within the rectangle. However, if other bands are close by, care has to be taken not to include intensity deriving from other bands.
2. Copy this rectangle and place it on the second lane of the same site. Always use identical rectangles to quantify the bands of the same site.

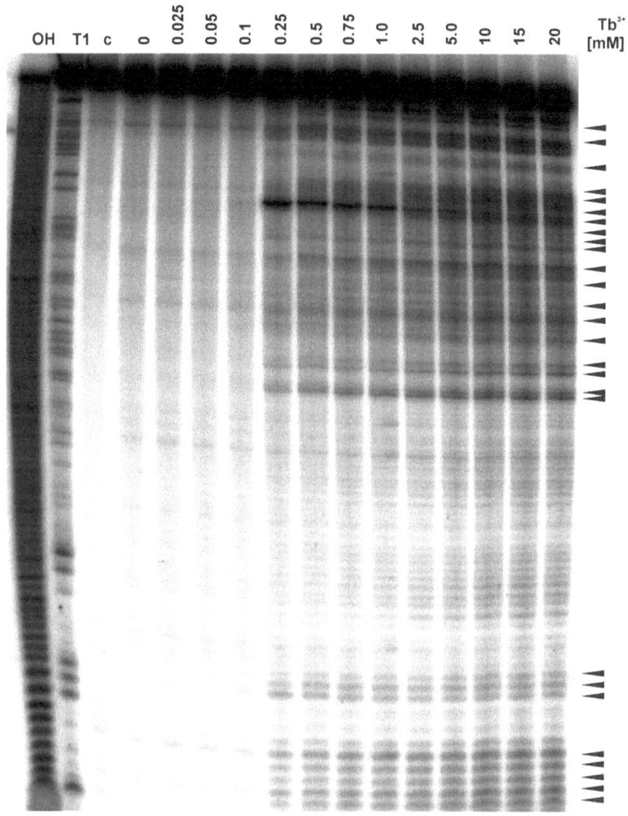

Fig. 4 Tb(III)-mediated footprinting of the *btuB* riboswitch. Footprint of the ^{32}P-labeled *btuB* riboswitch incubated with 0–20 mM Tb(III) after folding with 20 mM $MgCl_2$. *Lanes 1* and *2* indicate alkaline hydrolysis and RNase T1 ladder, respectively. *Lane 3, c,* is the control to check for nonspecific degradation of the RNA used in the experiment. *Lane 4* represents a second control to detect degradation of RNA under the employed reaction conditions but in the absence of Tb(III). The Tb(III) concentration increases from *lanes 5 to 16*. Tb(III) mediated cleavage of the RNA at specific nucleotides is indicated by *arrows* and can be categorized into weak, intermediate and strong cleavage sites

3. For background correction choose two sites within the cleavage pattern that do not change in intensity. Take two single bands or areas to get references R1 and R2 (*see* also Subheading 3.5.1). It is important to determine these two sites for all lanes of interest.

3.5.1 Evaluation of the In-Line Probing Experiments

1. To adjust the intensity of the different lanes calculate the relative intensity of each lane by the average of the two regions R1 and R2 (Fig. 3a, b).

$$A = \frac{(\mathrm{R1} + \mathrm{R2})}{2} \quad (1)$$

2. Out of these averages A, determine the most intensive lane X.
3. Divide all averages A by X to get a correction factor R for each lane.

$$R = \frac{A}{X} \tag{2}$$

4. Adjust the intensities of each band I by dividing with the correction factor R of the corresponding lane. This procedure gives the corrected intensity I_{corr}.

$$I_{corr} = \frac{I}{R} \tag{3}$$

5. Plot the intensities I_{corr} of the single bands versus the ligand concentration (or what is applicable) to quantify the intensity changes at each cleavage site using Origin9 or a respective program.
6. Fit the plot to a suitable fitting curve leading to the K_a values for the interaction between the RNA and its ligand (Fig. 3c, *see* **Note 21**).

3.5.2 Evaluation of Terbium(III) Cleavage Experiments

1. To correct for the background and to adjust the intensity of each band, perform **steps 1–4** as described in Subheading 3.5.1.
2. Determine I_{corr} for the corresponding site at 0 mM $TbCl_3$. This corrected cleavage intensity corresponds to $I_{corr(0)}$.
3. The final relative cleavage intensity I_R (compared to the absence of Tb(III)) of each band is calculated by taking the extent of cleavage at 0 mM $TbCl_3$ at the corresponding cleavage site into account. I_R is calculated by subtracting $I_{corr(0)}$ from I_{corr} at each applied $TbCl_3$ concentration:

$$I_R = I_{corr} - I_{corr(0)}. \tag{4}$$

4. The relative cleavage intensity of zero indicates a similar extent of cleavage as the control. Therefore only positive I_R values should be taken into account while I_R values of zero should be disregarded.
5. Classify the final calculated relative cleavage intensities I_R into categories from weak to (very) strong Tb(III) cleavage sites depending on the lowest and the highest observed cleavage intensity.
6. Compare the analyzed data to complementary experiments to confirm the accordance with Mg(II) binding sites (*see* **Note 22**) [17, 27–30]. A competition experiment can be set up in a similar way as the terbium cleavage assay, but here the Mg(II) concentration is increased in the presence of a constant Tb(III) concentration. Sites that involve Mg(II) binding exhibit a decrease in the cleavage intensity by Tb(III) with increasing Mg(II) concentration.

4 Notes

1. The percentage of the polyacrylamide gel solution has to be adapted depending of the length of the RNA analyzed (Fig. 2). For a 200 nt long RNA we use a 10 % gel solution. For smaller sequences the percentage has to be increased (e.g., for a 60 nt RNA use a polyacrylamide percentage of 18 %). The Long Ranger gel solution gives a higher resolution, although also other gel solutions can be used that have a ratio of acrylamide/bisacrylamide = 29:1.
2. The best resolution of the gels is achieved if the gel is packed the evening before. After polymerization of the gel, wet towels and plastic foil are wrapped on the sides of the plates to prevent the gel from drying, while the combs are left inside and removed only the next day. The gel is then kept at 4 °C to fully polymerize overnight. For RNA of 100–200 nt in length, the dimensions of the gels needed for in-line probing experiments are normally 30 cm × 45 cm and can hold 20 pockets of 1 cm width. For terbium cleavage experiments with large RNAs (>200 nt) we advise to use gel plates of 90 cm in length to allow good resolution at each nucleotide.
3. For small volumes it is advised to use formamide loading buffer over urea loading buffer as urea loading buffer tends to crystallize easily. Urea loading buffer can be rapidly brought back into solution by shortly heating in the microwave.
4. The concentrations of the $MgCl_2$ and KCl solutions have to be adapted to the RNA sequence. For the *btuB* riboswitch we use a final concentration of 20 mM $MgCl_2$ and 100 mM KCl to reach a complete folding and therefore prepare a 200 mM $MgCl_2$ and a 1 M KCl solution respectively.
5. The ratio between labeled and unlabeled RNA varies depending on the sequence and the labeling method. For the 5′ end-labeling of the *btuB* RNA, the ratio normally lies between 3 and 20. If desired, additional unlabeled RNA can be added to the reaction volume, although we do not advise to do so for the regular experiments.
6. Refolding of the aptamer region of the *btuB* riboswitch for in-line probing experiments: Dilute the RNA to a desired concentration in ultrapure water containing 100 mM KCl. Heat the sample at 90 °C for 1 min. Cool the sample at room temperature for 1 min. Add 1 M $MgCl_2$ and 1 M Tris–HCl, pH 8.3 (at 22 °C) to reach a concentration of 20 mM and 50 mM, respectively. Incubate at 37 °C for 15 min.
7. If many reactions need to be done contemporaneously (e.g., for a titration) it is easiest and safest to add the labeled RNA at the end.

8. Addition of unlabeled RNA reduces the non-specific cleavage by Tb(III) and also helps to precipitate the labeled RNA.
9. The pH of the Tb(III) cleavage reaction should be maintained between 7 and 7.5 to form sufficient amounts of the $Tb(OH)(aq)^{2+}$ species responsible for the cleavage.
10. The temperature and the time required for RNA folding after addition of $MgCl_2$ should be adjusted according to the system under investigation.
11. We recommend the Tb(III) cleavage reactions to be carried out at 4 °C (on ice bath) since it helps reducing the background degradation of the RNA although the cleavage reaction can also be carried out at higher temperatures (25–45 °C) and needs to be optimized carefully for an individual system.
12. For the cleavage reaction with Tb(III), the incubation time varies depending upon the target RNA under investigation and should be optimized for single-hit cleavage reactions where 90 % of the full-length RNA should remain uncleaved.
13. Instead of 3 M sodium acetate, pH 5.2, sodium chloride can be added to the samples to a final concentration of 200 mM to facilitate the precipitation of cleavage products.
14. Avoid excessive drying of the pellet in a SpeedVac evaporator as it makes the pellet difficult to redissolve in loading buffer.
15. The amount of RNase T1 can vary depending on the RNA used, e.g., higher amounts are needed for very stable and GC rich sequences.
16. The incubation time needed for both ladders varies depending on the RNA sequence and should be optimized before.
17. The power of 15 W is recommended only when the first and last two pockets are not loaded with samples (e.g., only 16 pockets are used out of the possible 20). To reach a uniform running-pattern, we clamp aluminum plates to the front of the mounted gels. In this way the generated heat is distributed evenly over the whole surface preventing the so-called "smiley-effect." In the absence of aluminum plates, it is advised to run the gel at lower Watt (around 8 W) for longer times.
18. To preheat the gel dryer, for us it works best to set the dryer to 90 °C and switch it on 10 min prior to demounting the gels.
19. Gels of up to 12 % acrylamide stick perfectly on Whatman chromatography paper (0.18 mm). Gels of higher percentage (e.g., 18 %) need to be transferred to the plastic foil first. In a second step they are covered with the Whatman chromatography paper and then put into the dryer with the paper pointing down.

20. Care should be taken while drying 15–18 % polyacrylamide gels as excessive drying can curl up the gels.
21. All experiments should be done at least three times and the binding constants given as the averaged value (arithmetic or weighted mean, as applicable).
22. For example, cleavage studies can be performed also with other metal ions, e.g., Mn(II), Pb(II), Zn(II), any Ln(III), or Mg(II) itself [27–30]. Note that some differences most likely will occur due to the different coordination properties. With Mg(II), the bands are usually only very weak, but help to verify real binding sites in combination with the Tb(III) data.

Acknowledgements

Financial support by the European Research Council (ERC Starting Grant to RKOS), the Swiss National Science Foundation, the Swiss State Secretariat for Education, Research, and Innovation (COST Action CM1105), and the University of Zurich is gratefully acknowledged.

References

1. Das R, Kwok LW, Millett IS et al (2003) The fastest global events in RNA folding: electrostatic relaxation and tertiary collapse of the *Tetrahymena* ribozyme. J Mol Biol 332:311–319
2. Woodson SA (2005) Metal ions and RNA folding: a highly charged topic with a dynamic future. Curr Opin Chem Biol 9:104–109
3. Shiman R, Draper DE (2000) Stabilization of RNA tertiary structure by monovalent cations. J Mol Biol 302:79–91
4. Batey RT, Rambo RP, Doudna JA (1999) Tertiary motifs in RNA structure and folding. Angew Chem Int Ed Engl 38:2326–2343
5. Costa M, Michel F (1995) Frequent use of the same tertiary motif by self-folding RNAs. EMBO J 14:1276–1285
6. Cech T, Golden BL (1999) Building a catalytic active site using only RNA. In: Gesteland RF, Cech TR, Atkins JF (eds) The RNA world. Cold Spring Harbor Laboratory Press, Cold Springs Harbor, NY, pp 321–349
7. Oivanen M, Kuusela S, Lonnberg H (1998) Kinetics and mechanisms for the cleavage and isomerization of the phosphodiester bonds of RNA by bronsted acids and bases. Chem Rev 98:961–990
8. Soukup GA, Breaker RR (1999) Relationship between internucleotide linkage geometry and the stability of RNA. RNA 5:1308–1325
9. Roth A, Breaker RR (2009) The structural and functional diversity of metabolite-binding riboswitches. Annu Rev Biochem 78:305–334
10. Pyle AM (2002) Metal ions in the structure and function of RNA. J Biol Inorg Chem 7:679–690
11. Freisinger E, Sigel RKO (2007) From nucleotides to ribozymes—a comparison of their metal ion-binding properties. Coord Chem Rev 251:1834–1851
12. Schnabl J, Sigel RK (2010) Controlling ribozyme activity by metal ions. Curr Opin Chem Biol 14:269–275
13. Sigel H, Sigel RKO (2013) Metal ion interactions with nucleic acids and their constituents. In: Reedijk J, Poepelmeier K (eds) Comprehensive Inorganic Chemistry. 2nd edn. Elsevier Ltd., Oxford, UK. doi: 10.1016/B978-0-08-097774-4.00317
14. Ciesiolka J, Marciniec T, Krzyzosiak W (1989) Probing the environment of lanthanide binding sites in yeast tRNA(Phe) by specific metal-ion-promoted cleavages. Eur J Biochem 182:445–450
15. Walter NG, Yang N, Burke JM (2000) Probing non-selective cation binding in the hairpin ribozyme with Tb(III). J Mol Biol 298: 539–555
16. Sigel RK, Pyle AM (2003) Lanthanide ions as probes for metal ions in the structure and

catalytic mechanism of ribozymes. Met Ions Biol Syst 40:477–512

17. Sigel RK, Vaidya A, Pyle AM (2000) Metal ion binding sites in a group II intron core. Nat Struct Biol 7:1111–1116
18. Saito H, Suga H (2002) Outersphere and innersphere coordinated metal ions in an aminoacyl-tRNA synthetase ribozyme. Nucleic Acids Res 30:5151–5159
19. Harris DA, Tinsley RA, Walter NG (2004) Terbium-mediated footprinting probes a catalytic conformational switch in the antigenomic hepatitis delta virus ribozyme. J Mol Biol 341:389–403
20. Nahvi A, Barrick JE, Breaker RR (2004) Coenzyme B_{12} riboswitches are widespread genetic control elements in prokaryotes. Nucleic Acids Res 32:143–150
21. Gallo S, Oberhuber M, Sigel RK et al (2008) The corrin moiety of coenzyme B_{12} is the determinant for switching the *btuB* riboswitch of *E. coli*. Chembiochem 9:1408–1414
22. Beckert B, Masquida B (2011) Synthesis of RNA by *in vitro* transcription. Methods Mol Biol 703:29–41
23. Gallo S, Furler M, Sigel RKO (2005) *In vitro* transcription and purification of RNAs of different size. Chimia 59:812–816
24. Hilario E (2002) End labeling procedures. An overview. Methods Mol Biol 179:13–18
25. Hilario E (2004) End labeling procedures: an overview. Mol Biotechnol 28:77–80
26. Huang Z, Szostak JW (1996) A simple method for 3′-labeling of RNA. Nucleic Acids Res 24:4360–4361
27. Erat MC, Sigel RK (2011) Methods to detect and characterize metal ion binding sites in RNA. Met Ions Life Sci 9:37–100
28. Pechlaner M, Sigel RK (2012) Characterization of metal ion-nucleic acid interactions in solution. Met Ions Life Sci 10:1–42
29. Hertweck M, Mueller MW (2001) Mapping divalent metal ion binding sites in a group II intron by Mn(2+)- and Zn(2+)-induced site-specific RNA cleavage. Eur J Biochem 268:4610–4620
30. Dorner S, Barta A (1999) Probing ribosome structure by europium-induced RNA cleavage. Biol Chem 380:243–251
31. Sambrook J, Russel DW (eds) (2001) Molecular cloning: a laboratory manual. Cold Spring Harbor Laboratory Press, Cold Spring Harbor, NY
32. Sigel RK, Freisinger E, Lippert B (2000) Effects of N7-methylation, N7-platination, and C8-hydroxylation of guanine on H-bond formation with cytosine: platinum coordination strengthens the Watson-Crick pair. J Biol Inorg Chem 5:287–299

Chapter 9

Chemical Probing of RNA in Living Cells

Michael Wildauer, Georgeta Zemora, Andreas Liebeg, Verena Heisig, and Christina Waldsich

Abstract

RNAs need to adopt a specific architecture to exert their task in cells. While significant progress has been made in describing RNA folding landscapes in vitro, understanding intracellular RNA structure formation is still in its infancy. This is in part due to the complex nature of the cellular environment but also to the limited availability of suitable methodologies. To assess the intracellular structure of large RNAs, we recently applied a chemical probing technique and a metal-induced cleavage assay in vivo. These methods are based on the fact that small molecules, like dimethyl sulfate (DMS), or metal ions, such as Pb^{2+}, penetrate and spread throughout the cell very fast. Hence, these chemicals are able to modify accessible RNA residues or to induce cleavage of the RNA strand in the vicinity of a metal ion in living cells. Mapping of these incidents allows inferring information on the intracellular conformation, metal ion binding sites or ligand-induced structural changes of the respective RNA molecule. Importantly, in vivo chemical probing can be easily adapted to study RNAs in different cell types.

Key words In vivo chemical probing, DMS, RNA structure, RNA folding, Metal-induced cleavage, Metal ions, Lead cleavage

1 Introduction

Exploring RNA folding paradigms in vitro has been a major objective in RNA biology [1–10]. RNA in general folds in a hierarchical manner: first monovalent ions are required to shield the negatively charged RNA backbone, allowing counterion-mediated condensation and assembly of secondary structure. By binding to distinct metal ion pockets within RNA, Mg^{2+} ions allow further compaction of the molecule and the formation of long-range, tertiary interactions. Although in vitro studies are inevitable for characterizing RNA structure and folding pathways, the in vitro refolding conditions are very different from the intracellular environment [11, 12]. Especially temperature and ion concentrations, two very important factors in RNA folding, can vary considerably.

Christina Waldsich (ed.), *RNA Folding: Methods and Protocols*, Methods in Molecular Biology, vol. 1086,
DOI 10.1007/978-1-62703-667-2_9, © Springer Science+Business Media, LLC 2014

Furthermore, the directionality and velocity of transcription and translation as well as the presence of *trans*-acting factors, such as proteins, can influence RNA folding in vivo [11, 12]. However, there are only few techniques available for determining the structure of RNAs in the cell. As a result little is known about RNA folding in vivo and the contribution of the cellular environment to RNA structure formation.

Chemical probing is a powerful method, which has been used most extensively to map RNA structure in vivo [13–25]. It is a simple, well-adaptable, and inexpensive approach. Dimethyl sulfate (DMS), the most widely used chemical probing reagent, rapidly penetrates the cell and all of its compartments, modifying N1 of adenines and N3 of cytosines. DMS only methylates proton accepting ring nitrogens, if these atoms are not engaged in hydrogen bonding (e.g., Watson–Crick base pairing or sheared AA base pairs) and if they are solvent accessible, while a reduced solvent exposure or binding of a protein results in protection from DMS modification (Fig. 1). Occasionally, certain uridines and guanines are reactive to DMS at their N3 or N1 position, respectively, if these are stabilized in an enol-tautomer due to a specific local environment [26, 27]. The sites of modification are mapped by reverse transcription, as the bulky methyl group at the Watson–Crick face of adenines and cytosines leads to termination of the primer extension. Thus, reverse transcription of the modified RNA pool results in a variety of cDNAs of different length, which can be resolved on a standard denaturing polyacrylamide gel. Information on the RNA structure is obtained by analyzing the DMS modification pattern in the context of secondary structure maps derived from phylogenetic or bioinformatic studies [13–25]. The experimental framework for the Pb^{2+}-induced cleavage assay is not fundamentally different to the DMS approach with one notable exception: in contrast to DMS, which is a nucleobase-specific probe, Pb^{2+} is able to displace Mg^{2+} ions in their binding pockets and induces strand scission of the RNA backbone in its vicinity [28–31]. These cleavage events can be mapped by reverse transcription as well.

Employing in vivo DMS chemical probing we recently provided the first structure-based insights into DEAD-box protein-facilitated RNA folding in living cells [18]. In yeast mitochondria efficient splicing of all group I and group II introns is dependent on the DEAD-box helicase Mss116p [32]. To monitor Mss116p-induced conformational changes within the ai5γ group II intron in yeast, we mapped the ai5γ structure in different genetic backgrounds (Fig. 2). While the intron adopts the native conformation in the wild-type yeast strain, ai5γ appears to be largely unfolded in the *mss116*-knockout strain, as most of the secondary structure elements, but none of the tertiary interactions are

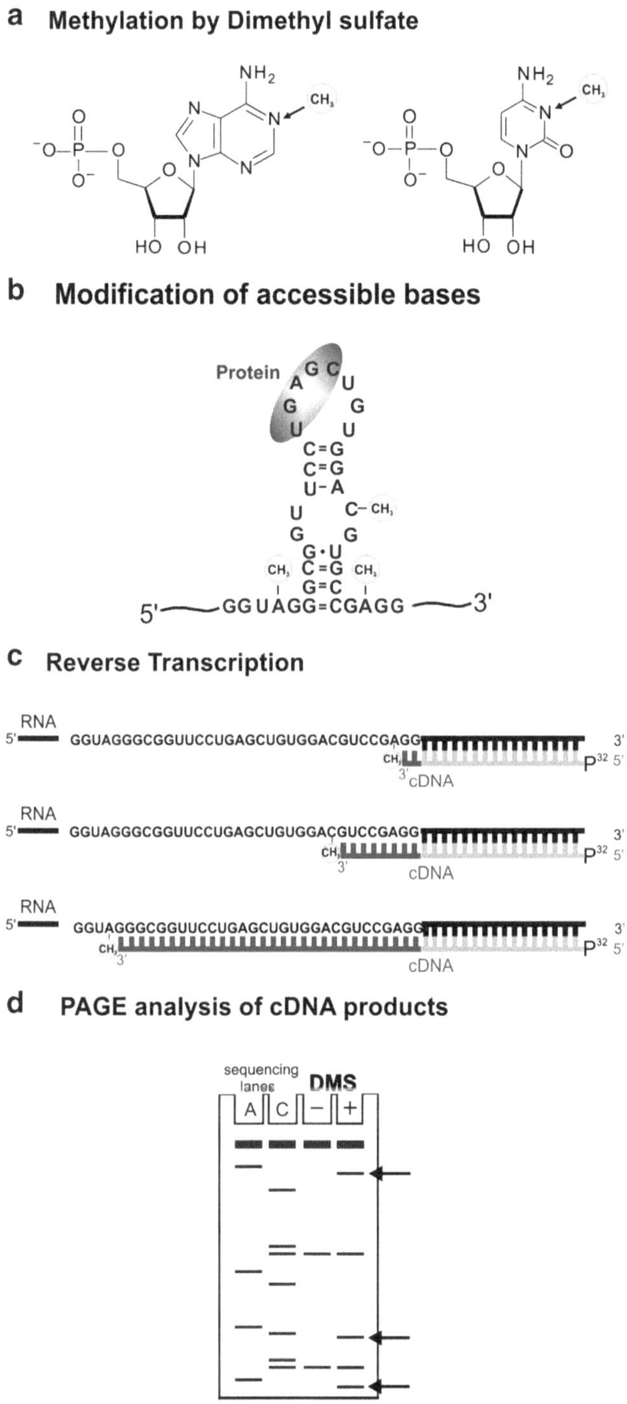

Fig. 1 Schematic representation of in vivo chemical probing using DMS. (**a**) The N1 atom of adenines and the N3 atom of cytosines are methylated by DMS, (**b**) if these atoms are not involved in H-bonding or protected by a protein. (**c**) The modified residues can be detected by reverse transcription using 5′end-labeled, gene-specific DNA primers (*light gray*), as the extension by the reverse transcriptase terminates when the enzyme encounters the bulky methyl group. This reaction generates a family of radio-labeled cDNAs of different length (*dark gray*). (**d**) The pool of cDNA is separated by a standard denaturing PAGE. A and C denote the sequencing **Fig. 1** (continued) lanes which are used to determine the position of the modification. The −lane is the RT stop control, showing natural stops occurring during reverse transcription, for which RNA extracted from cells that were not treated with the DMS was used. In the +lane RNA has been reverse transcribed that was extracted from cells treated with DMS. Comparing these two lanes reveals the sites of DMS modifications (indicated by an *arrow*)

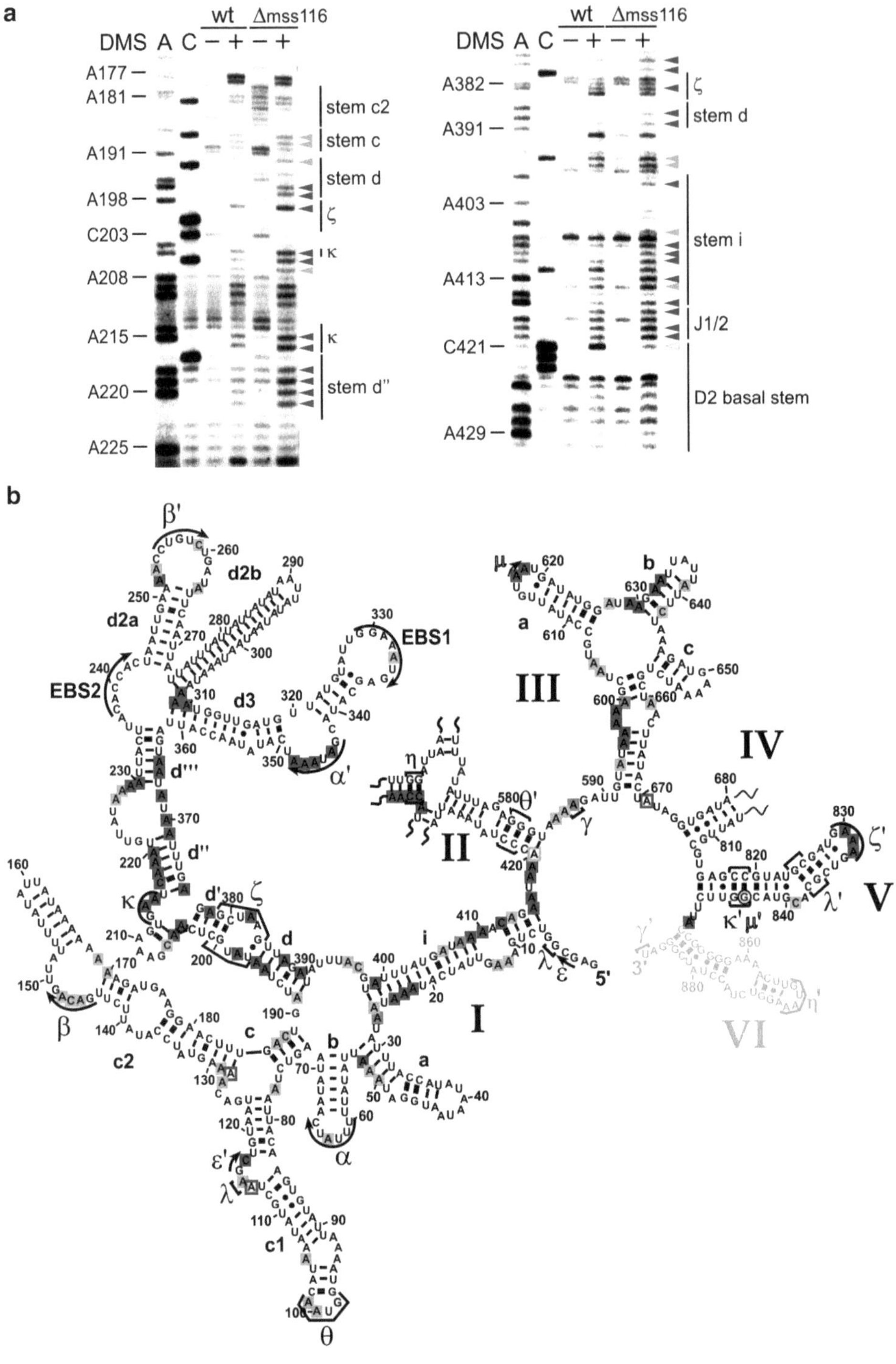

Fig. 2 Mss116p-induced conformational changes within the ai5γ intron in vivo. (**a**) The κ–ζ element depends on Mss116p for folding. Representative gel showing the modification intensity of nucleotides in the 5′ part (*left panel*) and 3′ part (*right panel*) of the D1 core structure composed of stems d, d′ and d″, in which the κ region, the ζ receptor and the coordination loop are embedded, in the wt and *mss116*-knockout strain. The *arrow heads* indicate residues, the accessibility of which changes due to the absence of Mss116p (*filled arrow heads*

formed [18]. In brief, most of the Mss116p-induced structural changes are observed within domain D1 (Fig. 2); thus Mss116p appears to facilitate the formation of this largest domain, which is the scaffold for docking of other intron domains. Based on the chemical probing data we proposed that Mss116p assists the ordered assembly of the ai5γ intron in vivo and is critical for folding of the RNA at an early step along the pathway [18]. In light of the fact that metal homeostasis plays a crucial role in yeast mitochondrial intron splicing [33] and that Mss116p lowers the Mg^{2+} requirement for intron folding in vitro [34, 35], we were also interested in assessing the impact of Mss116p on the formation of metal ion binding pockets within the ai5γ intron in vivo. Using the Pb^{2+}-induced cleavage assay we determined the metal ion binding sites within the ai5γ intron in vivo (Fig. 3) and observed so far that these correlate nicely with those identified previously in vitro [36], suggesting that Mss116p may not significantly influence the formation of metal ion binding sites in vivo (*Wildauer and Waldsich, unpublished*). Here we provide a detailed description of both the in vivo DMS chemical probing technique and the lead-induced cleavage assay for mapping RNA structure in eukaryotic cells.

2 Materials

Prepare all solutions with RNase-free ultrapure water (deionized water with a sensitivity of ≥18 MΩ cm at 25 °C) unless indicated otherwise. Sterilize all solutions by autoclaving or filtering (membrane pore size 0.2 μm) and store them at room temperature if not indicated otherwise. Use RNase-free glassware and plasticware for the experiments.

Fig. 2 (continued) represent an increase in accessibility, while open ones highlight bases with reduced accessibility). Strong changes in accessibility are displayed in dark gray (>2-fold), while smaller changes are shown in *light gray* (1.5 to 2-fold). These values were derived from normalized gel plots. In the –lane natural stops of the reverse transcriptase are seen. In the +lane the in vivo DMS pattern is shown. Notably, comparing these two lanes reveals the DMS-induced stops of the reverse transcriptase and thus accessible residues (N1-A, N3-C). Importantly, the – and +lanes are shown for both the wt and *mss116*-knockout strain. Comparing *lanes 4* and *6* reveals the altered DMS modification pattern and thus conformational changes within the ai5γ intron due to the absence of Mss116p. (**b**) Differential summary map: residues, the modification intensity of which changes in the absence of Mss116p (Δ*mss116* strain), are highlighted. This map is based on normalized plots, which had been derived from the modification gels. The *closed squares* indicate an increase in accessibility, while *open squares* represent bases with reduced accessibility in the absence of Mss116p. The A and C residues whose modification remains unaltered (i.e., equally modified or protected in both strains) are not highlighted. This figure has been adapted from [18], with permission

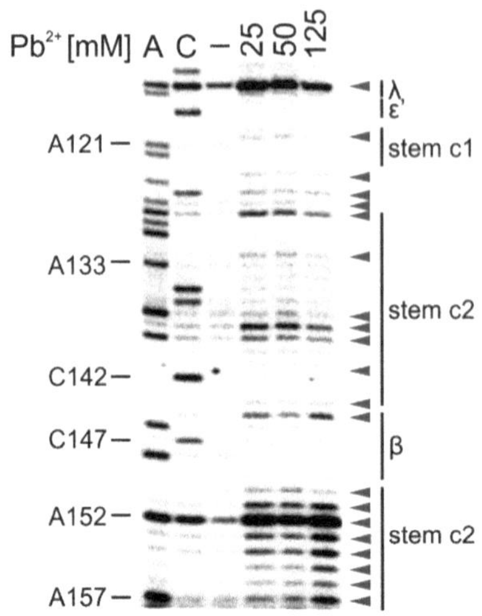

Fig. 3 Mapping metal ion binding sites within the ai5γ intron in yeast mitochondria. Pb^{2+}-induced cleavage sites (indicated with a *gray arrow*) were mapped via primer extension. A and C denote sequencing lanes; –lane is an RT stop control to detect natural stops of the extension (RNA, which was extracted from untreated cells, was reverse transcribed). Lanes labeled 25–125 indicate the concentration of $Pb(OAc)_2$ [mM] used to modify the intracellular RNA

2.1 Culturing Human Cells

1. Growth medium DMEM–FBS–PS for HEK 293 cells: Dulbecco's modified Eagle's medium (DMEM) with high glucose and supplemented with 10 % (v/v) fetal bovine serum (FBS) and 1 % (v/v) penicillin/streptomycin. Store at 4 °C.
2. Washing buffer for adherent HEK 293 cells: 1× Dulbecco's phosphate-buffered saline (1× DPBS) without Ca and Mg.
3. Detachment reagent for adherent HEK 293 cells: 1× Trypsin–EDTA. Store at 4 °C.

2.2 Growing Yeast Cells

1. YP growth medium: dissolve 1 % (w/v) yeast extract, 2 % (w/v) peptone from meat, pancreatic digest, in ddH_2O and autoclave. Add carbon source directly before use (*see* **Note 1**).
2. 10 % (w/v) glucose.
3. 10 % (w/v) raffinose (*see* **Note 2**).
4. TM buffer: 10 mM Tris–HCl, pH 7.5, 1 mM $MgCl_2$.

2.3 DMS Chemical Probing

1. 10.7 M dimethyl sulfate (DMS; *see* **Note 3**). Store at 4 °C.
2. 14.3 M β-mercaptoethanol. Store at 4 °C.
3. Isoamyl alcohol.

2.4 Pb^{2+}-Induced Cleavage Assay

1. 0.5 M EDTA, pH 8.0.
2. 50 mM lead(II) acetate.

2.5 RNA Isolation from Mammalian Cells

1. Isol-RNA lysis reagent or comparable product. Store at 4 °C.
2. Chloroform.
3. Isopropanol.
4. 75 % (v/v) ethanol.

2.6 RNA Isolation from Yeast

1. AE buffer: 50 mM NaOAc, pH 4.5, 10 mM EDTA, pH 8.0.
2. 10 % (w/v) SDS.
3. Water-saturated phenol, pH 4.5. Store at 4 °C.
4. PCI (phenol:chloroform:isoamyl alcohol, 25:24:1). Store at 4 °C.
5. CI (chloroform:isoamyl alcohol, 24:1).
6. Ethanol/0.3 M NaOAc, pH 5.0.
7. RNase-free DNase I (2 U/μL) and supplied 10× DNase I reaction buffer. Store at −20 °C.
8. RNase inhibitor (40 U/μL). Store at −20 °C.
9. 0.5 M EDTA, pH 8.0.

2.7 5′ End-Labeling of DNA Primers

1. 10 μM stocks of gene-specific DNA primers. Store at −20 °C.
2. [γ-^{32}P]-ATP (10 μCi/μL, 6,000 Ci/mmol). Store at 4 °C.
3. T4 polynucleotide kinase (PNK, 10 U/μL). Store at −20 °C.
4. 10× PNK buffer (supplied with enzyme). Store at −20 °C.
5. 0.5 M EDTA, pH 8.0.
6. Glycogen (10 mg/mL). Store at −20 °C.
7. Ethanol/0.3 M NaOAc, pH 5.0.

2.8 Primer Extension

2.8.1 Reverse Transcription with Transcriptor Reverse Transcriptase

1. 4.5× hybridization buffer: 225 mM K-HEPES, pH 7.0, 450 mM KCl. Store at 20 °C.
2. Transcriptor reverse transcriptase (20 U/μL; Roche), or comparable product. Store at −20 °C.
3. 5× reaction buffer supplied with Transcriptor reverse transcriptase (Roche). Store at −20 °C.
4. 10 mM dNTP mix. Store at −20 °C.
5. 10 mM ddTTP. Store at −20 °C.
6. 10 mM ddGTP. Store at −20 °C.
7. 0.1 M DTT (dithiothreitol). Store at −20 °C.
8. RNase inhibitor (40 U/μL). Store at −20 °C.

2.8.2 Reverse Transcription with AMV Reverse Transcriptase

1. 4.5× hybridization buffer: 225 mM K-HEPES, pH 7.0, 450 mM KCl. Store at −20 °C.
2. 10× extension buffer: 1.3 M Tris–HCl, pH 8.0, 0.1 M $MgCl_2$, 0.1 M DTT. Store at −20 °C.
3. 2.5 mM dNTP mix. Store at −20 °C.

4. 1 mM ddTTP. Store at −20 °C.
5. 1 mM ddGTP. Store at −20 °C.
6. AMV reverse transcriptase (10 U/μL). Store at −20 °C.

2.8.3 Solutions for Both Reverse Transcription Protocols

1. 1 M NaOH.
2. 1 M HCl.
3. 0.5 M EDTA, pH 8.0.
4. Glycogen (10 mg/mL). Store at −20 °C.
5. Ethanol/0.3 M NaOAc, pH 5.0.
6. 10× TBE buffer: 0.89 M Tris-base, 0.89 M boric acid, 20 mM EDTA.
7. Loading buffer: 7 M urea, 25 % (w/v) sucrose, 0.025 % (w/v) bromophenol blue, 0.025 % (w/v) xylene cyanol in 1× TBE.

2.9 Denaturing Polyacrylamide Gel Electrophoresis (PAGE)

1. Vertical gel electrophoresis apparatus (adjustable).
2. Glass plate sets: 42 cm (L) × 20 cm (W).
3. Metal plate: 26.5 cm (L) × 20 cm (W).
4. Combs and spacers, 0.4 mm thickness.
5. High-voltage power supply.
6. Whatman 3MM paper.
7. Saran wrap.
8. 10× TBE buffer: 0.89 M Tris-base, 0.89 M boric acid, 20 mM EDTA.
9. *N,N,N′,N′*- tetramethylethylenediamine (TEMED). Store at 4 °C.
10. 10 % (w/v) ammonium persulfate (APS) (*see* **Note 4**). Store at 4 °C.
11. 8 % denaturing acrylamide solution: 7 M urea, 8 % (v/v) acrylamide/bisacrylamide (19:1) in 1× TBE. Store at 4 °C.
12. Phosphorimager exposure cassette and screen (GE Healthcare) or comparable product.
13. Phosphorimager (e.g., STORM 820, GE Healthcare) or comparable equipment.

3 Methods

3.1 Modification of RNA In Vivo

3.1.1 DMS Modification of RNA in HEK Cells

1. Grow HEK cells to confluency in a 100 mm Ø dish (approx. 8.8×10^6 cells) in DMEM–FBS–PS medium at 37 °C with 5 % CO_2.
2. Discard the medium and wash the cells with 5 mL 1× DPBS.
3. Add 1 mL 1× Trypsin–EDTA and incubate at 37 °C with 5 % CO_2 for 2–3 min.

4. Add 9 mL DMEM–FBS–PS medium and transfer the cell suspension to 15 mL centrifugation tubes.
5. Centrifuge the cells at 1,200 × *g* for 5 min (4 °C).
6. Remove the supernatant and add 1 mL pre-warmed (37 °C) DMEM–FBS–PS medium.
7. Add DMS to a final concentration of 50 mM (vortex briefly). Incubate the cells at 37 °C for 2 min (*see* **Notes 3** and **5**).
8. Stop the DMS reaction by adding 50 μL 14.3 M β-mercaptoethanol and 50 μL isoamyl alcohol. Vortex strongly and centrifuge at 6,000 × *g* for 2 min (*see* **Notes 6** and 7).
9. Remove the supernatant carefully and resuspend the cells in 1 mL cold 1× DPBS and add another 50 μL β-mercaptoethanol.
10. Vortex briefly and centrifuge at 1,200 × *g* for 5 min.
11. Discard the supernatant and proceed with RNA isolation (Subheading 3.2.1).

3.1.2 DMS Modification of RNA in Yeast

1. Grow an overnight culture of the desired yeast strain at 30 °C in growth medium (e.g., YPD).
2. Inoculate 100 mL growth medium (e.g., YP + 2 % (w/v) raffinose, *see* **Notes 1** and **2**) with 5 mL of the overnight culture. Grow the culture at 30 °C to an OD_{600} of 1.0.
3. Harvest the cells (2 × 30 mL) by centrifugation at 4,500 × *g* for 5 min.
4. Discard the supernatant and resuspend the cell pellet in 1 mL pre-warmed (30 °C) growth medium (e.g., YPD).
5. Centrifuge the samples at 6,000 × *g* for 2 min. Carefully discard the supernatant.
6. Add DMS to a final concentration of 50 mM to one of the samples and vortex (*see* **Notes 3** and **5**).
7. Immediately incubate both samples at 30 °C for 2 min.
8. Stop the DMS reaction by adding β-mercaptoethanol to a final concentration of 0.7 M and 50 μL isoamyl alcohol. Vortex strongly and centrifuge both samples at 6,000 × *g* for 2 min (*see* **Notes 6** and 7).
9. Discard the supernatant and resuspend the pellet in 1 mL growth medium (e.g., YPD).
10. Add another 50 μL β-mercaptoethanol to the DMS-treated sample, vortex, and repeat the centrifugation step (6,000 × *g*, 2 min).
11. Discard the supernatant and freeze the pellet at −80 °C for at least 20 min but not longer than overnight.
12. Proceed with RNA isolation (Subheading 3.2.2).

3.1.3 Pb^{2+}-Induced Cleavage Assay of RNA in Yeast

1. Perform **steps 1–3** as described in Subheading 3.1.2.
2. Discard the supernatant and resuspend the cell pellet in 0.5 mL pre-warmed (30 °C) growth medium (e.g., YPD).
3. Add 500 μL of 50 mM $Pb(OAc)_2$ (prepare freshly) to the sample, yielding a final volume of 1 mL and a final concentration of 25 mM $Pb(OAc)_2$ (*see* **Notes 5** and **8**). To the second sample, add 500 μL ddH_2O as control.
4. Incubate the samples at 30 °C for 10 min with moderate shaking to prevent cells from pelleting at the bottom.
5. Stop the reaction by adding 125 μL of 0.5 M EDTA, pH 8.0 (2.5× molar excess relative to the [Pb^{2+}]) (*see* **Note 6**).
6. Pellet the cells by centrifugation at 4,500 × *g* for 5 min and carefully remove the supernatant.
7. Freeze the cells at −80 °C for at least 20 min to overnight.
8. Proceed with RNA isolation (Subheading 3.2.2).

3.2 Isolating Total RNA from Eukaryotic Cells

3.2.1 RNA Isolation from HEK Cells

1. Add 6 mL Isol-RNA lysis reagent to the cell pellet and resuspend it rapidly.
2. Keep the tube at room temperature for 5 min.
3. Add 1.2 mL chloroform and vortex vigorously for 15 s.
4. Repeat **step 2**.
5. Centrifuge at 12,000 × *g* for 15 min (4 °C).
6. Transfer the aqueous phase to a new tube.
7. Add 3 mL isopropanol and vortex 15 s. Incubate the tube at room temperature for 10 min. Centrifuge at 12,000 × *g* for 10 min (4 °C).
8. Carefully discard the supernatant.
9. Add 1 mL 75 % (v/v) ethanol.
10. Centrifuge at 7,500 × *g* for 5 min (4 °C).
11. Remove the supernatant completely and air-dry the RNA pellet for 5 min.
12. Dissolve the RNA in 20 μL ddH_2O (*see* **Note 9**). Store the RNA at −20 °C.

3.2.2 RNA Isolation from Yeast

1. Resuspend the frozen cells (*see* **Note 10**) in 600 μL AE buffer and 100 μL 10 % (w/v) SDS.
2. Split the cell suspension into two 1.5 mL tubes containing 700 μL water-saturated phenol, pH 4.5, preheated to 65 °C.
3. Vortex strongly and freeze the tubes in liquid nitrogen (ensure that the content is frozen completely). Thaw the samples at 65 °C, followed by vortexing for 30 s.
4. Repeat the freeze–thaw cycle twice.

5. Shake the samples at 65 °C for 4 min.
6. Centrifuge the samples at 18,000 × *g* for 5 min.
7. Transfer the aqueous phase to a new tube and add 700 μL PCI.
8. Vortex strongly for 30 s and centrifuge the samples at 18,000 × *g* for 5 min.
9. Transfer the aqueous phase to a new tube containing 700 μL CI.
10. Mix the sample thoroughly by vortexing for 30 s and centrifuge again at 18,000 × *g* for 5 min.
11. Transfer the aqueous phase to a new 2.0 mL tube. Precipitate the sample with 2.5× volumes of ethanol/0.3 M NaOAc, pH 5.0, and keep it at −20 °C for at least 60 min.
12. Centrifuge the samples at 18,000 × *g* for 30 min (4 °C) and discard the supernatant. Dry the pellet for 5 min at room temperature and then resuspend it in 100 μL ddH_2O (*see* **Note 9**).
13. Remove residual DNA by adding the following: 17.5 μL RNase-free DNase I (2 U/μL), 0.5 μL RNase inhibitor (40 U/μL), and 12 μL 10× DNase I reaction buffer. Incubate the samples on ice for 60 min.
14. Adjust the volume to 200 μL by adding 70 μL ddH_2O and perform a phenol extraction. Add 200 μL PCI, vortex vigorously and centrifuge at 18,000 × *g* for 5 min. Transfer the aqueous phase to a new tube and add 200 μL CI. After vortexing repeat the centrifugation step and transfer the aqueous phase to a new tube.
15. Precipitate the samples by adding 2 μL 0.5 M EDTA, pH 8.0, and 2.5× volumes of ethanol/0.3 M NaOAc, pH 5.0. Incubate the sample at −20 °C for at least 60 min.
16. Centrifuge samples at 4 °C, 18,000 × *g* for 30 min and discard the supernatant. Dry the pellet for 5 min at room temperature.
17. Resuspend the pellet in 10 μL ddH_2O (*see* **Note 9**).

3.3 5′ End-Labeling of Gene-Specific DNA Primers

1. Mix the reagents in the following order: 10 μL ddH_2O, 2 μL 10× T4 PNK buffer, 1 μL 10 μM DNA primer, 6 μL [γ-^{32}P]-ATP (10 μCi/μL, 6,000 Ci/mmol), and 1 μL T4 PNK (10 U/μL) (*see* **Note 11**).
2. Incubate the sample at 37 °C for 30–40 min.
3. Add 1 μL 0.5 M EDTA, pH 8.0, to the sample.
4. Incubate the sample at 95 °C for 1 min.
5. Immediately, place the sample on ice for 2 min.
6. Add 2 μL glycogen (10 mg/mL) and 2.5× volumes of ethanol/0.3 M NaOAc, pH 5.0. Keep the sample at −20 °C for at least 60 min.

7. Centrifuge the sample at 4 °C and 18,000×g for 30 min. Carefully remove the supernatant. Dry pellet for 5 min at room temperature and dissolve it in 40 μL ddH_2O.

3.4 Mapping the Sites of Modification or Strand Scission by Reverse Transcription

3.4.1 Reverse Transcription Using Transcriptor Reverse Transcriptase

1. Set up the annealing reaction by mixing 2.5 μL RNA (20–40 μg), 1 μL 0.25 μM ^{32}P-labeled primer (*see* **Note 12**), and 1 μL 4.5× hybridization buffer.
2. Incubate the sample at 95 °C for 1 min.
3. Immediately place the sample on ice for 2 min.
4. Add 15.5 μL extension mix consisting of 4 μL 5× reaction buffer, 2 μL 10 mM dNTPs, 1 μL 0.1 M DTT, 0.5 μL RNase inhibitor (40 U/μL), 0.5 μL Transcriptor reverse transcriptase (20 U/μL), and 5.5 μL ddH_2O. Incubate the samples at 50 °C for 60 min. To generate A and C sequencing lanes add 2 μL of a 10 mM ddTTP or ddGTP solution to the sample in addition to the extension mix, respectively (*see* **Notes 13–15**).
5. Degrade the RNA by adding 3 μL 1 M NaOH and incubate at 50 °C for 45 min.
6. Neutralize the pH with 3 μL 1 M HCl.
7. Precipitate the cDNAs by adding 2 μL glycogen (10 mg/mL), 2 μL 0.5 M EDTA, pH 8.0 and 2.5× volumes of ethanol/0.3 M NaOAc, pH 5.0. Incubate the samples at −20 °C for 1 h to overnight.
8. Centrifuge the sample at 4 °C and 18,000×g for 30 min. Discard the supernatant and resuspend the pellet in 8 μL loading buffer.

3.4.2 Reverse Transcription Using AMV Reverse Transcriptase

1. Set up the annealing reaction as described in **steps 1–3** in Subheading 3.4.1.
2. Add 2.2 μL extension mix consisting of 1.08 μL ddH_2O, 0.67 μL 10× extension buffer, 0.3 μL 2.5 mM dNTP mix, 0.15 μL AMV reverse transcriptase (10 U/μL) and incubate the sample at 42 °C for 1 h. To generate A and C sequencing lanes add 0.75 μL of 1 mM ddTTP or 1 mM ddGTP solution to the extension mix, respectively (*see* **Notes 13–15**).
3. Degrade the RNA by adding 1.5 μL 1 M NaOH and incubate the sample at 42 °C for 60 min.
4. Neutralize the pH with 1.5 μL 1 M HCl.
5. Precipitate the cDNAs by adding 2 μL glycogen (10 mg/mL), 1 μL 0.5 M EDTA, pH 8.0 and 2.5× volumes of ethanol/0.3 M NaOAc, pH 5.0. Incubate the sample at −20 °C for 1 h to overnight.
6. Centrifuge the sample at 4 °C and 18,000×g for 30 min. Discard the supernatant, dry the pellet at room temperature for 5 min and resuspend the pellet in 8 μL loading buffer.

3.5 Resolving the cDNA Pool on a Denaturing PAGE

1. Assemble the glass plates.
2. Mix 50 mL 8 % denaturing acrylamide solution with 500 μL 10 % (w/v) APS (*see* **Note 4**) and 50 μL TEMED.
3. Pour the gel solution immediately between the two glass plates. Avoid introducing air bubbles. Insert a suitable comb and let the gel polymerize for at least 60 min.
4. Place the gel in the electrophoresis apparatus and pre-run the gel for 30–60 min at 40 W (*see* **Note 16**).
5. Denature the samples at 95 °C for 1 min.
6. Rinse the wells thoroughly with 1× TBE buffer.
7. Load 4 μL of the sample and run the gel at 40 W for approximately 2 h or until bromophenol blue dye has reached the bottom of the gel.
8. Disassemble the gel by removing the top glass plate and transfer it to Whatman 3MM paper. Cover the gel with saran wrap.
9. Dry the gel in a vacuum dryer at 80 °C for 90 min.
10. Expose the gel to a phosphorimager screen overnight and scan the screen using a STORM 820 or comparable equipment.
11. Analyze the gel using ImageQuant or comparable software (*see* **Notes 17–19**).

4 Notes

1. Growth medium containing a carbon source gets easily contaminated. Therefore sugar is added directly prior use.
2. Using raffinose instead of glucose as a carbon source leads to proliferation of yeast mitochondria and in turn to an increased yield of mitochondrial RNA.
3. Care should be taken when working with DMS. DMS is a corrosive, toxic, and potentially carcinogenic substance that is readily absorbed through skin. Therefore DMS should be handled in the hood wearing gloves that do not permit penetration of organic chemicals.
4. The 10 % (w/v) APS solution can be kept at 4 °C for up to 2 weeks. Alternatively, store aliquots of the solution at −20 °C for up to 6 months.
5. The DMS concentration used in the modification reaction has to be optimized for each target RNA prior to experiments. To ensure single-hit conditions (one or less DMS modification event per RNA molecule). The amount of full-length cDNA of samples treated with 25–200 mM DMS should be compared to the RT stop control in order to decide which DMS concentration correlates best with single-hit conditions for the

target RNA. Single hit conditions have to be achieved for the Pb^{2+}-induced cleavage assay as well. Again, this is done by performing a concentration and/or time series; useful parameters for such series range from 10 mM to 150 mM $Pb(OAc)_2$ as final concentration and incubation times from 5 to 120 min. Like for DMS probing, the amount of full-length cDNA should be comparable between the RT stop control and the modified sample.

6. The DMS reaction has to be stopped efficiently before proceeding with total RNA extraction, to avoid extended modification during the RNA preparation. If this is not the case, the majority of the RNA bases will be modified by DMS during the denaturing conditions of the RNA preparation. To confirm efficient quenching, a stop control has to be carried out in the following manner: the indicated amount of β-mercaptoethanol has to be added prior to DMS and incubated for 2 min at 30 or 37 °C (in case of yeast or HEK cells, respectively). If no DMS pattern is observed for the stop control, this implies that the β-mercaptoethanol has successfully quenched the DMS reactivity and as a consequence it cannot methylate the RNA during subsequent RNA preparation. A stop control has to be prepared for Pb^{2+}-induced cleavage assay as well by adding the 2.5× molar excess of 0.5 M EDTA, pH 8.0, prior to $Pb(OAc)_2$ to the sample and subsequent incubation at 30 °C for 2 min. If no cleavage pattern is observed, this indicates that the amount of EDTA is sufficient to chelate all Pb^{2+}-ions present in the sample.

7. The cells treated with DMS are sometimes difficult to pellet. To avoid losing parts of the pellet, it is suggested to increase the centrifugation force or to increase the time of centrifugation.

8. $Pb(OAc)_2$ does precipitate in normal growth medium (e.g., YPD). The amount of precipitate correlates with the concentration of the stock solution added to the medium. We therefore propose to dilute the sample with the $Pb(OAc)_2$ stock only 1:2 (e.g., mix 500 μL cells suspension with 500 μL 50 mM $Pb(OAc)_2$ to achieve a final concentration of 25 mM) to reduce precipitation as much as possible. Furthermore, $Pb(OAc)_2$ has a strong preference to form precipitants under the following conditions: (1) the presence of chloride ions in the buffer and (2) as a function of the pH value of the buffers used. The lower the pH, the more precipitate is formed.

9. RNA is very sensitive to degradation. Therefore it is advisable to follow some general considerations when working with RNA: avoid as much as possible RNase contamination by always wearing gloves and by using consumables of "RNase-free" grade. Store the RNA stock at −20 °C and avoid repeated freeze–thaw cycles (aliquot, if necessary). In addition, when handling RNA,

keep it on ice if not stated otherwise. Prepare solutions with RNase-free water or DEPC-treated water. Also, the chemicals have to be of high quality and should be used for RNA only. If affordable, a separate set of pipettes should be used for working with RNA. In case of a severe RNAse contamination, it is helpful to clean the equipment and bench thoroughly with "RNaseZap" (Ambion) or a comparable product.

10. When working with yeast mitochondrial RNA, the best results in terms of quality and yield of mtRNA were obtained using the RiboPure Yeast Kit from Ambion. Using the method described in Subheading 3.2.2 we obtained RNA of good quality and high yield compared to all other protocols tested to extract total RNA from yeast.

11. We recommend gel-purification of the DNA primers before end-labeling. The purification is necessary to remove any organic remnants left from oligo synthesis. These could impair the T4 polynucleotide kinase during labeling and may result in a low signal-to-noise ratio. We purify DNA primers (18–30 nucleotides) on a 20 % denaturing polyacrylamide gel run with 25 W. Purification is time-consuming and should therefore be done ahead of time, at least 1 day prior to the labeling reaction.

12. Working with total RNA can be problematic due to mispriming events during reverse transcription. It is therefore advisable to prepare sequencing lanes from in vitro transcribed RNA and compare them with those obtained from in vivo isolated RNA, thereby confirming that the primer used does only bind to the RNA of interest. Otherwise the obtained DMS pattern might be misleading and/or inconclusive. Simple comparison of the band pattern in the sequencing lanes and the gene sequence is often insufficient.

13. The cDNA synthesis can be accomplished by different reverse transcriptases and is influenced by several parameters, like RNA sequence, highly structured RNA elements, low amount of target RNA, reaction temperature and ionic concentrations, among others. AMV reverse transriptase is the most commonly used enzyme. For highly structured or GC-rich RNA, a very sensitive and thermostable enzyme like Transcriptor reverse transcriptase should be used instead of AMV reverse transcriptase. Protocols for both enzymes are provided.

14. Include a stop control during reverse transcription. This is absolutely necessary for distinguishing between so-called natural stops of the reverse transcriptase enzyme and those induced by DMS modification or Pb^{2+}-cleavage, respectively. Secondary structures that form after denaturing of the RNA as well as G/C sequence stretches can block the reverse transcriptase enzyme, thereby terminating the extension. Also, RNAs tend to have "hot spots" for breakage due to a strain on

the backbone. Along this line, partial degradation by RNases results in RT stops as well. By reverse transcribing RNA from cells that were not treated with DMS or Pb^{2+}, these natural stops of different origin can be easily detected.

15. To generate sequencing lanes of good quality, the ddNTP concentration has to be adjusted for the respective RNA. If the concentration is inadequate, sequencing lanes tend to either give a very faint signal in the upper part of the gel or some nucleotides remain undetected (even though the flanking residues are well detectable).
16. Pre-run the gel for at least 30 min before loading the samples to reach optimal temperature (45–55 °C). A temperature indicator attached to the glass plates is recommended to ensure that the optimal temperature is applied. Higher temperature might damage the glass plates and cause bands to smear and therefore reduces the resolution of the gel. On the other hand uneven heat distribution during gel electrophoresis causes the "smile effect." Attaching a metal plate to the glass ensures a better heat distribution and thus improves the gel quality.
17. If the cDNA sample has a high salt content, this results in narrowing of the lanes during the gel electrophoresis. To avoid this problem, the cDNA pellet should be washed with 70 % (v/v) ethanol before resuspending it in loading buffer. The resolution of the gel is also drastically reduced, if the cDNA sample is precipitated for longer than overnight. Alternatively, it is possible to overcome this problem by increasing the percentage of the gel and the running time, as the salt front will run out of the gel.
18. A low signal-to-noise ratio usually points to the need for optimization of different steps. First, check the ^{32}P-labeling efficiency by comparing the amount of radioactivity of the supernatant versus the pelleted DNA oligo in the labeling step. Notably, the T4 PNK enzyme and the PNK buffer are both very sensitive to changes in temperature. Aliquoting the buffer helps avoiding frequent freeze–thaw cycles. Furthermore, it is advisable to use the radioactive stock no later than at its calibration date (the ^{32}P-isotope has a short half-life of 14.3 days). If no improvement is observed, several other parameters can be adjusted: the choice of reverse transcriptase used (AMV vs. Transcriptor reverse transcriptase); the amount of RNA and primer used for reverse transcription reaction; also the primer binding site and its melting temperature should be reevaluated along with the annealing and extension temperature, respectively; dNTP concentration and/or the preparation date of the stock.
19. Quantification is highly dependent on loading equal amounts of sample on each lane. In case of uneven loading, the second half of the sample can be used for a rerun with adjusted sample amount.

Acknowledgement

This work was funded by the Austrian Science Foundation (FWF; grants Y401 and P21017 to C.W.).

References

1. Brion P, Westhof E (1997) Hierarchy and dynamics of RNA folding. Annu Rev Biophys Biomol Struct 26:113–137
2. Herschlag D (1995) RNA chaperones and the RNA folding problem. J Biol Chem 270:20871–20874
3. Pyle AM, Fedorova O, Waldsich C (2007) Folding of group II introns: a model system for large, multidomain RNAs? Trends Biochem Sci 32:138–145
4. Schroeder R, Barta A, Semrad K (2004) Strategies for RNA folding and assembly. Nat Rev Mol Cell Biol 5:908–919
5. Sosnick TR, Pan T (2003) RNA folding: models and perspectives. Curr Opin Struct Biol 13:309–316
6. Treiber DK, Williamson JR (1999) Exposing the kinetic traps in RNA folding. Curr Opin Struct Biol 9:339–345
7. Treiber DK, Williamson JR (2001) Beyond kinetic traps in RNA folding. Curr Opin Struct Biol 11:309–314
8. Woodson SA (2000) Compact but disordered states of RNA. Nat Struct Biol 7:349–352
9. Woodson SA (2005) Structure and assembly of group I introns. Curr Opin Struct Biol 15:324–330
10. Woodson SA (2010) Compact intermediates in RNA folding. Annu Rev Biophys 39:61–77
11. Schroeder R, Grossberger R, Pichler A et al (2002) RNA Folding *in vivo*. Curr Opin Struct Biol 12:296–300
12. Zemora G, Waldsich C (2010) RNA folding in living cells. RNA Biol 7:634–641
13. Ares M Jr, Igel AH (1990) Lethal and temperature-sensitive mutations and their suppressors identify an essential structural element in U2 small nuclear RNA. Genes Dev 4:2132–2145
14. Balzer M, Wagner R (1998) Mutations in the leader region of ribosomal RNA operons cause structurally defective 30 S ribosomes as revealed by *in vivo* structural probing. J Mol Biol 276:547–557
15. Climie SC, Friesen JD (1988) *In vivo* and *in vitro* structural analysis of the rplJ mRNA leader of *Escherichia coli*. Protection by bound L10-L7/L12. J Biol Chem 263:15166–15175
16. Doktycz MJ, Larimer FW, Pastrnak M et al (1998) Comparative analyses of the secondary structures of synthetic and intracellular yeast MFA2 mRNAs. Proc Natl Acad Sci USA 95:14614–14621
17. Harris ME, Pace NR (1995) Analysis of the tertiary structure of bacterial RNase P RNA. Mol Biol Rep 22:115–123
18. Liebeg A, Mayer O, Waldsich C (2010) DEAD-box protein facilitated RNA folding *in vivo*. RNA Biol 7:803–811
19. Mereau A, Fournier R, Gregoire A et al (1997) An *in vivo* and *in vitro* structure-function analysis of the *Saccharomyces cerevisiae* U3A snoRNP: protein-RNA contacts and base-pair interaction with the pre-ribosomal RNA. J Mol Biol 273:552–571
20. Moazed D, Robertson JM, Noller HF (1988) Interaction of elongation factors EF-G and EF-Tu with a conserved loop in 23S RNA. Nature 334:362–364
21. Senecoff JF, Meagher RB (1992) *In vivo* analysis of plant RNA structure: soybean 18S ribosomal and ribulose-1,5-bisphosphate carboxylase small subunit RNAs. Plant Mol Biol 18:219–234
22. Waldsich C, Grossberger R, Schroeder R (2002) RNA chaperone StpA loosens interactions of the tertiary structure in the *td* group I intron *in vivo*. Genes Dev 16:2300–2312
23. Waldsich C, Masquida B, Westhof E et al (2002) Monitoring intermediate folding states of the *td* group I intron *in vivo*. EMBO J 21:5281–5291
24. Wells SE, Hughes JM, Igel AH et al (2000) Use of dimethyl sulfate to probe RNA structure *in vivo*. Methods Enzymol 318:479–493
25. Zaug AJ, Cech TR (1995) Analysis of the structure of *Tetrahymena* nuclear RNAs *in vivo*: telomerase RNA, the self-splicing rRNA intron, and U2 snRNA. RNA 1:363–374
26. Brunel C, Romby P (2000) Probing RNA structure and RNA-ligand complexes with chemical probes. Methods Enzymol 318:3–21
27. Ehresmann C, Baudin F, Mougel M et al (1987) Probing the structure of RNAs in solution. Nucleic Acids Res 15:9109–9128
28. Erat MC, Sigel RK (2011) Methods to detect and characterize metal ion binding sites in RNA. Met Ions Life Sci 9:37–100
29. Lindell M, Brannvall M, Wagner EG et al (2005) Lead(II) cleavage analysis of RNase P RNA *in vivo*. RNA 11:1348–1354

30. Lindell M, Romby P, Wagner GH (2002) Lead (II) as a probe for investigating RNA structure *in vivo*. RNA 8:534–541
31. Sigel RK, Pyle AM (2003) Lanthanide ions as probes for metal ions in the structure and catalytic mechanism of ribozymes. Met Ions Biol Syst 40:477–512
32. Huang HR, Rowe CE, Mohr S et al (2005) The splicing of yeast mitochondrial group I and group II introns requires a DEAD-box protein with RNA chaperone function. Proc Natl Acad Sci USA 102:163–168
33. Gregan J, Kolisek M, Schweyen RJ (2001) Mitochondrial Mg(2+) homeostasis is critical for group II intron splicing *in vivo*. Genes Dev 15:2229–2237
34. Halls C, Mohr S, Del Campo M et al (2007) Involvement of DEAD-box proteins in group I and group II intron splicing. Biochemical characterization of Mss116p, ATP hydrolysis-dependent and -independent mechanisms, and general RNA chaperone activity. J Mol Biol 365:835–855
35. Solem A, Zingler N, Pyle AM (2006) A DEAD protein that activatesintron self-splicing without unwinding RNA. Mol Cell 24:611–617
36. Sigel R, Vaidya A, Pyle A (2000) Metal ion binding sites in a group II intron core. Nat Struct Biol 7:1111–1116

Chapter 10

A Chemogenetic Approach to Study the Structural Basis of Protein-Facilitated RNA Folding

Olga Fedorova

Abstract

Large RNA molecules play important roles in all aspects of cellular metabolism ranging from mRNA splicing and protein biosynthesis to regulation of gene expression. In order to correctly perform its function in the cell, an RNA molecule must fold into a complex tertiary structure. Folding of many large RNAs is slow either due to formation of stable misfolded intermediates or due to high contact order or instability of obligate folding intermediates. Therefore many RNAs use protein cofactors to facilitate their folding in vivo. Folding of the yeast mitochondrial group II intron ai5γ to the native state under physiological conditions is facilitated by the protein cofactor Mss116. This chapter describes the use of Nucleotide Analog Interference Mapping (NAIM) to identify specific substructures within the intron molecule that are directly affected by the protein.

Key words NAIM, DEAD-box protein, Ribozyme, Group II intron, RNA folding

1 Introduction

In order to be able to correctly perform their biological function, RNA molecules must fold into complex tertiary structures within the cell. Many large RNAs fold slowly due to either high contact order [1–3], instability of on-pathway intermediates [4–7], or formation of stable misfolded intermediates ("kinetic traps") [8–10]. In order to be able to correctly fold in vivo, these RNA molecules require help of protein cofactors to either destabilize misfolded structures or to stabilize obligate transient intermediates. The "destabilizing" protein cofactors include RNA helicases [11–13], which use the energy of ATP hydrolysis to unwind misfolded structure of the RNA and give it a second chance to fold correctly [14, 15]; and RNA chaperones, which transiently bind and destabilize misfolded structures in an ATP-independent fashion [13, 16, 17]. Other cofactor proteins perform the stabilizing function: RNA annealers facilitate the formation of otherwise unstable secondary

Christina Waldsich (ed.), *RNA Folding: Methods and Protocols*, Methods in Molecular Biology, vol. 1086,
DOI 10.1007/978-1-62703-667-2_10, © Springer Science+Business Media, LLC 2014

structure of the RNA [13, 18, 19], other proteins selectively bind and stabilize the native state [20] or on-pathway intermediates [21–24].

The well-studied ai5γ group II intron from yeast folds slowly to the native state via one or more transient on-pathway intermediates [4, 7, 25]. In vitro folding of this RNA to its native state requires non-physiological conditions such as 42 °C and high concentrations of monovalent ions and magnesium (500 mM and 100 mM, respectively). We have recently shown that folding of the ai5γ intron under near-physiological conditions in the absence of any protein cofactors occurs under kinetic control, and the large energy barrier between unfolded and compact intermediate states makes it prohibitively slow (24–26 h) [26]. Folding of the ai5γ intron under near-physiological conditions in vitro and in vivo is greatly facilitated by the DEAD-box protein Mss116 [22, 23, 27–30]. The protein promotes the formation of an on-pathway compact intermediate in the intronic Domain 1 [22, 28]. This process has been described using two different models. According to the first model, Mss116 uses its helicase activity to disrupt misfolded secondary structure within the intron RNA [27, 30–32]. The second model suggests that the protein binds and stabilizes weak on-pathway intermediates, thus lowering the energy barrier between the unfolded and compact states [22–24, 28, 33]. Multiple studies have amassed experimental evidence in support of both models. Thus, Mss116-related proteins Cyt-19 and Ded-1 both unwind misfolded structural elements within a different large RNA, the *Tetrahymena* group I intron in the presence of ATP, which supports the first model [32, 34]. At the same time Mss116-facilitated compaction of the ai5γ intron is ATP-independent, which is inconsistent with the unwinding function of the protein [22, 28]. Other studies indicate that the folding pathway of the ai5γ group II intron is smooth and does not involve escape from kinetic traps, which also supports the second model [2, 4, 7, 25, 35]. Nevertheless, based on the current experimental evidence, neither model can be completely ruled out.

Most of the studies describe how Mss116 affects either self-splicing of the ai5γ intron or the global compaction of the intron RNA. However, the information on the RNA substructures that are specifically affected by the protein remains scarce. We have recently employed a direct chemogenetic approach (Nucleotide Analog Interference Mapping, or NAIM) to identify the regions within the intron RNA, which are directly affected by the protein

[36]. In this experiment, Mss116-facilitated compaction of the intron RNA was used as a selection step. First, RNA was transcribed in the presence of a nucleotide analog phosphorothioate, which produced an RNA library containing 1–2 modified nucleotides per molecule. The library was then radioactively labeled either at the 3′ or at the 5′ end and subjected to Mss116-promoted formation of the compact near-native state under near-physiological conditions (Fig. 1a). The unfolded and compact species were separated using native gel electrophoresis and subjected to cleavage of the phosphorothioate linkages by iodine and analysis of the samples on a sequencing gel (Fig. 1a). Analysis of the iodine cleavage patterns then revealed nucleotide positions where modifications interfered with or promoted compaction, causing the respective bands to be underrepresented or overrepresented in the population of the compact species (Fig. 1a).

In order to determine which RNA helices are stabilized or destabilized by the protein in the course of the intron compaction, we have chosen nucleotide analogs that stabilize or destabilize base pairing (Fig. 1b, c). Analogs like 2,6-diaminopurine (2,6-DAP) stabilize A–U base pair by adding an extra hydrogen bond, and analogs like inosine destabilize G–C pairs by reducing the number of hydrogen bonds from 3 to 2 (Fig. 1b, c). N6-methyl adenosine (N6MeA) destabilizes base pairing as well by hindering hydrogen-bonding with the complementary uridine residue (Fig. 1b, c). If duplex-destabilizing analogs (N6MeA and inosine) are located within RNA helices that must be disrupted by Mss116, they will be beneficial for the helicase activity of the protein. Thus, these analogs will assist protein-facilitated compaction of the intron RNA and exhibit enhancement effects (Fig. 1c). Conversely, 2,6-DAP residues located in the same duplexes will exhibit interference effects (Fig. 1c). The opposite is expected if the protein needs to stabilize certain RNA helices: 2,6-DAP will exhibit enhancement and inosine and N6MeA will exhibit interference effects (Fig. 1c).

Surprisingly, we found that Mss116 has a drastic effect on a small substructure within intronic Domain 1, which is called "the folding control element" [35]. Within this substructure, destabilizing analogs exhibited pronounced interference effects, while stabilizing analogs showed substantial enhancement effects [36]. These results suggest that the protein selectively stabilizes this substructure. The data are consistent with the previous NAIM results suggesting that the folding control element is the most unstable region within the intron [35], and its formation is the rate-limiting step in the intron folding [7].

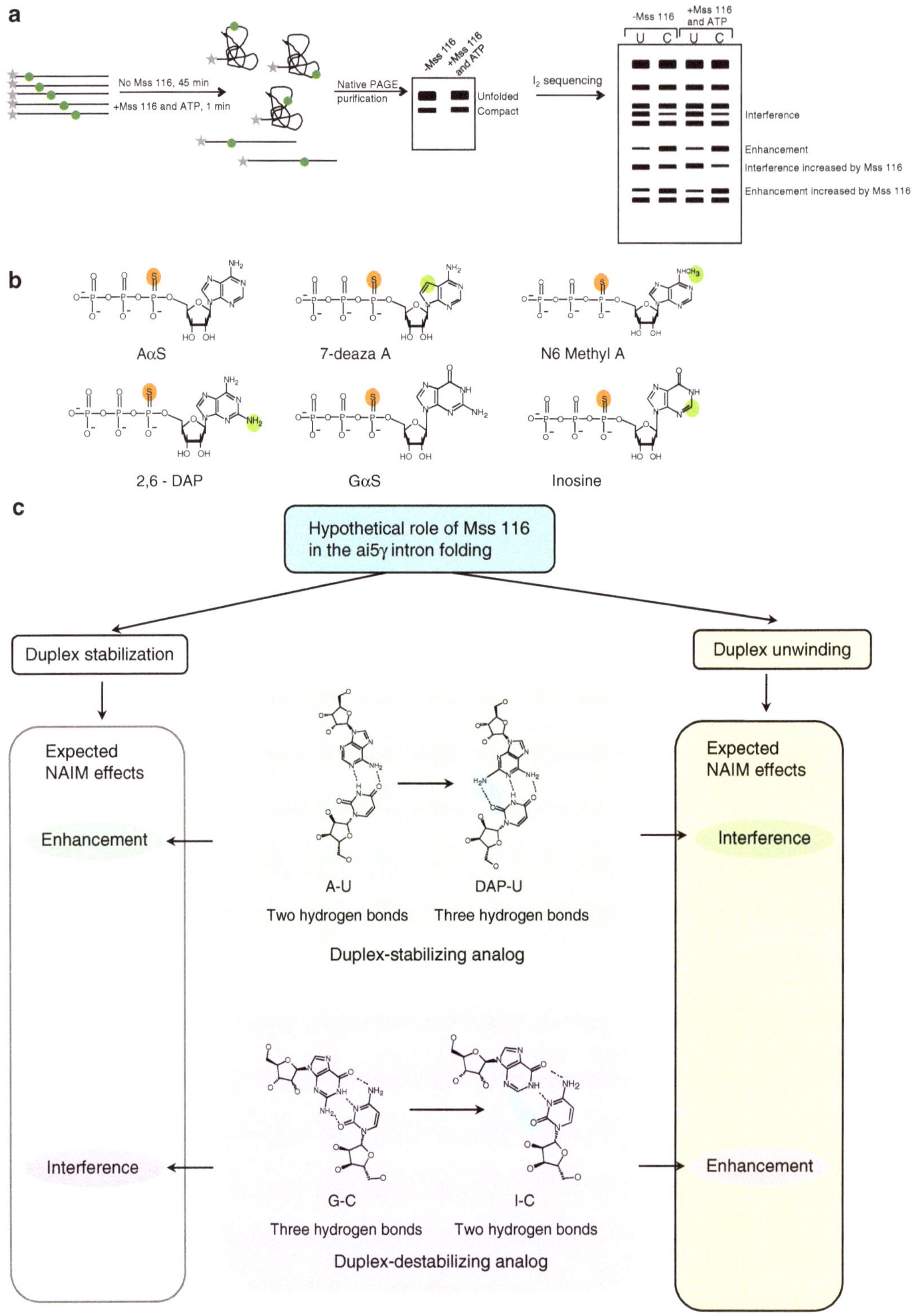

Fig. 1 Nucleotide Analog Interference Mapping (NAIM) to study protein-facilitated RNA folding. (**a**) A general schematic of the NAIM experiment, in which the selection step is RNA folding in the presence or in the absence of Mss116; (**b**) nucleotide analog phosphorothioates used in the NAIM experiment to identify functional groups important for protein-assisted group II intron folding; (**c**) a schematic of expected NAIM effects consistent with the destabilizing or stabilizing function of the protein. Reprinted with permission from: Fedorova, O., and Pyle, A. M. (2012) The brace for a growing scaffold: Mss116 protein promotes RNA folding by stabilizing an early assembly intermediate, *J. Mol. Biol.* **422**, 347–365 (Elsevier)

2 Materials

1. RNase-free water must be used for preparation of all solutions. It is preferable to use water deionized by Milli Q water purification system (Millipore). All solutions are then further sterilized using Nalgene filter units. Distilled water autoclaved in the presence of diethylpyrocarbonate may also be used for preparation of solutions, although this procedure may introduce contaminants. It is important that RNase-free glassware and disposable plastics are used for all experiments.
2. DNA templates for transcription (plasmids linearized by an appropriate restriction enzyme) at 1 μg/μL in TE buffer (*see* below). Stored at −20 °C.
3. T7 RNA polymerase. We express and purify T7 RNA polymerase in-house, which generally yields a ~2 μg/μL stock. Alternatively, T7 RNA polymerase may be purchased from for example Ambion (Life Technologies).
4. RNase inhibitor (40 U/μL). Stored at −20 °C.
5. 30 mM solutions of ATP, GTP, CTP, and UTP in 10 mM Tris–HCl, final pH 7.0, stored at −20 °C. These solutions can be made from powder or from premade NTP solutions of higher concentrations. It is important to adjust pH of NTP solutions to 7.0, in order to make them withstand several freeze-thaw cycles.
6. 10× solutions of nucleotide analog phosphorothioates (AαS, GαS, N6Me AαS, 2,6-DAPαS, IαS, 7-deaza AαS (7dAαS); Glen Research) stored at −20 °C.
7. 1 M Tris–HCl stock solution, pH 7.5 (at 25 °C), used for preparation of transcription buffer and TE buffer.
8. 1 M MOPS–NaOH, pH 6.0 (*see* **Note 1**), for making RNA elution and storage buffers.
9. 1 M DTT, aliquoted into 1.7 mL Eppendorf tubes and stored at −20 °C.
10. 0.5 M EDTA–NaOH, pH 8.0, stored at 4 °C.
11. TE buffer for DNA template storage: 10 mM Tris–HCl, pH 7.5, 1 mM EDTA, stored at 4 °C.
12. 10× transcription buffer: 400 mM Tris–HCl, pH 7.5, 150 mM $MgCl_2$, 20 mM spermidine, 50 mM DTT. Store at −20 °C in 1 mL aliquots.
13. RNA storage buffer ($M_{10}E_1$): 10 mM MOPS, pH 6.0, 1 mM EDTA. Stored at 4 °C.
14. RNA elution buffer: 20 mM MOPS, pH 6.0, 1 mM EDTA, 400 mM NaCl. Stored at 4 °C.
15. 5 M NaCl for RNA precipitation. Stored at 4 °C.

16. Calf intestine alkaline phosphatase (CIP, 1 U/μL). Stored at 4 °C.
17. 10× CIP buffer (supplied with CIP enzyme). Stored at 4 °C.
18. T4 Polynucleotide kinase (PNK, 10,000 U/mL). Stored at −20 °C.
19. Klenow Fragment (3′–5′ exo$^-$, 5,000 U/mL). Stored at −20 °C.
20. Proteinase K (20 mg/mL). Stored at 4 °C.
21. G-25 spin column or an equivalent product.
22. 40 % acrylamide (29:1 acrylamide/bisacrylamide ratio). Stored at 4 °C.
23. TEMED (N,N,N′,N′-tetramethyl ethylenediamine). Stored at 4 °C.
24. 10 % APS (ammonium persulfate) solution. Stored at 4 °C no longer than 3 weeks.
25. 10× TBE buffer for electrophoresis: 0.89 M Tris base, 0.89 M boric acid, 20 mM EDTA. Autoclaved and stored at room temperature.
26. 2× denaturing gel-loading buffer: 83 mM Tris–HCl, pH 7.5, 8 M urea, 1.7 mM EDTA, 17 % sucrose, 0.05 % each of bromophenol blue and xylene cyanol. Heat at 65 °C to dissolve urea, aliquot hot buffer into Eppendorf tubes (1 mL per tube, store at room temperature). Heat up at 65 °C before adding to a sample (urea will precipitate over time, and heating is needed to redissolve it).
27. Denaturing gel solutions: 5, 8 or 15 % acrylamide (29:1 acrylamide/bisacrylamide ratio), 7 M urea, 1× TBE. Filter through Nalgene filter unit (0.2 μm) and store at 4 °C no longer than 2 months (*see* **Note 2**).
28. 20× native gel running buffer stock (THE): 680 mM Tris, 1320 mM HEPES, 2 mM EDTA. Store at 4 °C. Use to prepare 1× native gel running buffer and native gel solutions.
29. 1× native gel running buffer: 34 mM Tris, 66 mM HEPES, 0.1 mM EDTA, 3 mM $MgCl_2$. Store at 4 °C.
30. 6 % native gel solution: 6 % acrylamide (29:1 acrylamide/bisacrylamide ratio), 1× THE, 3 mM $MgCl_2$. Store at 4 °C.
31. 4× native gel-loading buffer: 80 % (v/v) glycerol, 0.05 % each bromophenol blue and xylene cyanol. Store at 4 °C.
32. [α-^{32}P]-dCTP (3,000 Ci/mmol, 20 μCi/μL). Store at −20 °C.
33. [γ-^{32}P]-ATP (6,000 Ci/mmol, 150 μCi/μL). Store at −20 °C.
34. 1 M MOPS–KOH, pH 7.0 (*see* **Note 1**). This is a stock solution, which is used for the selection step (folding in the presence or in the absence of Mss116) at the final concentration of 80 mM.

35. 1 M $MgCl_2$ stock solution (*see* **Note 3**). After the stock solution of 1 M $MgCl_2$ is prepared, it is recommended to aliquot it into Eppendorf tubes and store at −20 °C.
36. 2 M KCl stock solution. Aliquot into Eppendorf tubes and store at −20 °C.
37. 10 mM iodine solution in ethanol. Store at room temperature no longer than 3 weeks. Protect from light.
38. 10× annealing buffer for 3′ end-labeling: 140 mM Tris–HCl, pH 7.5, 400 mM NaCl, 2 mM EDTA. Store at −20 °C.
39. 20× $MgCl_2$/DTT mix for 3′ end-labeling: 140 mM $MgCl_2$, 20 mM DTT. Store at −20 °C.
40. Phosphorimager and exposure cassette with screen (GE Healthcare) or equivalent equipment.

3 Methods

3.1 Preparation of an RNA Transcript Containing Nucleotide Analog Phosphorothioates

1. Thaw on ice, vortex and briefly spin down 10× transcription buffer, 1 M DTT solution, 30 mM ATP, GTP, UTP and CTP stocks, respective 10× nucleotide analog phosphorothioate stock and a DNA template solution.
2. In a 1 mL of a total volume, combine 20 μg of the DNA template, 100 μL of the 10× transcription buffer, 10 μL of 1 M DTT, 32 μL of each 30 mM NTP stock, 100 μL of the 10× nucleotide analog phosphorothioate stock, 5 μL of RNase inhibitor (40 U/μL), RNase-free water and 30–50 μL of T7 RNA polymerase.
3. Incubate the resulting mixture at 37 °C for 4–5 h.
4. Prior to gel-purification, centrifuge the transcription in order to precipitate inorganic pyrophosphate (*see* **Note 4**). Transfer the supernatant into a 14-mL culture tube, add 40 μL of 0.5 M EDTA, pH 8.0, 50 μL of 5 M NaCl, and 10 mL of ethanol and incubate at −80 °C for 1 h or overnight for precipitation.
5. Centrifuge ethanol-precipitated RNA, redissolve the pellet in water, mix with an equal volume of the urea-loading buffer and purify on a denaturing polyacrylamide gel. Visualize the band corresponding to the RNA transcript by UV shadowing and excise the band from the gel.
6. Crush the band into a 14 mL tube, add three volumes of the elution buffer, freeze the mixture on dry ice and incubate on a shaking platform at 4 °C for 4–5 h.
7. Isolate the supernatant from acrylamide and precipitate the RNA by adding three volumes of ethanol to the supernatant. Incubate at −80 °C for 1 h or overnight.

8. Centrifuge the RNA and dispose of the supernatant. Redissolve the pellet in the RNA storage buffer (15–20 μL) and determine the concentration using the spectrophotometer.

3.2 5′ End-Labeling of the RNA Transcript

In order to map the 5′ end of the RNA using NAIM, it is necessary to first 5′ end-label the RNA transcript. Since the transcribed RNA always contains a triphosphate at its 5′ end, it must be dephosphorylated using the following protocol:

1. Mix 100 pmol of RNA in 43 μL of ddH_2O with 5 μL of 10× CIP buffer and 2 μL of CIP (1 U/μL).
2. Incubate the reaction mixture at 37 °C for 1 h, then phenol-extract and ethanol-precipitate the RNA.
3. Centrifuge the precipitated RNA, redissolve the pellet in the RNA storage buffer, and store at −20 °C. After dephosphorylation, the RNA is 5' end labeled as follows:
4. Combine 30 pmol of dephosphorylated RNA with 1 μL of 10× PNK buffer, 1 μL of [γ-^{32}P]-ATP (6,000 Ci/mmol, 150 μCi/μL) and 1 μL of T4 PNK (10 U/μL) in 10 μL of a total volume.
5. Incubate the reaction mixture for 1 h at 37 °C.
6. Remove the unreacted radioactive material by passing the reaction mixture through a G-25 spin column or an equivalent product from a different manufacturer. This step greatly reduces the danger of the radioactive contamination of the work area and facilitates subsequent cleanup of the bench area and electrophoresis equipment.
7. Add 10 μL of 2× denaturing gel-loading buffer and purify the labeled RNA on a denaturing polyacrylamide gel.

3.3 3′ End-Labeling of the RNA Transcript

During transcription T7 RNA polymerase tends to add one or more untemplated adenosine residues at the 3′ end of the transcript, thus resulting in RNA transcripts with heterogeneous 3′ ends. Splint-guided addition of a single [α-^{32}P]-dCTP is a preferred method for 3′ end-labeling of such transcripts. This method allows one to selectively label only the transcript of the correct length using the following protocol:

1. Design a 12–13 nt DNA splint oligo, which is complementary to the 3′ end of the RNA except for its 5′ terminal 3 nucleotides, which are usually 5′ CCG. The third nucleotide from the 5′ end is a G, because it should be complementary to [α-^{32}P]-dCTP, and the two upstream C residues are added to ensure that only one labeled nucleotide is added to the RNA.
2. Mix 40 pmol of the RNA, 200 pmol of the DNA splint oligo and 1 μL of 10× annealing buffer in 10 μL of total volume.

3. Denature the RNA by heating the mixture at 95 °C for 30 s, then anneal the RNA and the splint by slowly cooling the mixture to 42 °C.
4. Add 1 μL of 20× $MgCl_2$/DTT mix, 5 μL ddH_2O, 2 μL of [α-^{32}P]-dCTP (3,000 Ci/mmol, 20 μCi/μL), and 2 μL of Klenow DNA polymerase exo- (5 U/μL), and incubate for 2 h at 37 °C. Remove the unreacted radioactive nucleotide by using G-25 spin columns and purify the labeled RNA on a denaturing polyacrylamide gel.

3.4 Selection Step: RNA Folding in the Absence of Mss116

Although this chapter is dedicated to NAIM study of protein-facilitated RNA folding, it is necessary to carry out in parallel the same experiments without the protein and compare the results side by side. This allows one to identify NAIM effects that are specifically caused by the protein (increase or decrease of already existing interference or enhancement effects, appearance of new effects or disappearance of some of the effects observed without the protein). Folding of the RNA in the absence of the protein is carried out using the following protocol:

1. Denature 5′- or 3′- end-labeled RNA (2 nM or 20 nM final concentration, respectively) at 95 °C for 1 min in 80 mM MOPS–KOH, pH 7.0.
2. Cool the RNA to 30 °C and add KCl and $MgCl_2$ to final concentrations of 100 mM and 8 mM, respectively. The final volume of the reaction mixture is 80 μL.
3. Incubate the reaction mixture at 30 °C for 40–45 min or until 15–20 % of the compact population is formed.
4. Add 27 μL of 4× native gel-loading buffer and load the mixture on a 6 % native gel containing 3 mM $MgCl_2$. Since the gel contains $MgCl_2$, it is critical to run it in the cold room at 4 °C to prevent overheating.
5. Use phosphorimager or equivalent equipment to visualize bands corresponding to unfolded and compact species (Fig. 2a, b). For some RNAs you may also be able to isolate folding intermediates (Fig. 2).
6. Cut respective bands out of the gel, elute the samples, and redissolve in the RNA storage buffer.

3.5 Selection Step: RNA Folding in the Presence of Mss116

Folding of the RNA in the presence of the protein is carried out using the following protocol:

1. Denature 5′- or 3′- end-labeled RNA (2 nM or 20 nM final concentration, respectively) at 95 °C for 1 min in 80 mM MOPS–KOH, pH 7.0.
2. Cool the RNA to 30 °C and add KCl and $MgCl_2$ to final concentrations of 100 mM and 8 mM, respectively. Add Mss116

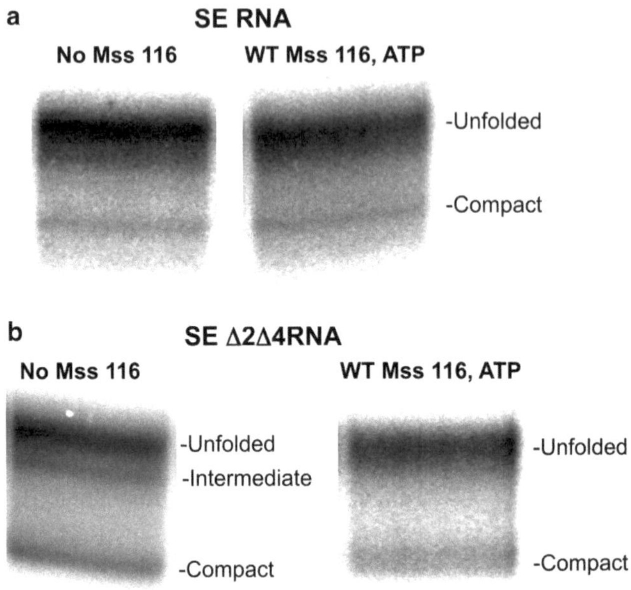

Fig. 2 Representative native gels showing the formation of the compact species (NAIM selection step) in the presence and in the absence of Mss116 by the SE RNA (**a**) and SE Δ2Δ4 RNA (**b**). Both RNAs were transcribed in the presence of 2,6 DAP. SE RNA construct contains the full-length ai5γ intron and short 5′ and 3′ exons (28 nucleotides and 15 nucleotides, respectively). SE Δ2Δ4 construct was created from SE construct by shortening nonessential domains 2 and 4 to short stem-loops [36]. Reprinted with permission from: Fedorova, O., and Pyle, A. M. (2012) The brace for a growing scaffold: Mss116 protein promotes RNA folding by stabilizing an early assembly intermediate, *J. Mol. Biol.* **422**, 347–365 (Elsevier)

to the final concentration of 20 nM (*see* **Note 5**). The final volume of the reaction mixture is 80 μL.

3. Incubate the reaction mixture at 30 °C for 1–3 min or until 15–20 % of the compact population is formed.
4. Add Proteinase K (20 mg/mL) solution to the final concentration of 6 mg/mL (*see* **Note 6**). Incubate at 30 °C for 30 s.
5. Add 27 μL of 4× native gel-loading buffer and load the mixture on a 6 % native gel containing 3 mM $MgCl_2$. Since the gel contains $MgCl_2$, it is critical to run it in the cold room at 4 °C to prevent overheating.
6. Use phosphorimager or equivalent equipment to visualize bands corresponding to unfolded and compact species. For some RNAs you may also be able to isolate folding intermediates (Fig. 2).
7. Cut respective bands out of the gel, elute the samples, and redissolve in 5 μL of the RNA storage buffer.

3.6 Iodine Cleavage, Sequencing, and Data Analysis

In order to identify atoms and functional groups that interfere with or facilitate the group II intron RNA folding in the presence or in the absence of Mss116, after the selection step gel-purified folded and unfolded RNAs are subjected to iodine cleavage at the phosphorothioate tag positions. This is carried out according to the following protocol:

1. Denature the RNA (previously dissolved in 5 μL of the RNA storage buffer) at 95 °C for 1 min and chill on ice.
2. Add 1 μL of freshly prepared 10 mM iodine solution in ethanol, incubate at 37 °C for 3 min (*see* **Note 7**).
3. Add 6 μL of denaturing gel-loading buffer.
4. Determine the amount of the radioactive material in each sample by using a liquid scintillation counter and load equal amounts of radiolabeled unfolded and folded (compact) RNA samples onto sequencing gels (*see* **Note 8**). This allows one to easily visualize interference and enhancement effects after radio-analytic imaging of the gels (Fig. 3).
5. Dry the gels, expose to a phosphorimager screen (usually overnight), scan and quantify the intensities of bands corresponding to various nucleotide positions (depending on the analogs used) in unfolded and compact RNA samples. Correct for background if necessary.
6. Calculate effects by using the following formula: $\text{Effect} = \text{N}\alpha\text{S}_{\text{unfolded}}/\text{N}\alpha\text{S}_{\text{compact}}$. Calculated effects above one indicate analog interference, and lower than one indicate analog enhancement.
7. Adjust resulting values for variations in sample loading onto the gel: calculate the average interference value for all positions that are within two standard deviations from the mean and then divide each interference (or enhancement) value by that average. Resulting adjusted values are considered interferences if higher than 2 (*see* **Note 9**), and enhancements if lower than 0.5.

4 Notes

1. The pH of MOPS buffers tends to get lower upon dilution. This has to be taken into account when making more concentrated stocks of this buffer. It is recommended to take an aliquot of the concentrated solution, dilute it to the desired final concentration, check the pH of the diluted solution and then adjust the pH of the concentrated stock accordingly.
2. Acrylamide solutions stored for long periods of time form aromatic compounds, which absorb UV light at 254 nm. Gels

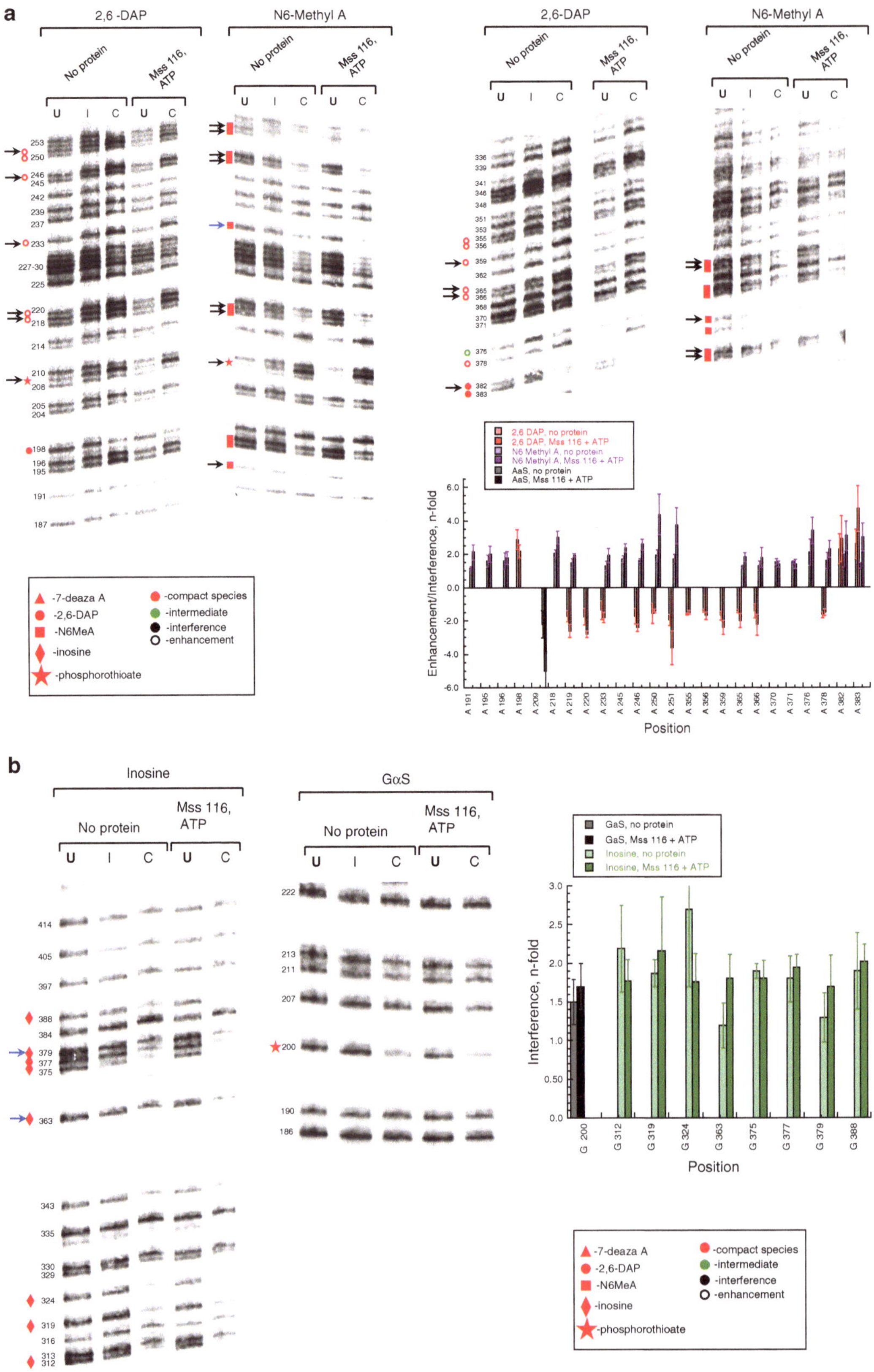
a
2,6-DAP
N6-Methyl A
No protein
Mss 116, ATP
U I C U C
-7-deaza A
-2,6-DAP
-N6MeA
-inosine
-phosphorothioate
-compact species
-intermediate
-interference
-enhancement
2,6 DAP, no protein
2,6 DAP, Mss 116 + ATP
N6 Methyl A, no protein
N6 Methyl A, Mss 116 + ATP
AaS, no protein
AaS, Mss 116 + ATP
Enhancement/Interference, n-fold
Position
b
Inosine
GαS
No protein
Mss 116, ATP
U I C U C
GaS, no protein
GaS, Mss 116 + ATP
Inosine, no protein
Inosine, Mss 116 + ATP
Interference, n-fold
Position

prepared from these solutions cannot be used for UV shadowing of nucleic acids.

3. It is important to use the highest purity magnesium chloride, because traces of transition metal ions will induce cleavage of the RNA. Magnesium chloride is hygroscopic, therefore weighing out the calculated amount of the powder will likely result in lower concentration of $MgCl_2$ solution. Therefore it is more practical to calculate the amount of $MgCl_2$ required to make a 2 M solution, determine its concentration (which will be lower than 2 M) and then dilute it to 1 M with an appropriate amount of water. After dissolving the calculated amount of $MgCl_2$ in water, the concentration of the stock solution should be determined by using density versus concentration plots [37].
4. In some transcription protocols it is suggested to add inorganic pyrophosphatase to the transcription mixture in order to hydrolyze inorganic pyrophosphate. However, in our hands this does not affect the transcription yield.
5. Always use fresh aliquots of Mss116. It cannot be refrozen after thawing, because this results in drastic loss of activity.
6. Without Proteinase K treatment the samples tend to stay in the wells and do not enter the gel.
7. Depending on the RNA sample, iodine cleavage conditions may need optimization. Iodine final concentration in the reaction mixture may be varied from 0.1 to 1.5 mM. We get best results when the cleavage reaction is carried out at 37 °C, but room temperature is used often as well. In addition, the incubation times may be varied between 1 and 5 min.
8. Although all the interference and enhancement effects are calculated using appropriate software (like ImageQuant from GE Healthcare), it always helps to see them on the sequencing

Fig. 3 Dissecting Mss116-facilitated compaction of the ai5γ intron at atomic resolution. Representative sequencing gels (*top panel*) showing functional groups interfering with compaction at adenosine (**a**) and guanosine (**b**) positions in the "folding control element". On each gel, *lane U* corresponds to Unfolded species, *I* to Extended Intermediate species and *C* to Compact species. *Symbols on the left of each gel* indicate nucleotide positions that interfere with (*closed symbols*) or enhance (*open symbols*) the formation of the compact (*red*) or intermediate (*green*) species. *Black arrows* indicate positions where NAIM effects are increased in the presence of Mss116. Positions where NAIM effects appear only in the presence of Mss116 are denoted by *blue arrows*. Bar graphs on the *bottom* (**a**) and on the *right* (**b**) show average value and standard deviation for interference and enhancement effects observed in the "folding control element" upon formation of the compact state in the absence and in the presence of Mss116. (**c**) A schematic of functional groups exhibiting increased interference or enhancement effects in the presence of Mss116. Reprinted with permission from: Fedorova, O., and Pyle, A. M. (2012) The brace for a growing scaffold: Mss116 protein promotes RNA folding by stabilizing an early assembly intermediate, *J. Mol. Biol.* **422**, 347–365 (Elsevier)

gel by eye. Adjustment of the amount of the radioactive samples prior to loading allows one to see effects as low as 1.5–2-fold.

9. In some cases the cut-off can be lowered to 1.5-fold for interference effects if they are reproducible in at least three independent experiments.

Acknowledgements

I would like to thank Prof. A.M. Pyle for guidance and support and members of the Pyle lab for helpful discussions. I would like to apologize to all our colleagues whose work could not be extensively cited in this chapter. O.F. is a Research Specialist at Howard Hughes Medical Institute.

References

1. Sosnick TR (2008) Kinetic barriers and the role of topology in protein and RNA folding. Protein Sci 17:1308–1318
2. Swisher JF, Su LJ, Brenowitz M et al (2002) Productive folding to the native state by a group II intron ribozyme. J Mol Biol 315:297–310
3. Heilman-Miller SL, Woodson SA (2003) Perturbed folding kinetics of circularly permuted RNAs with altered topology. J Mol Biol 328:385–394
4. Su LJ, Waldsich C, Pyle AM (2005) An obligate intermediate along the slow folding pathway of a group II intron ribozyme. Nucleic Acids Res 33:6674–6687
5. Ramaswamy P, Woodson SA (2009) Global stabilization of rRNA structure by ribosomal proteins S4, S17, and S20. J Mol Biol 392:666–677
6. Mayerle M, Bellur DL, Woodson SA (2011) Slow formation of stable complexes during coincubation of minimal rRNA and ribosomal protein S4. J Mol Biol 412:453–465
7. Waldsich C, Pyle AM (2008) A kinetic intermediate that regulates proper folding of a group II intron RNA. J Mol Biol 375:572–580
8. Treiber DK, Williamson JR (1999) Exposing the kinetic traps in RNA folding. Curr Opin Struct Biol 9:339–345
9. Pan J, Woodson SA (1998) Folding intermediates of a self-splicing RNA: mispairing of the catalytic core. J Mol Biol 280:597–609
10. Pan T, Sosnick TR (1997) Intermediates and kinetic traps in the folding of a large ribozyme revealed by circular dichroism and UV absorbance spectroscopies and catalytic activity. Nat Struct Biol 4:931–938
11. Karginov FV, Uhlenbeck OC (2004) Interaction of *Escherichia coli* DbpA with 23S rRNA in different functional states of the enzyme. Nucleic Acids Res 32:3028–3032
12. Diges CM, Uhlenbeck OC (2001) *Escherichia coli* DbpA is an RNA helicase that requires hairpin 92 of 23S rRNA. EMBO J 20:5503–5512
13. Rajkowitsch L, Chen D, Stampfl S et al (2007) RNA chaperones, RNA annealers and RNA helicases. RNA Biol 4:118–130
14. Henn A, Cao W, Hackney DD et al (2008) The ATPase cycle mechanism of the DEAD-box rRNA helicase, DbpA. J Mol Biol 377:193–205
15. Pan C, Russell R (2010) Roles of DEAD-box proteins in RNA and RNP folding. RNA Biol 7:667–676
16. Woodson SA (2010) Taming free energy landscapes with RNA chaperones. RNA Biol 7:677–686
17. Mayer O, Waldsich C, Grossberger R et al (2002) Folding of the *td* pre-RNA with the help of the RNA chaperone StpA. Biochem Soc Trans 30:1175–1180
18. Yang Q, Jankowsky E (2005) ATP- and ADP-dependent modulation of RNA unwinding and strand annealing activities by the DEAD-box protein DED1. Biochemistry 44:13591–13601
19. Vogel J, Luisi BF (2011) Hfq and its constellation of RNA. Nat Rev Microbiol 9:578–589
20. Weeks KM (1997) Protein-facilitated RNA folding. Curr Opin Struct Biol 7:336–342

21. Bokinsky G, Nivon LG, Liu S et al (2006) Two distinct binding modes of a protein cofactor with its target RNA. J Mol Biol 361:771–784
22. Fedorova O, Solem A, Pyle AM (2010) Protein-facilitated folding of group II intron ribozymes. J Mol Biol 397:799–813
23. Solem A, Zingler N, Pyle AM (2006) A DEAD protein that activates intron self-splicing without unwinding RNA. Mol Cell 24:611–617
24. Zingler N, Solem A, Pyle AM (2010) Dual roles for the Mss116 cofactor during splicing of the ai5gamma group II intron. Nucleic Acids Res 38:6602–6609
25. Steiner M, Karunatilaka KS, Sigel RK et al (2008) Single-molecule studies of group II intron ribozymes. Proc Natl Acad Sci USA 105:13853–13858
26. Fedorova O, Waldsich C, Pyle AM (2007) Group II intron folding under near-physiological conditions: collapsing to the near-native state. J Mol Biol 366:1099–1114
27. Halls C, Mohr S, Del Campo M et al (2007) Involvement of DEAD-box proteins in group I and group II intron splicing. Biochemical characterization of Mss116p, ATP hydrolysis-dependent and -independent mechanisms, and general RNA chaperone activity. J Mol Biol 365:835–855
28. Karunatilaka KS, Solem A, Pyle AM et al (2010) Single-molecule analysis of Mss116-mediated group II intron folding. Nature 467:935–939
29. Huang HR, Rowe CE, Mohr S et al (2005) The splicing of yeast mitochondrial group I and group II introns requires a DEAD-box protein with RNA chaperone function. Proc Natl Acad Sci USA 102:163–168
30. Potratz JP, Del Campo M, Wolf RZ et al (2011) ATP-dependent roles of the DEAD-box protein Mss116p in group II intron splicing *in vitro* and *in vivo*. J Mol Biol 411:661–679
31. Del Campo M, Tijerina P, Bhaskaran H et al (2007) Do DEAD-box proteins promote group II intron splicing without unwinding RNA? Mol Cell 28:159–166
32. Del Campo M, Mohr S, Jiang Y et al (2009) Unwinding by local strand separation is critical for the function of DEAD-box proteins as RNA chaperones. J Mol Biol 389:674–693
33. Liebeg A, Mayer O, Waldsich C (2010) DEAD-box protein facilitated RNA folding *in vivo*. RNA Biol 7:803–811
34. Bhaskaran H, Russell R (2007) Kinetic redistribution of native and misfolded RNAs by a DEAD-box chaperone. Nature 449:1014–1018
35. Waldsich C, Pyle AM (2007) A folding control element for tertiary collapse of a group II intron ribozyme. Nat Struct Mol Biol 14:37–44
36. Fedorova O, Pyle AM (2012) The brace for a growing scaffold: Mss116 protein promotes RNA folding by stabilizing an early assembly intermediate. J Mol Biol 422:347–365
37. Haynes WH (2012) Handbook of chemistry and physics, 93rd edn. CRC Press, Boca Raton, FL

Chapter 11

Applying UV Crosslinking to Study RNA–Protein Interactions in Multicomponent Ribonucleoprotein Complexes

Andrey S. Krasilnikov

Abstract

Ribonucleoprotein complexes (RNPs) play crucial roles in a wide range of biological processes. Here, we describe experimental approaches to the UV crosslinking-based identification of protein-binding sites on RNA, using multicomponent *Saccharomyces cerevisiae* RNPs of the RNase P/MRP family as an example. To identify the binding sites of a protein component of interest, a hexahistidine affinity tag was fused to that protein. Then RNase P/MRP RNPs were purified from yeast cells that had expressed the protein component of interest with the fused tag, subjected to UV crosslinking, and disassembled to separate the non-covalently-bound components. The protein component of interest was isolated under denaturing conditions using the hexahistidine tag as a purification handle. Provided that the isolated protein formed UV-induced crosslinks with the RNA component of the studied RNP, the isolation of the protein resulted in the co-isolation of the covalently bound RNP RNA. The isolated protein was enzymatically degraded, and the UV crosslinked RNA was purified. The locations of the crosslinks formed between the protein component of interest and the RNP RNA were identified by primer extension with a reverse transcriptase followed by gel electrophoresis; this procedure was repeated for all of the protein components of RNases P/MRP.

Key words Ribonucleoprotein complexes, RNA–protein interactions, UV crosslinking, Reverse transcription, Ribonuclease P, Ribonuclease MRP

1 Introduction

A wide range of key biological processes rely on complexes between RNA and proteins (RNPs). The presence of both RNA and protein components, often combined with their large size and overall complexity, makes RNPs challenging for structural studies. Crosslinking studies of RNA–protein interactions that reveal the positioning of protein components on RNA are among the main tools in the systematic analysis of the structural organization of RNPs. Here, we describe the procedures that we applied to identify protein-binding sites on the RNA components of *S. cerevisiae*

Christina Waldsich (ed.), *RNA Folding: Methods and Protocols*, Methods in Molecular Biology, vol. 1086,
DOI 10.1007/978-1-62703-667-2_11, © Springer Science+Business Media, LLC 2014

RNase P and RNase MRP RNPs [1]. RNases P/MRP are essential multicomponent ribonucleoproteins involved in the metabolism of a variety of RNA molecules, including tRNA, rRNA, and others [2, 3]. In *S. cerevisiae*, RNase P has a large (369 nucleotides long) catalytic RNA component and nine essential proteins (ranging from 15 to 100 kDa in size) of unclear functions [2, 4]; RNase MRP has a 340 nucleotides long RNA moiety resembling that of RNase P, and ten essential proteins, eight of which are shared with RNase P [2–6]. The overlap in the protein compositions of RNase P and RNase MRP and the similarity of their RNA moieties allowed us to study these closely related RNPs in parallel [1, 7, 8].

The presence of multiple protein components requires special approaches to crosslinking studies of RNA–protein interactions to discriminate between crosslinks formed by different proteins. The procedures described in this chapter were developed for RNase P/MRP RNPs [1]; however, they can easily be adapted for other multicomponent ribonucleoproteins. Briefly, we copurified RNases P/MRP RNPs from yeast strains that had hexahistidine tags fused to each of the protein components (one tagged protein per strain), subjected these RNPs to UV crosslinking (optimized to produce only a small—ideally about one per molecule—number of crosslinks), disassembled the complexes under denaturing conditions, and then used the hexahistidine tag as a handle to isolate the tagged protein component together with the crosslinked RNP RNA. To identify the locations of the UV-induced crosslinks between the tagged protein of interest and RNP RNA, we then enzymatically degraded the protein using proteinase K and purified the resulting RNA along with small fragments of the crosslinked protein still covalently attached. The resultant RNA was analyzed for the locations of the crosslinking sites using primer extension with a reverse transcriptase and primers that were specific to either RNase P or RNase MRP RNAs (Fig. 1). This procedure was repeated for each of the protein components and allowed us to generate a map of RNA–protein interactions in RNases P/MRP [1].

2 Materials

2.1 Preparation of Deproteinated Control RNA

All reagents must be RNase-free (*see* **Note 1**).

1. A refrigerated microcentrifuge capable of providing 15,000 × *g* at 4 °C (*see* **Note 2**).
2. A vortexer.
3. A freezer (−20 °C) and a deep freezer (−76 °C or below).
4. SpeedVac or a similar evaporator.
5. Purified RNP of interest (*see* **Notes 3** and **4**).
6. 3 M sodium acetate, pH 7.0. Store at 4 °C.

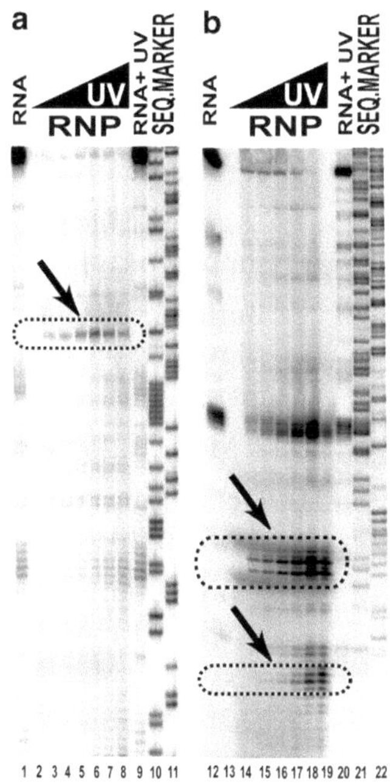

Fig. 1 Typical results of the primer extension analysis of RNase P/MRP RNA crosslinked to specific protein components [1]. (**a**) RNase MRP RNA crosslinked to protein Pop7; (**b**) RNase P RNA crosslinked to protein Pop4. *Lanes 1*, *12*: primer extension reactions for deproteinated RNAs (no UV irradiation); *lanes 2–8*, *13–19*: primer extension reactions for RNA extracted from UV-treated RNPs, the UV dose increasing from 0 to 640 mJ/cm^2; *lanes 9*, *20*: primer extension reactions for UV-irradiated deproteinated RNAs; *lanes 10*, *11*, *21*, *22*: sequencing markers (*see* **step 12** in Subheading 3.4). The *arrows* and *outlined regions* indicate the locations of the RNA–protein crosslinks

7. Polyacryl carrier (Molecular Research Center). Store at 4 °C.
8. Tris–HCl-saturated phenol, pH 8.0. Store at 4 °C.
9. Ethanol and 80 % (v/v) ethanol solution. Store at room temperature.
10. Annealing buffer: 5 mM Tris–HCl, pH 8.0, 10 mM KCl, 0.1 mM EDTA, pH 8.0. Store frozen.

2.2 UV Crosslinking and Isolation of the Tagged Protein of Interest

All reagents must be RNase-free.

1. XL-1500 UV crosslinker (Spectronics Corp.) or a similar UV (254 nm) crosslinker.
2. RNase-free glass cover slides. To remove possible contamination with ribonucleases, put slides in a clean glass container, cover the container with aluminum foil, and bake at 180 °C in a dry oven overnight.
3. A standard aluminum dry bath thermoblock.

4. A dry bath capable of maintaining 37 and 60 °C.
5. A styrofoam container, approximately 15 cm × 15 cm × 10 cm (W × L × H).
6. A microcentrifuge.
7. A gel box, glass plates, spacers, a comb, and a suitable power supply to run denaturing SDS-polyacrylamide protein gels.
8. Buffer 1: 20 mM Tris–HCl, pH 7.8, 150 mM NaCl, 2.5 mM $MgCl_2$, 0.05 mM EDTA, pH 8.0, 5 % (v/v) glycerol. Store frozen.
9. 1 M DTT. Prepare using RNase-free water. Keep frozen.
10. Denaturing buffer: 4 % (w/v) SDS, 100 mM Tris–HCl, pH 7.8, 20 % (v/v) glycerol, 50 mM DTT. Prepare fresh.
11. Binding buffer: 8 M urea, 15 mM Tris–HCl, pH 7.8, 500 mM NaCl, 10 mM Na-imidazole, pH 7.4, 1 % (v/v) Tween 20. Prepare fresh.
12. Ni-IDA resin (Macherey-Nagel). Equilibrate with the binding buffer. Keep at 4 °C.
13. Elution buffer: 4 M urea, 15 mM Tris–HCl, pH 7.8, 250 mM NaCl, 300 mM Na-imidazole, pH 7.4, 0.5 % (v/v) Tween 20. Prepare fresh.
14. Proteinase K. Dissolve dry enzyme in 20 mM Tris–HCl, pH 8.0, 1 mM $CaCl_2$ to 1 mg/mL. Store in aliquots at −76 °C. Do not refreeze.
15. Annealing buffer: 5 mM Tris–HCl, pH 8.0, 10 mM KCl, 0.1 mM EDTA, pH 8.0. Store frozen.
16. Components of denaturing SDS-polyacrylamide gels: 40 % acrylamide/bisacryamide (38:1) solution, tetramethylethylenediamine (TEMED), 10 % (w/v) ammonium persulfate solution, 1.5 M Tris–HCl, pH 8.8, 0.5 M Tris–HCl, pH 6.8, glycine (powder), 10 % (w/v) SDS. Prepare the gel mix immediately before use.
17. Silver staining kit. Follow the manufacturer's recommendations on the usage and storage.

2.3 Reverse Transcription

1. Primers for reverse transcription (*see* **Notes 5–8**). Dilute to 200 pmol/μL in water. Store frozen.
2. Dry baths capable of maintaining 42, 52, 65, and 85 °C.
3. A gel box, glass plates, spacers, a comb, and a suitable high-voltage (up to 2,000 V) power supply to run denaturing (sequencing) polyacrylamide gels.
4. Sequenase 2.0 DNA sequencing kit (Affymetrics) or a similar set of components for DNA sequencing (*see* **Note 9**). Follow manufacturer's recommendations on the use and storage.

5. A plasmid encoding the RNA component of the RNP of interest (*see* **Note 9**).
6. Phosphorimager or other means to visualize radioactive bands in a gel.
7. 0.5 M ethylenediaminetetraacetic acid (EDTA). Adjust pH to 8.0 with NaOH. Store at 4 °C.
8. SuperScript II reverse transcriptase kit. Store at −20 °C.
9. 10 mM solutions of dATP, dGTP, dTTP, and dCTP, pH 8.0. Store at −20 °C.
10. RNase inhibitor (20 U/μL). Store at −20 °C.
11. [α-^{32}P]-dATP (3,000 Ci/mmol, 10 mCi/mL).
12. Primer extension solution. Make a master mix: per 10 primer extension reactions, mix together 12 μL of the 5× SuperScript II reverse transcriptase buffer (from the kit), 4.2 μL of 100 mM $MgCl_2$, 1.8 μL of 100 mM DTT, 3 μL (each) of 10 mM solutions of dGTP, dTTP, dCTP, 0.6 μL of RNase inhibitor (20 U/μL, 1.5 μL of [α-^{32}P]-dATP (3,000 Ci/mmol, 10 mCi/mL), and 3 μL of SuperScript II reverse transcriptase (200 U/μL). Mix the components on ice; add the enzyme last. Prepare fresh.
13. A mixture of RNase T1 and RNase A (1 mg/mL each). Store at −20 °C. Dilute to 20 ng/μL (each) before use.
14. 10× Tris–borate–EDTA (TBE) buffer: 890 mM Tris-base, 890 mM boric acid, 20 mM EDTA, pH 8.4. Store at room temperature.
15. Loading buffer. To prepare, add to formamide: 10 % (v/v) of 10× TBE buffer, 4 % (v/v) of 500 mM EDTA, pH 8.0, trace amounts of xylene cyanol and bromophenol blue dyes. Store at −20 °C.
16. Components of denaturing polyacrylamide gels: 40 % acrylamide/bisacrylamide (19:1) solution, tetramethylethylenediamine (TEMED), 10× TBE, 10 % (w/v) ammonium persulfate solution, urea. Prepare the gel mix immediately before use.

3 Methods

3.1 Preparation of Deproteinated Control RNA

1. Take 2 pmol (*see* **Note 10**) of the RNP of interest, mix with water and 3 M sodium acetate, pH 7.0, to produce 100 μL of the RNP solution in 600 mM sodium acetate; small amounts of other buffers/salts (carryovers from the RNP storage buffer) are allowed.
2. Add 10 μL of polyacryl carrier.
3. Add 100 μL of phenol.

4. Vortex for 1 min, spin the sample at 15,000 × *g* at 4 °C for 5 min, then carefully transfer the aqueous phase to a new tube.
5. Repeat **steps 3** and **4** two more times.
6. Add 350 μL of cold (−20 °C) ethanol, mix, and incubate at −76 °C for at least 40 min.
7. Spin the sample at 15,000 × *g* at 4 °C for 15 min.
8. Remove supernatant. Do not disturb the pellet.
9. Add 300 μL of cold (−20 °C) 80 % (v/v) solution of ethanol, vortex briefly.
10. Spin the sample at 15,000 × *g* at 4 °C for 10 min.
11. Repeat **steps 8–10**.
12. Remove supernatant. Do not disturb the pellet.
13. Briefly spin the tube to collect the remaining liquid at the bottom of the tube; carefully remove the liquid.
14. Dry the pellet in a SpeedVac.
15. Dissolve the pellet in the annealing buffer and adjust the concentration to 0.25 pmol/μL. Store frozen at −76 °C.

3.2 UV Crosslinking

1. Fill a styrofoam container with ice; the ice level should be approximately 1 in. below the top of the container walls. Put a standard aluminum thermoblock from a dry bath (or any similar metal object) upside down into the styrofoam container, on ice.
2. Put the styrofoam container with the thermoblock into a UV crosslinker. The top of the thermoblock should be at a sufficient distance from the UV lamps of the crosslinker to ensure even illumination. Make sure that the walls of the container or any other objects are not casting shadows on the top of the thermoblock when the UV crosslinker is running (*see* **Note 11**). Make sure that the styrofoam container is not casting a shadow on the UV sensor mounted inside the crosslinker (*see* **Note 12**). Select a different styrofoam container if necessary (*see* **Notes 13** and **14**).
3. Give the thermoblock enough time to cool down.
4. Dilute the RNP of interest to 0.5 pmol/μL in buffer 1; approximately 100 μL of the diluted sample will be needed.
5. Put aside a 15 μL aliquot of the sample.
6. Take six RNase-free glass cover slides. Using a thin marker, mark one slide "control," and number the remaining slides "1" through "5"; make sure that the tops of the slides remain RNase-free. Put 15 μL of the RNP solution from **step 4** on each of the glass cover slides marked "1" through "5." Put 15 μL of the control RNA solution from **step 15** in Subheading 3.1 on the slide marked "control."

7. Carefully put the cover slides with the drops of the sample on the top of the cold thermoblock and into the UV crosslinker as described above.
8. Set the desired UV dosage to 40 mJ/cm^2 (the irradiation time will vary depending on the crosslinker; do not use time settings, but use UV dosage ("energy") settings instead). Irradiate the slides with the samples.
9. Carefully remove the slide numbered "1". Transfer the solution from this slide into a new microcentrifuge tube. This is the sample exposed to the 40 mJ/cm^2 of UV radiation.
10. Repeat **step 8**.
11. Carefully remove the slide numbered "2." Transfer the solution into a new microcentrifuge tube. This is the sample exposed to the 80 mJ/cm^2 of UV radiation.
12. Set the desired UV dosage to 80 mJ/cm^2; irradiate the remaining slides.
13. Carefully remove the slide numbered "3." Transfer the solution into a new microcentrifuge tube. This is the sample exposed to the 160 mJ/cm^2 of UV radiation.
14. Repeat **steps 12** and **13** using increasing irradiation dosages: 160 mJ/cm^2 (sample "4," the total dosage of 320 mJ/cm^2), then 320 mJ/cm^2 (sample "5," the total dosage 640 mJ/cm^2). The "control" slide should also be exposed to the total dosage of 640 mJ/cm^2 (*see* **Note 15**).

3.3 Isolation of the Tagged Protein of Interest

1. Add 20 μL of the denaturing buffer to the samples from **steps 9–14** in Subheading 3.2 and to the sample that was put aside in **step 5** in Subheading 3.2.
2. Incubate the samples for 10 min in a dry bath at 60 °C.
3. Add a fresh portion of DTT to increase its concentration by an additional 30 mM, repeat the 10 min incubation at 60 °C (*see* **Note 16**).
4. Repeat **step 3** two more times (*see* **Note 16**).
5. Transfer the samples into fresh tubes containing 1 mL of the binding buffer.
6. Incubate the samples with 50 μL of Ni-IDA resin (*see* **Note 17**) at room temperature for 2 h with intensive agitation (*see* **Note 18**).
7. Spin the tubes for 1 min at 500 × *g* at room temperature to collect the resin at the bottom of the tubes; remove the liquid.
8. Add 1 mL of the binding buffer, wash the resin by gentle vortexing (*see* **Note 19**).
9. Spin the tubes for 1 min at 500 × *g* at room temperature; carefully remove the binding buffer.

10. Repeat **steps 8** and **9** four more times.
11. Add 100 μL of the elution buffer to each of the sample tubes; vortex gently for 20 min at room temperature.
12. Spin the tubes for 1 min at 500 × *g* at room temperature to pellet the resin. Collect the elution buffer (avoid any resin particles), keep it on ice.
13. Repeat **steps 11** and **12**. Combine the elution buffer collected in **step 12**.
14. Take 20 μL aliquots from each of the samples. Keep the aliquots frozen for further analysis.
15. Add 300 μg of proteinase K to each of the samples, mix gently.
16. Incubate samples with proteinase K for 20 min at 37 °C, and then put them on ice.
17. Add 10 μL of polyacryl carrier, and 1/5 volume of 3 M sodium acetate pH 7.0, mix.
18. Add 200 μL of phenol.
19. Vortex for 1 min, spin the sample at 15,000 × *g* at 4 °C for 5 min, carefully transfer the aqueous phase to a new tube.
20. Repeat **steps 18** and **19** two more times.
21. Add 1 mL of cold (−20 °C) ethanol, mix, and incubate at −76 °C for at least 40 min.
22. Perform ethanol precipitation of RNA as described in **steps 6–14** in Subheading 3.1.
23. Store dry RNA samples at −76 °C.
24. Analyze the protein content of the aliquots that were put aside in **step 14** using silver-stained SDS-polyacrylamide gels; use 1 μL of the original RNP preparation (**step 4** in Subheading 3.2) loaded on the same gel as a reference (*see* **Note 20**).

3.4 Reverse Transcription

1. For each of the samples from **step 23** in Subheading 3.3: dissolve RNA in 2.5 μL of the annealing buffer supplemented with 100 pmol of the primer for reverse transcription. Due to the low volume of the buffer, multiple rounds of vortexing followed by short spins in a microcentrifuge are required to ensure RNA recovery.
2. Mix 1 μL of the control RNA from **step 15** in Subheading 3.1 with 100 pmol of the primer for reverse transcription; adjust the volume to 2.5 μL with the annealing buffer.
3. Anneal the primer to the samples from **steps 1** and **2** (above) by incubating the mix in a dry bath at 85 °C for 3 min, immediately followed by incubation on ice for 15 min (*see* **Note 21**).

Briefly spin the samples to recover any possible water condensate from the walls of the tubes.

4. Add 3 μL of the primer extension solution to each of the samples (*see* **Note 22**).
5. Incubate the samples in a dry bath at 42 °C for 5 min.
6. Quickly add 0.3 μL of 10 mM dATP to each of the samples, mix. Do not put the samples on ice.
7. Incubate the samples in a dry bath at 52 °C for 10 min.
8. Add 1 μL of 100 mM EDTA, pH 8.0, to each of the samples, mix.
9. Add 1 μL of RNase T1/RNase A mixture (20 ng/μL of each) to each of the samples, mix.
10. Incubate the samples in a dry bath at 52 °C for 5 min (*see* **Note 23**).
11. Add 15 μL of the loading buffer to each of the samples. After this step the samples can be stored at −76 °C for up to a few days. Prolonged storage will considerably decrease the quality of the results.
12. Follow the sequencing kit manufacturer's protocol to sequence a plasmid encoding the RNA component of the RNP of interest; use the same primer as in the reverse transcription (**steps 1** and **2**). This sequence will serve as a marker to identify the locations of the crosslinks (*see* **Note 24**).
13. Prepare a 5 % denaturing (sequencing) polyacrylamide gel.
14. Pre-run the gel (*see* **Note 25**). Adjust the power output of the power supply so that the temperature of the glass plates is about 50 °C (*see* **Note 26**).
15. Denature 5 μL of the samples from **step 11** and the sequence ladder from **step 12** by incubating them in a dry bath at 65 °C for 5 min.
16. Load 5 μL of the denatured samples that originated from **step 11**, and 1 μL of the denatured sequence ladder that originated from **step 12** on the gel (*see* **Notes 27** and **28**) and run the gel (*see* **Note 29**).
17. Stop the gel, disassemble the glass plates, transfer the gel on a sheet of chromatography paper, and dry it.
18. Visualize the radioactive bands in the gel using a phosphorimager or X-ray film. Sample gels obtained for RNase P/MRP protein components Pop1, Pop4, and Pop7 [1] are shown in Fig. 1.
19. Interpret the results (*see* **Note 30**) and make corrections to the experimental procedure if necessary (*see* **Note 31**).

4 Notes

1. When possible, prepare solutions using RNase-free stock components and certified RNase-free plasticware. When the initial components of the solution are not available in an RNase-free form, treat the prepared solution with 0.1 % (v/v) of diethylpyrocarbonate (DEPC) overnight at 37 °C with agitation, followed by autoclaving. Note that DEPC cannot be used to treat reagents that do not tolerate autoclaving (e.g., MES) or can interact with DEPC (e.g., Tris). Store prepared RNase-free solutions frozen when possible. It is advisable to keep RNase-free solutions sterile as bacterial growth will introduce ribonucleases.
2. Alternatively, a non-refrigerated centrifuge kept in a cold room or a cold chromatography cabinet can be used. However, during prolonged operation of non-refrigerated centrifuges, the rotor may become noticeably warmer than the environment.
3. The exact procedure for the purification depends on the RNP being studied. Methods based on the tandem-affinity tag (TAP-tag) purification [9] are commonly used to purify RNPs; the TAP-tag purification procedure is described in detail in ref. [10].
4. Purified RNP should have a hexahistidine tag fused to the analyzed protein. Both N- and C-terminal fusions are acceptable for the described procedure; however, the tolerance of a particular RNP to the introduction of N- or C-terminally fused tags may vary. For yeast RNPs, standard genetic manipulation techniques [11] allow for an easy introduction of tags.
5. Expect that a primer extension will allow to efficiently cover a ~100–200 nucleotides long region of the template RNA that starts about 10 nucleotides from the 3′ end of the primer (the extent of the coverage will depend on the quality of the template RNA, the presence of strong secondary structure elements, and the location of the primer). Accordingly, for long RNAs, multiple primers are likely to be needed. In addition, to obtain reliable results, it is highly advisable to at least duplicate the results for each part of RNA using distinct primers. In our analysis of RNases P/MRP [1, 8], we used sets of primers that were positioned about 50 nucleotides apart.
6. The presence of strong secondary structure elements in the template RNA may interfere with the efficient annealing of primers. To counteract this effect, we used long (30–40 nucleotides) primers [1, 8]. Note that due to the nature of the primer extension analysis, the 3′ terminus of the template RNA cannot be analyzed. To minimize this "dead area" in the analysis of the 3′ terminal region, one is forced to use the

shortest primers possible. The size of the shortest acceptable primer depends on the properties of the template RNA and should be determined experimentally. For yeast RNase P, such a "minimal" primer for the 3′region was 11 nucleotides long, while for the similar RNase MRP it was 17 nucleotides long [8]. These primers provided acceptable quality of the primer extension in the immediate vicinity of their 3′ termini, but longer reads required considerably longer primers as described above.

7. It is our experience that the best results are obtained when the 3′ end of the primer anneals to an expected single-stranded part of the template (as opposed to a part of a stable double helix).
8. Best results are obtained using primers that have been gel-purified. Gel-purification of synthesized oligonucleotides is a standard option in commercial DNA synthesis; alternatively, oligonucleotides can be gel-purified in-house.
9. If the plasmid encoding the RNA component of the RNP of interest is not available, alternative sequence ladder can be produced (*see* **Note 24**). In this case, the DNA sequencing kit is not required.
10. At the 260 nm wavelength, RNA absorbs UV light about two orders of magnitude stronger than proteins. This fact can be used to estimate concentrations of RNPs. Provided that the molecular weight of the RNA moiety is known, one can estimate the molar concentration of the complex based on the 260 nm absorption, disregarding the contribution of proteins. It should be noted that this simple approach will not work for complexes that have a very high (more than 10 or so) protein-to-RNA ratio (by mass) or if a noticeable amount of contaminating nucleic acids is present. An alternative (more time- and material-consuming) approach is to estimate the amount of RNA of interest per given volume of sample by comparing the intensity of the corresponding band on an ethidium bromide-stained denaturing polyacrylamide gel and that of a series of a sequentially diluted reference RNA of a known concentration and a similar molecular weight. It is highly unadvisable to attempt loading the whole RNP on a denaturing gel as in some cases RNA may be lost in the wells of the gel. Instead, RNP RNA should be extracted with phenol prior to loading on a gel; care should be taken to avoid RNA loss during phenol extraction.
11. Powerful UV light generated by UV lamps of the crosslinker is dangerous. Make sure that your skin and, especially, eyes are not exposed to UV light.
12. A shadow casted on the sensor will result in a poor reproducibility of the results. To check, set an arbitrary UV dosage, start

the crosslinker, and measure how long it takes to achieve that dosage with and without the styrofoam container in the crosslinker. The times should be nearly identical; a longer time when the container is present indicates a problem.

13. It is crucial that the intensity of the UV light is even over the top surface of the thermoblock. Uneven intensity will lead to poorly reproducible results.
14. Make sure that all UV lamps of the crosslinker are functional. It is not uncommon for older crosslinkers to have one or more of UV lamps burned out; such a crosslinker may function well enough for other applications, but it may not provide the even distribution of light required in this experiment.
15. The drops sitting on the glass slides will evaporate if kept too long. Any noticeable evaporation should be avoided; all manipulations involving the drops on slides should be performed as quickly as possible.
16. At these steps, all potential disulfide bonds joining protein components are to be disrupted. The rate of the reduction (if any reduction is required at all) may vary dramatically for different RNPs. In the case of RNases P/MRP [1], some of the disulfide bonds connecting individual protein components were remarkably difficult to disrupt. To overcome this problem, a prolonged incubation in the presence of DTT was required; since DTT was not stable under the conditions of incubation, fresh DTT had to be added in several rounds. Also note that residual DTT will interfere with the subsequent binding of the protein to Ni-IDA resin; accordingly, the amount of added DTT should be minimized when possible.
17. In principle, any type of Ni-affinity resin should work with this procedure. In our case, with yeast RNases P/MRP [1], the Ni-IDA resin was found to give better results (more efficient and specific binding).
18. It is convenient to attach the microcentrifuge tubes to the vortexer head with tape; special vortexer attachment can also be used if available.
19. Avoid overly intensive vortexing of solutions containing SDS as otherwise they will foam.
20. For the samples that were not exposed to UV irradiation (control) or exposed to a lower dosage, expect to observe a single protein band that should have the same mobility as the protein of interest in the reference (the original RNP) lane; the intensity of the band should be comparable to the intensity of the corresponding band in the reference lane. The absence of any protein bands in the control sample that was not subjected to UV irradiation (**step 5** in Subheading 3.2) indicates that the experiment has failed. In this case, aliquots of the sample from

all key steps (including aliquots of the resin) should be collected and analyzed. If the protein of interest fails to bind to the Ni-IDA resin, different binding conditions (e.g., a lower concentration of imidazole or a lower residual concentration of DTT) or different resin can be tried. The presence of multiple bands corresponding to protein components of the RNP other than the tagged protein of interest (in the control sample that was not subjected to UV irradiation (**step 5** in Subheading 3.2)) may indicate that the denaturation of the RNP (**steps 1–4** in Subheading 3.3) was not sufficiently stringent, that the resin wash (**steps 8–10** in Subheading 3.3) was not sufficient, that a significant quantity of resin was accidentally carried over in **steps 12** and **13** in Subheading 3.3, or that the resin binds some protein components of the studied RNP nonspecifically. In the latter case, more stringent binding conditions (a higher concentration of imidazole) or an alternative type of the resin should be tried. In the worst case, a different type of the protein tag may be required. Note that at higher UV irradiation dosages, additional protein bands (corresponding to cross-linked RNA–protein and protein–protein complexes) may start appearing; this should be noted, but should not be a cause for alarm at this stage.

21. To avoid evaporation of the sample during the 85 °C incubation, make sure that the entire surface of the tube is contacting the thermoblock of the dry bath; use an additional thermoblock heated to 85–95 °C to cover the top (lid) of the tube if any traces of condensation are observed. Maintaining an even temperature across the entire surface of the tube during high-temperature incubations essentially eliminates evaporation of the sample.
22. The primer extension solution contains a radioactive isotope. This and all subsequent steps that involve radioactive samples must be performed by trained authorized personnel following appropriate rules and regulations.
23. This step is required to remove long RNAs that can interfere with the subsequent denaturing gel electrophoresis. If gels show an abrupt drop in resolution (excessively smeared bands) above a certain level, increase concentrations of RNase A/RNase T1 in **step 9** in Subheading 3.4.
24. If the required plasmid is not available, direct Sanger-type sequencing of RNA can be used instead. To perform sequencing of RNA, use the deproteinated RNA component of the RNP of interest (Subheading 3.1) as the template. Perform primer annealing and primer extension as described in Subheading 3.4, but add dideoxyribonucleotides (to a final concentration of 0.5 mM in the reaction) to the deoxyribonucleotides in the primer extension (**step 4** in Subheading 3.4).

The addition of ddGTP will result in the appearance of the bands corresponding to the positions of Cs on RNA, etc., similar to the regular Sanger DNA sequencing. RNA sequencing performed according to this approach does not produce the quality of DNA Sanger sequencing results (**step 12** in Subheading 3.4), but is still very useful for identifying the locations of crosslinks.

25. When the gel is run at a constant wattage, the voltage will initially be dropping, then the voltage will stabilize and start going up. At this point one can stop pre-running the gel and load the samples.
26. To take instant readings of glass plates' temperatures, it is convenient to use a basic commercially available IR thermometer.
27. Use 4 μL of the sequence ladder if it was produced by the Sanger-type sequencing of RNA as described in **Note 24**.
28. Load samples immediately after pre-running. Do not allow the gel to cool down.
29. It is convenient to use the mobility of the dyes present in the loading buffer (bromophenol blue and xylene cyanol) to estimate the position of the region of interest in the gel as it allows one to optimize the run time. On a 5 % denaturing polyacrylamide gel, bromophenol blue, and xylene cyanol run approximately as 35 and 130 nucleotides long DNAs, respectively.
30. The method of the detection of UV-induced protein–RNA crosslinks described in this chapter relies on the appearance of shorter products of primer extension. However, similarly looking primer extension products may result from artifacts such as: (a) reverse transcriptase pausing on the residual elements of the RNA template secondary structure that have been present during primer extension; (b) partially degraded template RNA; (c) inter- and intramolecular RNA–RNA UV-induced crosslinks; (d) protein–RNA crosslinks formed by proteins other that the tagged protein of interest. The cases of (a), (b), and some of the (c) can be easily identified as these bands appear in the deproteinated control RNA lanes as well. It should be noted that while the deproteinated RNA control should allow one to identify most of the UV-induced RNA–RNA crosslinks, it cannot be excluded that due to the likely change of the RNA conformation upon deproteination, some of the RNA–RNA crosslinks may potentially appear only in the presence of the proteins, making it difficult to differentiate between the RNA–RNA crosslinks and the protein–RNA crosslinks of interest. However, when the results obtained for different tagged protein components of interest within the same RNP are compared, only the bands that correspond to the UV-induced crosslinks between the tagged protein of interest and RNA should be

unique for each tagged protein; the RNA–RNA crosslinks will not change no matter which protein has been tagged. This allows for a reliable differentiation between the RNA–RNA crosslinks and the protein–RNA crosslinks of interest. A similar approach should be used to identify the crosslinks that involve proteins other than the tagged protein of interest (the chances of the appearance of such crosslinks raise with the increase of the UV dosage as UV-induced protein–protein crosslinks start appearing); the bands representing the protein of interest-to-RNA crosslinks are generally expected to be unique for that particular protein.

31. It is very likely that the first attempt at running the gel will result in significantly different intensities of the bands in the control lanes, sequence ladders, and the actual experimental lanes as the efficiency of crosslinks may differ from RNP to RNP; in this case, the gel may have to be rerun with adjusted volumes of the samples/ladders loaded.

Acknowledgements

We would like to thank Robert Fagerlund for his help with the preparation of this chapter. This work was supported by NIH grant GM085149 to A.S.K.

References

1. Khanova E, Esakova O, Perederina A et al (2012) Structural organizations of yeast RNase P and RNase MRP holoenzymes as revealed by UV-crosslinking studies of RNA-protein interactions. RNA 18:720–728
2. Esakova O, Krasilnikov AS (2010) Of proteins and RNA: the RNase P/MRP family. RNA 16:1725–1747
3. Krasilnikov AS (2011) Ribonucleoprotein ribonucleases P and MRP. In: Nicholson AW (ed) Ribonucleases, vol 26, Nucleic acids and molecular biology. Springer, Berlin, Heidelberg, pp 319–342
4. Chamberlain JR, Lee Y, Lane WS et al (1998) Purification and characterization of the nuclear RNase P holoenzyme complex reveals extensive subunit overlap with RNase MRP. Genes Dev 12:1678–1690
5. Schmitt ME, Clayton DA (1994) Characterization of a unique protein component of yeast RNase MRP: an RNA-binding protein with a zinc-cluster domain. Genes Dev 8:2617–2628
6. Salinas K, Wierzbicki S, Zhou L et al (2005) Characterization and purification of *Saccharomyces cerevisiae* RNase MRP reveals a new unique protein component. J Biol Chem 280:11352–11360
7. Perederina A, Esakova O, Koc H et al (2007) Specific binding of a Pop6/Pop7 heterodimer to the P3 stem of the yeast RNase MRP and RNase P RNAs. RNA 13:1648–1655
8. Esakova O, Perederina A, Quan C et al (2008) Footprinting analysis demonstrates extensive similarity between eukaryotic RNase P and RNase MRP holoenzymes. RNA 14:1558–1567
9. Rigaut G, Shevchenko A, Rutz B et al (1999) A generic protein purification method for protein complex characterization and proteome exploration. Nat Biotechnol 17:1030–1032
10. Puig O, Caspary F, Rigaut G et al (2001) The tandem affinity purification (TAP) method: a general procedure of protein complex purification. Methods 24:218–229
11. Brachmann CB, Davies A, Cost GJ et al (1998) Designer deletion strains derived from *Saccharomyces cerevisiae* S288C: a useful set of strains and plasmids for PCR-mediated gene disruption and other applications. Yeast 14:115–132

Chapter 12

The Kinetics of Ribozyme Cleavage: A Tool to Analyze RNA Folding as a Function of Catalysis

Nora Zingler

Abstract

As catalytically active RNAs, ribozymes can be characterized by kinetic measurements similar to classical enzyme kinetics. However, in contrast to standard protein enzymes, for which reactions can usually be started by mixing the enzyme with its substrate, ribozymes are typically self-cleaving. The reaction has to be initiated by folding the RNA into its active conformation. Thus, ribozyme kinetics are influenced by both folding and catalytic components and often enable indirect observation of RNA folding. Here, I describe how to obtain quantitative ribozyme cleavage data via denaturing polyacrylamide gel electrophoresis (PAGE) of radioactively labeled in vitro transcripts and discuss general considerations for subsequent kinetic analysis.

Key words RNA, Ribozymes, Denaturing PAGE, Body labeling, In vitro transcription, Kinetics, Nonlinear regression

1 Introduction

Enzymes must fold to a specific conformation to perform catalysis. This applies not only to proteins but also to catalytically active RNAs, the ribozymes. Aided by metal ions, ribozymes can adopt their active three-dimensional structure, the native conformation, by forming multiple secondary and tertiary interactions between bases and the sugar backbone. However, due to the limited repertoire of functional groups in RNA (4 bases vs. 20 amino acids), many alternative structures of similar stability exist and may have to be sampled before the most stable form is found. Thus, RNA folding is often slow, and stable misfolded structures (kinetic traps) can severely impede the RNA's biological function. Therefore, numerous protein cofactors have evolved that assist RNA folding in various ways [1]. To study the role of these cofactors in RNA folding, careful monitoring of the folding state of the RNA is necessary. Several methods for direct observation of the folding state exist. Native gel electrophoresis, analytical ultracentrifugation, and single-molecule

Christina Waldsich (ed.), *RNA Folding: Methods and Protocols*, Methods in Molecular Biology, vol. 1086,
DOI 10.1007/978-1-62703-667-2_12, © Springer Science+Business Media, LLC 2014

studies give information about the compaction status, and biochemical structural probing and crosslinking reveal secondary and tertiary interactions within the RNA [2–6]. However, these techniques can be laborious and time-consuming and may often be replaced by the comparatively simpler measurement of catalytic activity of the correctly folded ribozyme. If folding is slower than catalysis (i.e., rate-limiting), these measurements provide an independent approach to determine the rate constant of folding. An added advantage is that catalytic activity is an unequivocal indication that the observed molecules are actually folded into the native conformation. An example of experiments that showed that kinetic measurements can indeed complement more direct techniques is discussed in Swisher et al. 2002 [7]: The D135 ribozyme, a shortened derivative of the ai5γ group II intron, was characterized by various methods which yielded comparable rate constants for the formation of tertiary contacts within the RNA and ribozyme activity. Similarly, investigations on the role of the protein Mss116 in folding of a self-splicing ai5γ RNA flanked by short exons (SE construct) resulted in similar rate constants in both compaction analysis via native gels [8] and kinetic experiments [9] (Fig. 1).

In order to analyze these details, it is vital to obtain high precision data of the reaction kinetics. Here, I will describe the experimental setup to generate reliable time courses of protein-assisted splicing using the mitochondrial ai5γ intron from *Saccharomyces cerevisiae*, and the basics of kinetic data analysis are discussed. This protocol can easily be adapted to study different aspects of ribozyme folding by varying the reaction conditions.

2 Materials

2.1 General

1. Gloves! When working with RNA, always wear gloves. Change the gloves after every potential contamination.
2. Only use certified RNase-free chemicals.
3. Prepare all buffers or washing solutions with RNase-free water (DEPC-treated or ultrafiltrated) and RNase-free reagents.
4. Before use, filter all solutions through a 0.2 μm bottletop filter or syringe filter (*see* **Note 1**).

2.2 In Vitro Transcription and Splicing

1. *Hin*dIII (20 U/μL). Store at −20 °C.
2. 10× *Hin*dIII buffer. Store at −20 °C.
3. 10× transcription buffer: 220 mM $MgCl_2$, 400 mM Tris–HCl, pH 8.0, 20 mM spermidine, 100 mM NaCl, 0.1 % Triton X-100. Store at −20 °C.
4. NTP set (100 mM each, pH 7–7.5; separate solutions, not a mix). Store at −20 °C.

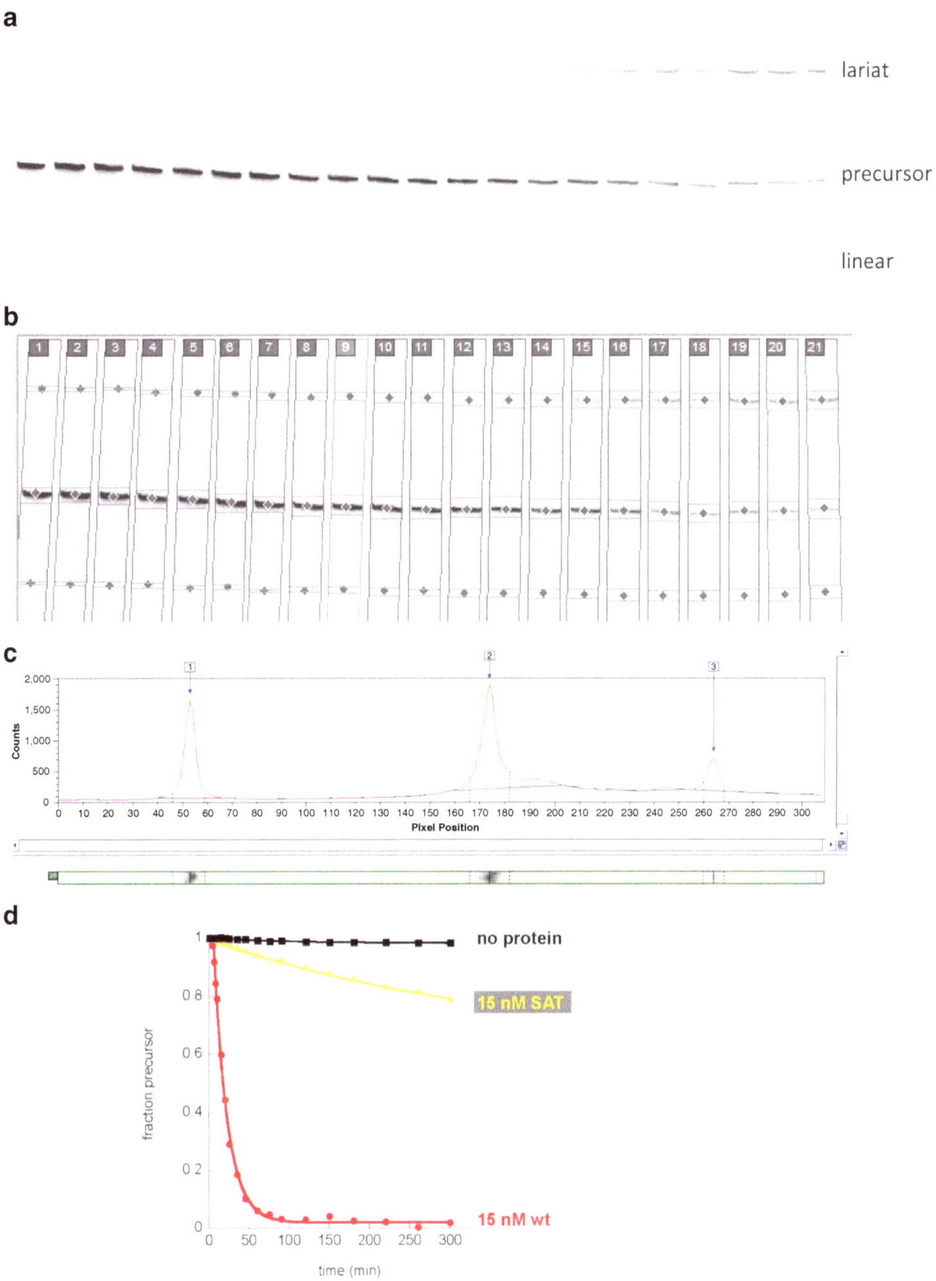

e

	rate constant k_{obs}	t_{lag}	endpoint B
pJD20/no protein	0.01113	-	0.98236
pJD20/15 nM Mss116 wt	0.05828	5.5	0.02142
pJD20/15 nM Mss116 SAT mutant	0.00321	6.0	0.67368

Fig. 1 Kinetic analysis of Mss116-assisted ai5γ (pJD20) splicing with 21 data points. (**a**) Phosphorimager screen image, (**b**) same image as in panel A after lane creation and peak picking. (**c**) Example lane illustrating peak area integration and background subtraction. A screenshot of one lane from a denaturing PAGE gel (shown sideways in the *lower panel*) with corresponding peak intensities (*upper panel*) is shown. Background correction using the rolling ball method is depicted in *pink*, the limits for area integration of the peaks are given as *black broken lines*. (**d**) Diagram plotting the fraction of precursor molecules of three splicing reactions (without protein, with wild-type Mss116, and with the Mss116 SAT mutant) against time [9]. (**e**) Parameters resulting from fitting the data points from panel D to the formula $A\,e^{(-k_{obs}(t-t_{lag}))} + B$

5. Radioactively labeled NTP (e.g., [α-^{32}P]-UTP: 10 mCi/mL, specific activity 3,000 Ci/mmol).
6. 1 M dithiothreitol (DTT). Store at −20 °C.
7. RNase inhibitor (40 U/μL). Store at −20 °C.
8. T7 RNA polymerase (preferentially the mutant P266L, *see* **Note 2**). Store at −20 °C.
9. pJD20 [10, 11], a template DNA for runoff transcription of the ai5γ intron (*see* **Note 3**). Store at −20 °C.
10. Purified Mss116 protein: purify wild-type Mss116 and its SAT mutant as reported [12].
11. 100 mM ATP. Store at −20 °C.
12. 1 M MOPS-KOH, pH 7.5. Store at 4 °C.
13. 1 M KCl.
14. 1 M $MgCl_2$.

2.3 Denaturing Polyacrylamide Gel Electrophoresis

1. 10× TBE buffer: 890 mM Tris-base, 890 mM boric acid, 20 mM EDTA disodium salt, pH is 8.3. Autoclave at 121 °C for 20 min.
2. 5 % acrylamide gel solution: 7 M urea, 5 % high quality acrylamide/bisacrylamide (29:1) (*see* **Note 4**) in 1× TBE buffer. Heat to 65 °C for 10 min or until the urea is dissolved, filter through a 0.2 μm bottletop filter and store in the dark at room temperature for up to 1 month.
3. 10 % ammonium persulfate (APS). Store at −20 °C.
4. *N*,*N*,*N*′,*N*′-tetramethyl-ethylenediamine (TEMED). Store at 4 °C.
5. Quench buffer: 3.6 % sucrose, 0.18× TBE, 0.04 % xylene cyanol, 72 % formamide (deionized), 50 mM EDTA, pH 8.0. Store at 4 °C.
6. Elution buffer: 0.3 M NaCl, 10 mM MOPS, pH 6.0 (*see* **Notes 5** and **6**), 1 mM EDTA, pH 8.0.
7. Scalpel blades.
8. 0.2 μm syringe filters.
9. Storage buffer: 10 mM MOPS, pH 6.0, 1 mM EDTA, pH 8.0 (*see* **Notes 5** and **6**).

2.4 Lab Equipment and Software

1. High-voltage power supply that allows a voltage output of at least 1,200 V.
2. Large vertical gel chambers and glass plates.
3. Scintillation counter.
4. Gel dryer.
5. Phosphorimager or X-ray film and developing equipment.

6. Excel or other table calculation program.
7. Curve fitting software.
8. Kinetics simulation and nonlinear regression program.

3 Methods

3.1 Limiting In Vitro Transcription

In order to visualize the reaction products, the RNA has to be labeled. Since fluorescent dyes might interfere with folding due to their size, radiolabeling is the method of choice for folding experiments (*see* **Note 7**).

1. Digest the template DNA pJD20 with *Hin*dIII (20 U/μL), the restriction enzyme used to define the transcriptional stop.
2. Analyze an aliquot of the digested plasmid by agarose gel electrophoresis. If remnants of undigested plasmid are visible, add more restriction enzyme and increase the incubation time.
3. Purify the DNA by phenol/chloroform/isoamyl alcohol extraction and ethanol precipitation.
4. Assemble the transcription reaction: 4 μg restricted DNA template, 5 μL 10× transcription buffer, 0.5 μL 1 M DTT, 8.7 mM ATP, 3.5 mM GTP, 3.0 mM CTP, 0.6 mM UTP (*see* **Note 8**), 7.5 μL [α-^{32}P]-UTP (10 mCi/mL), 40 U RNase inhibitor, 3 μL T7 RNA polymerase mutant P266L (homemade [13]), ad 50 μL RNase-free water.
5. Incubate the reaction for 1–6 h at 37 °C (*see* **Note 9**).
6. Centrifuge at high speed (≥11,000×*g*) in tabletop centrifuge for 2 min to remove pyrophosphate.
7. Pipette the supernatant to a new reaction tube.
8. Add an equal volume of quench buffer to the supernatant (*see* **Note 10**).

3.2 Denaturing Polyacrylamide Gel Electrophoresis

Both purification of the RNA (*see* **Note 11**) and separation of the reaction products in subsequent steps is performed on polyacrylamide gels that contain urea. Under these denaturing conditions, secondary and tertiary structures are destroyed and the separation of the different RNA species depends on size alone.

1. Calculate the volume (x) of gel solution needed to pour a gel of the desired size. Gel thickness should be ~0.5 mm.
2. Assemble the glass plates. One of the glass plates should be silanized so that the gel does not stick to it when disassembling the gel.
3. Mix x mL of gel solution of the appropriate percentage with x/200 mL 10 % APS and x/2,000 mL TEMED.

4. Pour a polyacrylamide gel. Insert combs. For a 50 μL transcription reaction, the well should be 3–4 cm wide.
5. Allow the gel to polymerize for 45–60 min.
6. Remove combs and rinse wells thoroughly with RNase-free water.
7. Assemble the gel apparatus and pre-run gel for about an hour to get rid of the salt front and to equilibrate the temperature to 45–50 °C (can be measured with an adhesive gel thermometer). Set the power to a constant value of 15–25 W depending on the gel size (e.g., 15 W for a gel 25 cm long and 15 cm wide, 25 W for a gel 25 cm long and 30 cm wide). When starting the pre-run, the voltage should not exceed 1,000 V (it will rise later during the run).
8. Heat the sample to 95 °C for 1 min.
9. After switching off the current, wash the wells thoroughly with 1× TBE buffer using a syringe and then load sample on the gel.
10. Run the preparative gel at the same settings as the pre-run until xylene cyanol reaches the bottom of the gel (*see* **Note 12**).
11. Disassemble the gel sandwich and transfer the gel between two layers of Saran wrap (*see* **Note 13**).
12. Visualize the band with a phosphorimager or exposure to an X-ray film.
13. Cut out band (this is important to get rid of premature termination products) with a scalpel and place the gel piece in a 1.5 mL Eppendorf tube.
14. Reexpose the gel on phosphorimager or X-ray film to check that the band was cut out completely.
15. Crush the gel piece with a pipette tip.
16. Estimate the volume of the crushed gel slab and add three times as much elution buffer.
17. Quick-freeze on dry ice or in liquid nitrogen to break up the gel even more. Let it thaw and elute the RNA with shaking at 4 °C for 2–4 h.
18. Transfer the gel slurry into a syringe and filter through a 0.2 μm filter to remove all polyacrylamide.
19. Precipitate the RNA by adding 3 volumes of ethanol. There is no need to add extra sodium as the elution buffer already contains the correct salt concentration.
20. Incubate on dry ice for 5 min or at −80 °C for 1 h. Centrifuge in a chilled centrifuge at ≥11,000 × *g* for 30 min.
21. Remove supernatant and let the pellet dry briefly. Do not overdry the pellet, since this will make it harder to redissolve.
22. Dissolve the pellet in an appropriate volume (e.g., 10 μL) of storage buffer.

3.3 Determining the RNA Concentration

1. Measure 1 μL of the labeled transcript in a scintillation counter.
2. Calculate the amount of radioactive Us in the sample by taking into account the calibration date and the exponential decay law of radioactivity ($t_{1/2}$ of ^{32}P is 14.26 days).
3. Calculate the concentration of the RNA by taking into account the number of Us in the transcript and the ratio of hot and cold UTP used in the transcription reaction.

3.4 Ribozyme Reaction and Sampling

1. Incubate the body-labeled ai5γ precursor RNA (10 nM) in 40 mM MOPS, pH 7.5, at 90 °C for 1 min, and at 30 °C for 3 min.
2. Assemble the splicing mastermix for a final concentration of 40 mM MOPS-KOH, pH 7.5, 100 mM KCl, 11 mM $MgCl_2$, and 2 mM ATP and equilibrate it at 30 °C for 5 min.
3. Set up reactions in PCR strips (*see* **Note 14**) by adding 1 nM precursor RNA (*see* **Note 15**) to the splicing mix. Make sure to set up the reactions in a larger volume than needed for the samples, e.g., a 50 μL reaction volume in order to take 25 aliquots of 1 μL. This reduces any potential errors by evaporation of the buffer which might be an issue especially for the later time points.
4. Initiate folding (*see* **Note 16**) by adding 15 nM Mss116 protein, 15 nM Mss116 SAT mutant or an equivalent volume of protein storage buffer as negative control.
5. Take 1 μL aliquots at varying time points and directly add them to 3.5 μL quench buffer.
6. For preliminary tests or optimization experiments, 6–8 time points yield sufficient information. However, for the actual kinetics, take 20–30 time points to obtain a sufficiently detailed data set for curve fitting. Take more time points in the regions with the most drastic changes. Since the reaction most likely has exponential decay characteristics, tightly spaced time points should be taken in the beginning of the reaction. At later time points, the intervals should gradually widen. The last three or four time points should be in the plateau phase where the reaction has virtually stopped. For protein-assisted ai5γ splicing, a good regimen is 0, 1, 2, 3, 4, 6, 8, 10, 12.5, 15, 20, 25, 30, 35, 40, 50, 60, 75, 90, 105, 120, 140, 160, 180, 220, 260, and 300 min (*see* **Note 17**).
7. Heat samples to 95 °C for 1 min and load 2 μL on an 5 % analytical denaturing PAGE gel (0.4 mm spacers).
8. Run the gel until xylene cyanol has reached the bottom of the gel.
9. Disassemble the gel and dry it between Whatman 3MM paper and Saran wrap (*see* **Note 18**).
10. Expose the dried gel overnight to a phosphorimager screen (Fig. 1a).

3.5 Quantification

This section describes how to quantify the bands with ImageQuant TL, but other programs are likely to have similar functions. Older programs require drawing of boxes around individual bands, while newer versions automatically detect lanes and bands (Fig. 1b). Even in the newer versions, some manual adjustments are most likely necessary, but it is still much faster than drawing boxes.

1. Scan screen and save the image as a gel file.
2. Open gel file with ImageQuant TL or a similar software.
3. Detect lanes and adjust their shape if necessary.
4. Detect bands. Adjust the sensitivity of peak detection and thus choose a setting that gives optimal results, but often it is necessary to do some manual adjustments. Set peak marks also in regions where a precursor has already disappeared or a product has not been produced yet. This allows the detection of even small amounts of a reaction species which may not be visible by eye. It also simplifies systematic data analysis later on.
5. Correct for background: Use rolling ball background correction with a setting that does not eat away the integral area under the peak, but removes the background that can be seen throughout the lane (Fig. 1c) (*see* **Note 19**).
6. Make sure to adjust the band borders to the points where the peak actually rises from the background (Fig. 1c).
7. After background correction, measure the peak integrals and export them to an Excel sheet or a similar tabular format.
8. Since the peak intensities not only directly correlate with the number of molecules constituting the peak but also with the uridine-content of the RNA species, the peak intensities have to be corrected by dividing by the number of uridines in each reaction product.
9. Calculate the relative fractions of the precursor and products (*see* **Note 20**).
10. Repeat each kinetic measurement at least three times independently and make sure that the results are reasonably reproducible. If the variation is too high, go back to the experimental setup and find the source of variability.

3.6 Data Analysis

The ai5γ intron is a self-cleaving ribozyme and therefore follows unimolecular first-order kinetic equations (*see* **Note 21**). First-order reactions are only dependent on the concentration of a single educt (in this case, the precursor RNA). The rate of this first-order reaction is defined by the differential rate law $d[A]/dt = -k_{obs}[A]$, or its integrated exponential form $[A] = [A]_0 e^{(-k_{obs}\ t)}$. Surprisingly, these unimolecular reactions often follow a bi-exponential curve (or colloquially called a “double exponential”), they may include a lag of a few minutes before the first reaction products can be observed, and usually there is a fraction of molecules that do not react at all so that the decay curve plateaus at an endpoint. This behavior stems from individual molecules following parallel and/or consecutive folding pathways [3, 14] and sometimes ending up in "dead end" misfolded conformations that are slow to resolve (kinetic traps) [15] (*see* **Note 22**). By measuring how the reaction products evolve over time, it is thus possible to deduce the underlying folding steps and to determine the rate constants of both the folding steps and the reaction steps.

1. The classic way to extract kinetic rate constants from ribozyme time courses is semi-logarithmic plotting and determination of k_{obs} by the slope of the resulting line (e.g., [10]). However, nowadays fitting the data via regression is easier and arguably more exact (Fig. 1d).
2. There are a number of software applications that allow fitting of kinetic data, e.g., KaleidaGraph 3.5 (Synergy Software), Origin (OriginLab), or Mathematica (Wolfram Research).
3. The simplest way to model a unimolecular chemical reaction is exponential decay. Therefore, initial fitting should be conducted with the broadly applicable formula $A\,e^{(-k_{obs}(t-t_{lag}))}+B$, where A: amplitude, t_{lag}: length of the lag, B: endpoint. In addition to the exponential decay, this formula also takes into consideration the fraction of unreactive molecules (endpoint), and a lag that may indicate several non-observable consecutive reaction steps (e.g., folding steps, binding of protein/ligand) prior to the catalytic reaction (Fig. 1e).
4. In reactions with two populations following parallel pathways, a bi-exponential curve allows determination of the fast and slow rate constants and the relative amounts (amplitudes) of fast and slow population ($A_1\ e^{(-k_{1obs}\ t)}+A_2\ e^{(-k_{2obs}\ t)}+B$) (*see* **Note 23**).
5. Extracting folding information from these kinetic measurements often involves comparison of the parameters obtained under different conditions. In the example, the fitted variables show that the SAT mutation reduces the rate constant of Mss116-assisted splicing approximately 20-fold relative to the wild-type protein (Fig. 1e). Also, the mutant seems unable to assist folding of about two-thirds of the precursor RNA, indicating that two different folding populations may exist (*see* **Note 24**).
6. In more complicated cases where the differential equations are not that easy (or impossible) to derive, simulations can be applied. Data can be simulated by numerical integration (e.g., with Excel) and compared to the experimental data [16]. Alternatively, programs such as DYNAFIT may be used [17]. Here, a set of chemical equations is entered and subsequently used as a postulated theoretical model to fit the experimental

data. Thus, it can be tested whether the data fit best to an equilibrium or a unidirectional reaction, or whether there are any reaction intermediates. In multimolecular reactions, association and dissociation steps can be introduced.

7. The program DYNAFIT varies the rate constants for each of these reactions and performs multiple iterations to approach the experimental data as closely as possible. If by specifying different initial values, vastly different results are obtained each time, it is likely that this model is too complex for the data set and the approach has to be changed. Figure 2 shows an example of how certain reaction pathways can be excluded by applying data simulations. In protein-mediated ai5γ splicing, an intermediate can be observed (Fig. 2a). This intermediate most likely represents the lariat/3' exon species as its migration behavior on a gel matches that of the product generated by the first step of splicing of a mutant ai5γ intron [18]. By comparing the kinetic profile of the intermediate with simulations including a protein binding step and a varying number of folding steps, a simple folding regimen with just one direct folding step can be clearly excluded (Fig. 2b). This result agrees with the extensive lag phase that is observed in this reaction (*see* **step 3** in Subheading 3.6).

4 Notes

1. Using membrane material that has a high protein binding capacity (e.g., cellulose nitrate or polyamide) not only makes the solutions sterile, but has the added effect of removing any stray RNase molecules that might have made their way into the solution.
2. In vitro transcription is often limited by abortive initiation, i.e., the polymerase binds to the promoter and initiates transcription, but then does not enter the elongation phase and dissociates from the template. The P266L mutant of T7 RNA polymerase increases transcription efficiency by weaker binding to the promoter and more efficient promoter clearance [13].
3. When characterizing a different ribozyme, the template may be either a plasmid or a PCR product that contains a T7 promoter sequence (5′ TAATACGACTCACTATAgG 3′, with the lower case g indicating the transcriptional start site +1) upstream of the region to be transcribed. The transcription end is defined by a restriction enzyme recognition site or the end of the PCR product. The restriction enzyme should leave a blunt end or a 5′ overhang. 3′ overhangs can lead to template switch and thus long U-turn transcripts [19]. If there is concern about homogeneous 3′ ends of the transcript, a PCR template

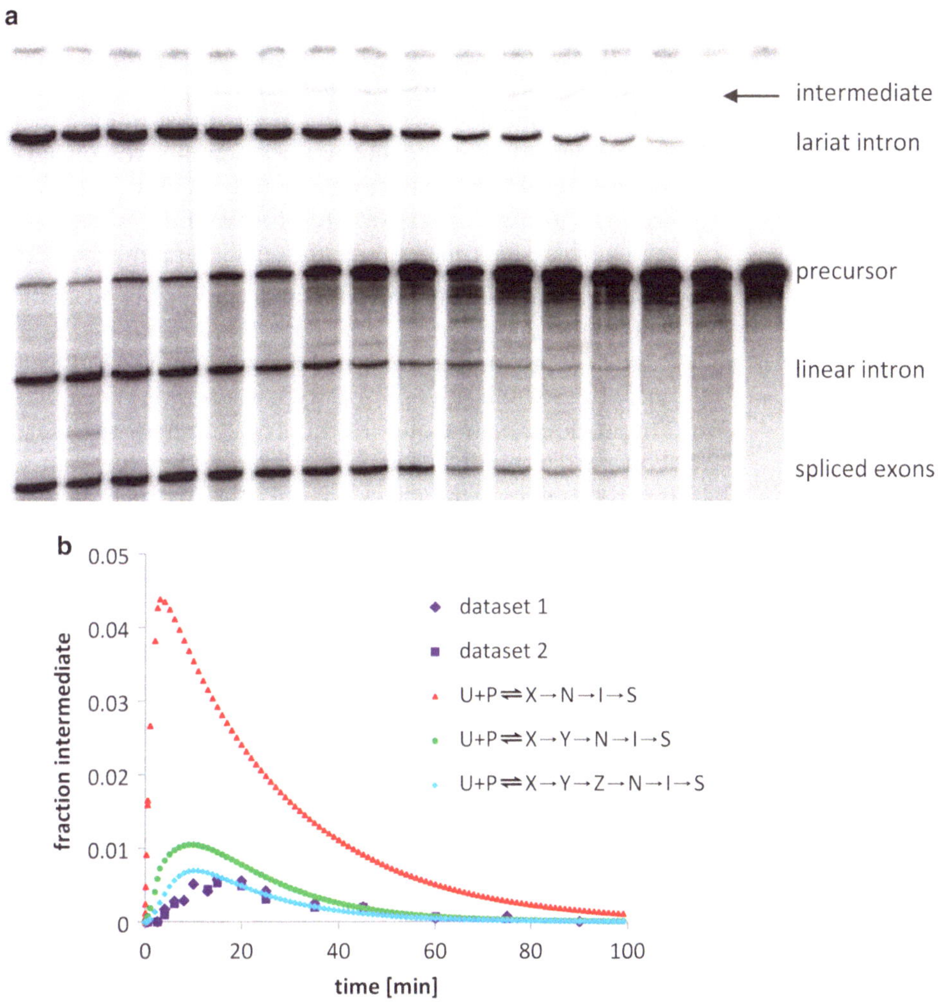

Fig. 2 Data simulation with DYNAFIT. (**a**) During Mss116-mediated ai5γ (pJD20) splicing, a splicing intermediate—most likely the lariat/3' exon species—can be observed (indicated by an *arrow*). (**b**) Comparison of the experimental evolution of the splicing intermediate (*purple symbols*) with three simulations that involve a protein binding step and 1 (*red triangles*), 2 (*green circles*), or 3 (*blue diamonds*) subsequent RNA folding steps (U, unfolded RNA; P, protein; N, native state of the RNA; X, Y, Z, folding intermediates; I, observed splicing intermediate; S, observed splicing products). The folding intermediates cannot be observed on the gel, but inferred from the kinetics of the evolution of the splicing intermediate. Direct formation of the native state after protein binding would lead to a dramatically different accumulation pattern of the splicing intermediate and can thus be excluded. Therefore, at least two distinct folding intermediates have to form consecutively before the native state is reached

is preferable. T7 RNA polymerase tends to add a few untemplated nucleotides to the transcript [20], but this can be reduced by using a PCR template that was created with a primer which introduces a 2′ OMe modification at the penultimate nucleotide of the template strand [21].

4. The percentage depends on the size of the RNA products that have to be resolved. Recommended polyacrylamide concentrations range from 4 to 5 % for fragments >150 nt to 20 % for oligonucleotides <30 nt. If fragments of vastly different sizes are expected to emerge during the reaction, two-concentration gels (e.g., 20 % in the bottom half and 5 % in the top half of the gel) can be used to get good resolution of the larger fragments and at the same time capture the smaller fragments.
5. The pH of MOPS buffers can change considerably upon dilution and is also temperature dependent. Thus, titrate the stock to a higher pH than needed in the reaction. A 1 M MOPS solution at pH 6.5 will yield a pH of ~6.0 at 40 mM. When preparing the stock, take an aliquot of the stock, dilute it to the working concentration, equilibrate the sample to the temperature used for the reaction, and check whether the pH is correct.
6. Buffers containing MOPS should be stored in the dark and discarded if they turn yellow.
7. To produce the radioactively labeled RNA, in vitro transcription is used. α-Labeled UTP is added to the in vitro transcription reaction so that the labels are statistically distributed throughout the RNA sequence (body labeling). Depending on the required RNA concentration in the experiment and the specific labeling efficiency required for sufficiently intense bands, either limiting or non-limiting transcription may be used. In non-limiting transcription, optimal amounts of all NTPs are provided, and the labeled NTP is supplied in addition. Here the yield is high, but the specific activity of the RNA is low, since only a small proportion of the incorporated nucleotides is labeled. To achieve higher labeling efficiencies, the ratio of labeled to unlabeled NTP must be increased. Since raising the concentration of the radioactive NTP is limited by the low concentration of commercially available stock solutions (e.g., 3.33 μM for [α-^{32}P]-UTP with 10 mCi/mL and 3,000 Ci/mmol), the ratio must be changed by lowering the concentration of cold NTP. Thus, a higher percentage of nucleotides will be labeled, but the overall efficiency of the transcription reaction will decrease.
8. The amounts of individual NTPs should be adapted to reflect the composition of the transcript. Due to salt inhibition of T7 RNA polymerase, total NTP concentration should not exceed 16 mM when using the sodium salts of the NTPs and can be increased up to 20 mM with the lithium salts. To establish limiting conditions, drop the concentration of cold UTP in the in vitro translation reaction while keeping all other components constant. As each transcript behaves differently, it is advisable to carry out a few small labeling reactions with different UTP

concentrations to optimize the yield. A good starting point is to use 10-fold and 50-fold less cold UTP than in the non-limiting transcription. Run an analytical polyacrylamide gel, dry it, visualize the bands, and choose the optimal dilution for a large-scale transcription.

9. After 15–30 min, the reaction will often turn cloudy due to precipitation of magnesium pyrophosphate. To avoid this and at the same time improve the yield, add pyrophosphatase (PPase) to the reaction. However, since the yield is usually considerable, this is not absolutely necessary.
10. Since the RNA will be purified via denaturing PAGE, it is not necessary to remove a plasmid template by DNase digestion. However, if a PCR template is used, the sizes will be so similar that a digestion step with certified RNase-free DNase is recommended.
11. It is very important not to skip the gel-purification step since components of the in vitro transcription reaction (e.g., magnesium ions bound to the RNA, template DNA) and premature termination products will be carried over when the RNA is merely purified by gel-filtration or precipitation.
12. Unincorporated nucleotides migrate much faster than the transcript. Thus, a considerable fraction of the radioactive label will end up in the bottom buffer reservoir. Make sure to collect and properly dispose of the buffer with extra caution.
13. Once the gel sandwich is opened, the crucial part is to transfer the gel to the Saran Wrap without breaking it. Cover the glass plate with the gel with Saran Wrap (do not forget to label the lanes!), put the glass plate upside down and slowly lift it on one edge to transfer the gel by gravity onto the Saran Wrap. Use a scalpel or razor blade to gently nudge the gel off the glass plate. After the gel is completely transferred to the Saran Wrap, cover it with another layer so that the gel is sandwiched. Make sure that the gel is completely covered with Saran Wrap, lest it may contaminate the film development apparatus or phosphorimager screen with radioactivity.
14. To increase the throughput and reproducibility, it is advisable to perform the reactions in PCR strips and use a multichannel pipette to start the reactions and to take samples. It is highly expedient to choose combs with wells that have the same spacing as the multichannel pipette in order to be able to load multiple samples directly from the strips to the gel.
15. Make sure that the concentration and specific activity of the labeled RNA is sufficient to allow detection of all reaction products but follow the ALARA concept (as low as reasonably achievable). RNA should be freshly transcribed and gel-purified to have low degradation background. Degradation may lead to

overestimation of weak bands and thus reduce the accuracy of the kinetic measurements.

16. In other experiments, folding initiation may be achieved by addition of magnesium ions, ATP, or a ligand.
17. It may be necessary to adapt the time points for specific purposes, e.g., when there is a pronounced lag at the beginning of the reaction. If the reaction is very fast and time points less than 10 s apart are needed, an alternate strategy such as quench-flow may be used. Invest the time to make a detailed pipetting scheme: start a new set of experiments as soon as the more densely spaced time points in the beginning of the time course are done. In the pipetting scheme mentioned above, a good starting point for the next set of experiments would be after 41 min.
18. For low percentage (5–10 %) gels, cover the gel with Whatman 3MM paper and gently press on it. Then peel off the gel attached to the Whatman 3MM paper and cover it with Saran Wrap. Higher percentage gels (12–20 %) do not stick to Whatman 3MM paper. Here, it is best to cover the gel with Saran wrap and proceed as described above (*see* **Note 12**). Then cover the gel with Whatman 3MM paper. For intermediate percentage gels, test whether the Whatman 3MM paper adheres to the gel. If so, peel off the gel, if not, remove the paper and use Saran wrap instead.
19. When drawing boxes around the bands, the background must be corrected by subtracting the intensity of a box of equal size in a region of the gel that has similar background but no bands.
20. Please note: by dividing the band intensity by the number of Us in the product, the intensities are basically converted to arbitrary molecular units. Thus when the relative fractions of the products are calculated, the stoichiometry must be taken into account. This offers an opportunity to check whether the quantification was correct. Consider this simple example: precursor A (containing 6 Us) is cleaved to yield products B (4 Us) and C (2 Us). The measured band intensities are A: 18, B: 28, and C: 14. The U-corrected arbitrary units are A: 3, B: 7, and C: 7, which agrees with the stoichiometry of the reaction and a 70 % conversion. However, the measured band intensities might be A: 18, B: 28, and C: 8. Now the U-corrected units are A: 3, B: 7, and C: 4, which does not agree with the stoichiometric model. This can occur if product C reacts further to form smaller products that run off the gel, or if a strong background decreases the signal-to-noise ratio to a point where reliable measurements are no longer possible. Note that if the intensities had not been corrected for U content, these problems would not be detected in the second example, and the conversion rate would be falsely calculated as 66 %.

21. There are some examples of ribozymes that catalyze the conversion of a substrate *in trans* or require a ligand for folding (the ribosome, RNase P, or riboswitches), but here the focus is on self-cleaving ribozymes.
22. If working with a new RNA species, set up the mechanistic and kinetic framework of the reaction first [22]. With well-studied ribozymes, the published frameworks can be used as starting point (see examples in ref. [16, 23–25]).
23. However, it is important not to overdo the number of variables: for example, all data sets that can be fitted by a single exponential equation can also be fitted by a double exponential. The fit will be at least incrementally better (i.e., the *R2-value* will increase), but the two rate constants will be almost identical and the amplitudes of the two phases may vary with each fitting. In such cases, use "Occam's Razor" and stick to the easiest model instead of assuming two different populations that happen to behave very similarly. Alternatively, one rate constant may be a small negative value: in this case, the program is almost certainly fitting the error in the curve not a real reverse reaction.
24. Discussing how to design other reaction conditions to address specific aspects of ai5γ splicing would go beyond the scope of this chapter, but more details on this particular example can be found in ref. [9].

Acknowledgements

I would like to thank O. Fedorova, C. Waldsich, A. Solem, and M. Roitzsch for useful tips and explanations of the techniques described here. M. Roitzsch and A. Vela optimized the transcription buffer to substantially improve yields. I am also grateful to A. Solem and M. Roitzsch for valuable comments on the manuscript.

References

1. Rajkowitsch L, Chen D, Stampfl S et al (2007) RNA chaperones, RNA annealers and RNA helicases. RNA Biol 4:118–130
2. Fedorova O, Waldsich C, Pyle AM (2007) Group II intron folding under near-physiological conditions: collapsing to the near-native state. J Mol Biol 366:1099–1114
3. Su LJ, Brenowitz M, Pyle AM (2003) An alternative route for the folding of large RNAs: apparent two-state folding by a group II intron ribozyme. J Mol Biol 334:639–652
4. de Lencastre A, Hamill S, Pyle AM (2005) A single active-site region for a group II intron. Nat Struct Mol Biol 12:626–627
5. Karunatilaka KS, Solem A, Pyle AM et al (2010) Single-molecule analysis of Mss116-mediated group II intron folding. Nature 467:935–939
6. Waldsich C, Pyle AM (2007) A folding control element for tertiary collapse of a group II intron ribozyme. Nat Struct Mol Biol 14:37–44
7. Swisher JF, Su LJ, Brenowitz M et al (2002) Productive folding to the native state by a group II intron ribozyme. J Mol Biol 315: 297–310
8. Fedorova O, Solem A, Pyle AM (2010) Protein-facilitated folding of group II intron ribozymes. J Mol Biol 397:799–813

9. Zingler N, Solem A, Pyle AM (2010) Dual roles for the Mss116 cofactor during splicing of the ai5gamma group II intron. Nucleic Acids Res 38:6602–6609
10. Daniels DL, Michels WJ Jr, Pyle AM (1996) Two competing pathways for self-splicing by group II introns: a quantitative analysis of *in vitro* reaction rates and products. J Mol Biol 256:31–49
11. Jarrell KA, Dietrich RC, Perlman PS (1988) Group II intron domain 5 facilitates a *trans-splicing* reaction. Mol Cell Biol 8: 2361–2366
12. Solem A, Zingler N, Pyle AM (2006) A DEAD protein that activates intron self-splicing without unwinding RNA. Mol Cell 24: 611–617
13. Guillerez J, Lopez PJ, Proux F et al (2005) A mutation in T7 RNA polymerase that facilitates promoter clearance. Proc Natl Acad Sci USA 102:5958–5963
14. Russell R, Herschlag D (1999) New pathways in folding of the *Tetrahymena* group I RNA enzyme. J Mol Biol 291:1155–1167
15. Treiber DK, Williamson JR (1999) Exposing the kinetic traps in RNA folding. Curr Opin Struct Biol 9:339–345
16. Diegelman-Parente A, Bevilacqua PC (2002) A mechanistic framework for co-transcriptional folding of the HDV genomic ribozyme in the presence of downstream sequence. J Mol Biol 324:1–16
17. Kuzmic P (1996) Program DYNAFIT for the analysis of enzyme kinetic data: application to HIV proteinase. Anal Biochem 237:260–273
18. Mikheeva S, Murray HL, Zhou H et al (2000) Deletion of a conserved dinucleotide inhibits the second step of group II intron splicing. RNA 6:1509–1515
19. Sambrook J, Russel DW (eds) (2001) Molecular cloning: a laboratory manual. Cold Spring Harbor Laboratory Press, Cold Spring Harbor, New York
20. Milligan JF, Groebe DR, Witherell GW et al (1987) Oligoribonucleotide synthesis using T7 RNA polymerase and synthetic DNA templates. Nucleic Acids Res 15:8783–8798
21. Kao C, Rudisser S, Zheng M (2001) A simple and efficient method to transcribe RNAs with reduced 3' heterogeneity. Methods 23:201–205
22. Fersht A (1985) Enzyme structure and mechanism, 2nd edn. W.H. Freeman, New York
23. Hertel KJ, Herschlag D, Uhlenbeck OC (1994) A kinetic and thermodynamic framework for the hammerhead ribozyme reaction. Biochemistry 33:3374–3385
24. Bergman NH, Johnston WK, Bartel DP (2000) Kinetic framework for ligation by an efficient RNA ligase ribozyme. Biochemistry 39:3115–3123
25. Pyle AM, Green JB (1994) Building a kinetic framework for group II intron ribozyme activity: quantitation of interdomain binding and reaction rate. Biochemistry 33:2716–2725

Chapter 13

RNA Catalytic Activity as a Probe of Chaperone-Mediated RNA Folding

Brant Gracia and Rick Russell

Abstract

For structured RNAs that possess catalytic activity, this activity provides a powerful probe for measuring the progress of folding and the effects of RNA chaperone proteins on the folding rate. The crux of this approach is that only the natively folded RNA is able to perform the catalytic reaction. This method can provide a quantitative measure of the fraction of native RNA over time, and it can readily distinguish the native state from all misfolded conformations. Here we describe an activity-based method measuring native folding of ribozymes derived from self-splicing group I introns, and we show how the assay can be used to monitor acceleration of native folding by DEAD-box RNA helicase proteins that function as general RNA chaperones. By measuring the amount of substrate that is converted to product in a rapid first turnover, we describe how to determine the fraction of the ribozyme population that is present in the native state. Further, we describe how to perform a two-stage or discontinuous assay in which folding proceeds in stage one and then solution conditions are changed in stage two to permit catalytic activity and block further folding. This protocol allows folding to be followed under a broad range of solution conditions, including those that do not support catalytic activity, and facilitates studies of chaperone proteins.

Key words Catalysis, Ribozyme, Folding, Helicase, Chaperone

1 Introduction

Although it was known as early as the 1960s that certain RNAs folded to specific structures, interest in RNA structure and folding began to grow dramatically in the 1980s when it was discovered that some RNAs possess catalytic activity. Since then, the field of RNA folding has continued to grow because of continued interest in these and other functional RNAs, as well as a general interest in the fundamental properties of biological macromolecules. Although the conceptual challenges of folding are shared between RNA and the other major folded macromolecule, protein, the basic physical properties of RNA lead to important characteristics of its folding processes. RNA secondary structures (helices) and tertiary contacts can form rapidly, and local structure can be independently

Christina Waldsich (ed.), *RNA Folding: Methods and Protocols*, Methods in Molecular Biology, vol. 1086,
DOI 10.1007/978-1-62703-667-2_13, © Springer Science+Business Media, LLC 2014

stable. With only four nucleotides in RNA, incorrect secondary and/or tertiary structure forms frequently, and the independent stability allows incorrect contacts to persist for long periods [1, 2]. This property can be restated as a folding free energy profile that is rugged, with deep valleys corresponding to stable misfolded forms, also termed kinetic traps [3]. Understanding how RNA folds correctly and avoids kinetic traps is universally important because structured RNAs are required in all cells.

DEAD-box proteins are ubiquitous in RNA-mediated processes and have been shown to resolve kinetic traps and promote folding to the native state [4, 5]. The mitochondrial DEAD-box proteins CYT-19 and Mss116p, from *Neurospora crassa* and *Saccharomyces cerevisiae*, respectively, have been shown to function as general RNA chaperones in vivo [6, 7]. Further, both of these proteins have been shown to facilitate native folding of group I and group II introns and their ribozyme derivatives in vitro [6, 8–11], primarily by resolving misfolded conformations [12–14]. Crystal structures and extensive biochemical analysis have indicated that DEAD-box proteins use cycles of ATP binding and hydrolysis to unwind RNA helices non-processively [15–19], and this unwinding is thought to underlie the ability of these proteins to function as RNA chaperones [13, 20].

A key method to probe native RNA folding is the catalytic activity assay [21–23]. In principle, this method can be applied to any RNA that possesses catalytic activity. Here we focus on ribozyme derivatives of group I introns, which lack exons and cleave an oligonucleotide substrate in a reaction that mimics the first step of the splicing reaction. The specific protocols shown were developed for use with the *Tetrahymena* ribozyme, which is derived from an intron found within the precursor rRNA that encodes the large ribosomal subunit RNAs. Folding of this ribozyme has been studied extensively by catalytic activity and other approaches [21, 24–29].

The operating principle of the catalytic activity assay is that only correctly folded, native RNA displays catalytic activity. Thus, the amount of native ribozyme can be determined from the amount of product produced rapidly when substrate is added in excess of the ribozyme. The "burst" of product formation is equal to the amount of native ribozyme if one of the products is released slowly, allowing the first round of cleavage to proceed to completion while preventing subsequent rounds. An example is shown in Fig. 1a, b for the *Tetrahymena* ribozyme. Upon cleavage of the substrate, the 3′ product is released fast while the 5′ product is released slowly because it forms a six base pair helix with the ribozyme that is additionally stabilized by tertiary contacts [30]. For some ribozyme constructs, both cleavage products are released slowly, allowing substrate cleavage and ligation to reach equilibrium [31]. In this case, the amount of product obtained in the burst is smaller than

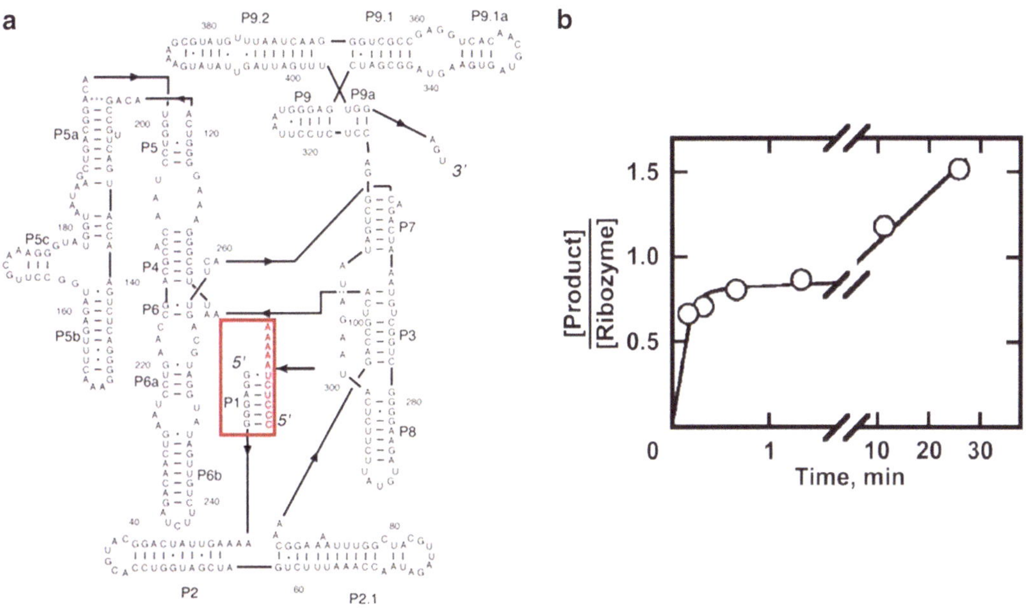

Fig. 1 *Tetrahymena* ribozyme secondary structure and oligonucleotide substrate cleavage reaction. (**a**) The standard "L-21" *Tetrahymena* ribozyme, with the oligonucleotide substrate highlighted in *red*. The *dark black arrow* to the right of the substrate indicates the cleavage site. The 3′ product (A_5) lacks base pairing with the ribozyme such that it dissociates fast after cleavage, while the 5′ product (CCCUCU) dissociates slowly, resulting in burst kinetics. (**b**) Cleavage reaction of pre-folded *Tetrahymena* ribozyme with excess substrate [22, 24]. The second slower phase is dominated by slow release of the 5′ product. It should be noted that an incomplete burst of 0.8–0.9 is characteristic of a homogenous population of native ribozyme, most likely due to a small fraction of damaged ribozyme which varies from preparation to preparation [24]. Reprinted from New pathways in folding of the *Tetrahymena* group I RNA enzyme, 291/5, Russell R. and Herschlag D., New pathways in folding of the *Tetrahymena* group I RNA enzyme, 1155–1167, Copyright (1999), with permission from Elsevier

the amount of native ribozyme, requiring a correction based on the equilibrium constant for cleavage and ligation [22].

Although the simplest experimental design for catalytic activity examines catalysis and folding concurrently (termed a continuous assay), a more complex two-stage or "discontinuous" assay can be useful for many applications including studying the effects of RNA chaperone proteins [22, 23]. As shown in Fig. 2a, the reaction is separated into two stages: folding and catalysis. By arresting folding at different times and performing cleavage reactions, the accumulation of native ribozyme can be observed as the RNA folds in the folding stage. As shown in Fig. 2b, c, plotting the burst amplitudes from individual cleavage reactions over folding time reveals the progress of native folding.

The discontinuous assay has been used to study the effects of CYT-19 on native folding of the *Tetrahymena* ribozyme when starting from a long-lived misfolded conformation [8, 12]. Conditions that give accumulation of misfolded ribozyme have

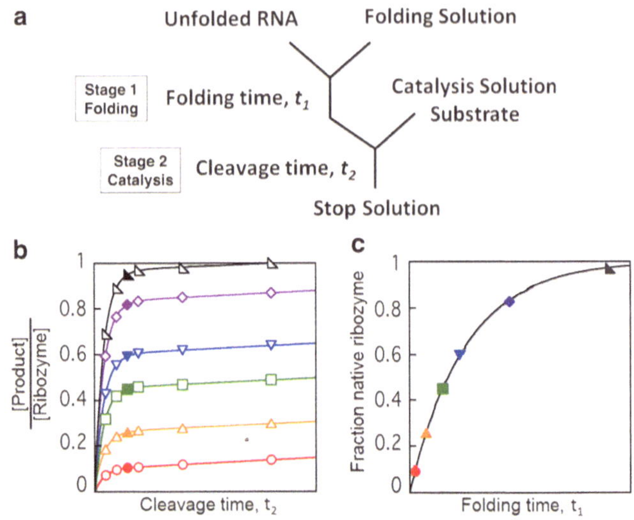

Fig. 2 Discontinuous catalytic activity assay. (**a**) Reaction scheme for the discontinuous assay. The assay is performed in two stages. Ribozyme folding occurs in the first stage and at different times (t_1) aliquots are transferred to the catalysis stage, where the fraction of native ribozyme is determined by cleavage of substrate over time (t_2). (**b**) Simulated data of cleavage reactions. The data follow substrate cleavage over time t_2. Each curve represents an aliquot removed at a different time t_1 during folding of the ribozyme, beginning with the *red curve* and ending with the *black curve* [23]. The solid curves simulate fits by an equation that includes an exponential term and a linear term [22]. The amplitudes of the exponential regions of the curves correspond approximately to the solid symbols. As indicated in the protocols and in the figure, at least six time points should be collected per cleavage reaction in initial experiments. However, after the substrate cleavage kinetics are well understood, it is possible to collect fewer time points. Even a single time point can be sufficient if it is collected at the time corresponding to the solid symbols. (**c**) The fraction of native ribozyme over folding time t_1 taken from burst amplitudes in panel (**b**). Fitting the folding progress to a single exponential, in this case, results in determination of a single rate constant for folding to the native state. Reprinted from Catalytic activity as a probe of native RNA folding, 468, Wan Y., Mitchell D., and Russell R., Catalytic activity as a probe of native RNA folding, 195–218, Copyright (2009), with permission from Elsevier

been previously established (10 min, 10 mM Mg^{2+}, 25 °C). In experiments starting from misfolded ribozyme and using CYT-19 in the folding stage at different concentrations, CYT-19 has been shown to increase the observed rate constant for folding to the native state. In this chapter, we present a protocol used to measure this folding transition from misfolded to native ribozyme and the effects of CYT-19 [12]. It should be noted that variations of this protocol have also been used to monitor effects of CYT-19 and related DEAD-box proteins on the native ribozyme and effects of DEAD-box proteins on folding when starting from unfolded

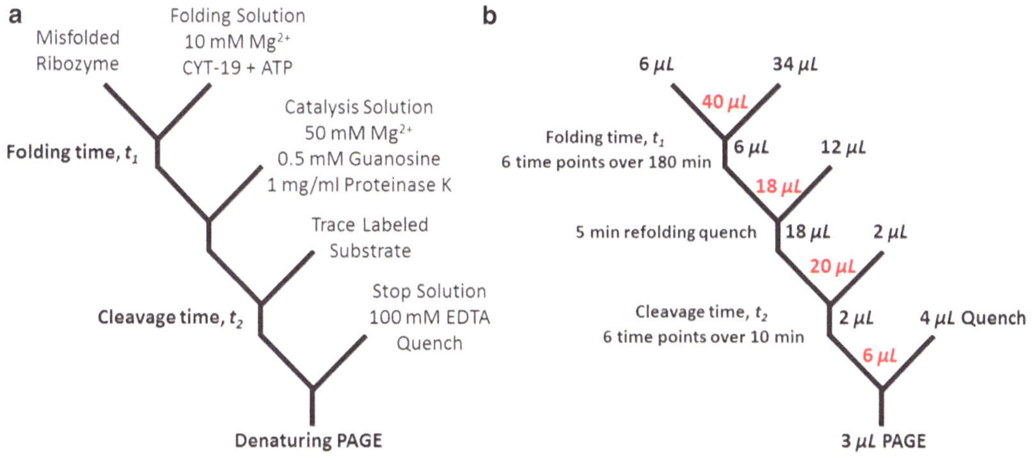

Fig. 3 Detailed reaction scheme for using the discontinuous assay to probe CYT-19 effect on folding. (**a**) Solution conditions for each stage of the assay. Ribozyme is pre-folded to a population of predominantly misfolded RNA and enters the folding stage with the addition of CYT-19 and ATP at 10 mM Mg^{2+}. Folding is quenched at different times t_1 (typically ranging from 5 to 180 min) by raising the magnesium concentration to 50 mM and adding proteinase K with a short incubation between stages (typically 5 min, but this time can vary, provided that it is short enough to prevent additional accumulation of native ribozyme). Trace-labeled substrate is added to initiate the reaction and the fraction of native ribozyme is determined by the amplitude of the "burst" of product formation during time t_2 (times for t_2 typically range from 15 s to 10 min). (**b**) Reaction volumes in this protocol. *Red* values indicate total reaction volumes after aliquots from each step are combined

conformations. It should also be noted that while this method has been used principally to monitor effects of DEAD-box proteins—i.e., ATP-dependent RNA helicases—that function as chaperones, it is expected to be equally applicable to other groups of RNA chaperone proteins [32, 33].

Figure 3a shows a detailed scheme for measurement of native ribozyme folding in the presence of CYT-19. A high concentration of magnesium (50 mM Mg^{2+}) is used as a folding quench because it dramatically slows refolding transitions from the misfolded conformation, effectively blocking further folding, and it promotes fast catalysis, allowing measurement of the fraction of native ribozyme at different times during the folding reaction [34]. Proteinase K can also be present in the catalysis solution to ensure that the DEAD-box chaperone protein does not remain active in the catalysis stage and perturb the reaction by promoting rapid exchange of the substrate. In Fig. 3b, we present typical reaction volumes, a typical number of time points for each stage, and total reaction volumes for this particular experiment. The oligonucleotide substrate is typically radio-labeled using T4 polynucleotide kinase (PNK) and [γ-^{32}P]-ATP, and time points from the cleavage reactions are analyzed using denaturing polyacrylamide gel electrophoresis (PAGE) to separate the substrate and product. Each cleavage reaction is expected to have a fast phase with a burst amplitude corresponding to the amount of native ribozyme and an observed rate constant for cleavage, and a slow phase limited by dissociation of one of the products from the ribozyme (the product that is released more slowly).

We present two protocols below that are used commonly in our laboratory. The first protocol describes measurement of the rate constant of ribozyme folding to the native state when the folding process is begun from unfolded ribozyme by adding Mg^{2+}, and the second protocol describes measurement of native folding starting from a population of predominantly misfolded ribozyme and including DEAD-box protein CYT-19 in the folding reaction.

2 Materials

To avoid RNA degradation, all solutions should be prepared using RNase-free water and conventional filter sterilization methods. Store all stock solutions at −20 °C (except the proteins, which should be stored at −80 °C or the temperature recommended by the manufacturer) and store all solutions on ice during the experiment. Folding and catalysis reactions are performed in siliconized or "slick" 500 μL microcentrifuge tubes (VWR), and solutions containing RNA are typically prepared in these tubes also. Ribozymes are transcribed using recombinant T7 RNA polymerase and run-off transcription, as defined by a restriction site, and purified using a QIAGEN RNeasy Mini or Midi kit as described [35].

2.1 Labeling 5′ Termini of Substrate Oligonucleotide

1. T4 polynucleotide kinase (PNK; 10 U/μL). Stored at −20 °C.
2. 5× T4 PNK buffer: 350 mM Tris–HCl, 25 mM $MgCl_2$, 12.5 mM dithiothreitol (DTT), pH 7.6. Stored at −20 °C.
3. [γ-^{32}P]-adenosine triphosphate ([γ-^{32}P]-ATP): 150 mCi/mL.
4. Oligonucleotide substrate: 5′ CCCUCUAAAAA 3′. Stored at −20 °C.
5. Tris–EDTA (TE) buffer: 1 mM EDTA and 10 mM Tris–HCl, pH 8.0.
6. Non-denaturing polyacrylamide gel: 20 % acrylamide/bisacrylamide (29:1), 1× TBE buffer. Stored at 4 °C.
7. X-ray film.
8. Film processor (e.g., KODAK X-omat 2000A).
9. Microcentrifuge.
10. Scintillation counter.

2.2 Catalytic Activity Solutions

1. Working stock solutions: 0.5 M Na-MOPS, pH 7.0, 10 μM *Tetrahymena* ribozyme, 50 mM $Mg(CH_3COO)_2$ (magnesium acetate), 0.5 M $MgCl_2$, 5 mM guanosine, 5 μM CYT-19, 20 mM ATP-Mg^{2+}.

2. Folding solution 1: 40 μL solution of 0.6 μM ribozyme, 10 mM $Mg(CH_3COO)_2$, 50 mM Na-MOPS, pH 7.0. Do not add 8 μL of 50 mM $Mg(CH_3COO)_2$ until ready to initiate the folding (*see* **Note 1**).
3. Catalysis solution 1: 80 μL solution of 78.3 mM $MgCl_2$, 833 μM guanosine, 50 mM Na-MOPS, pH 7.0 (*see* **Notes 2** and **3**).
4. Pre-folding solution: 20 μL solution of 0.6 μM ribozyme, 10 mM $Mg(CH_3COO)_2$, 50 mM Na-MOPS, pH 7.0. Do not add 4 μL of 50 mM $Mg(CH_3COO)_2$ until ready to initiate pre-folding (*see* **Note 1**).
5. 20 mM ATP-Mg^{2+}. Mix ATP with the same concentration of $Mg(CH_3COO)_2$ to achieve 20 mM ATP-Mg^{2+} (*see* **Note 4**).
6. Folding solution 2: 34 μL solution of 10 mM $Mg(CH_3COO)_2$, 0.25 μM CYT-19, 2 mM ATP-Mg^{2+}, 50 mM Na-MOPS, pH 7.0. Wait to add 2 μL of 5 μM CYT-19 until ready to initiate the reaction (15 s–1 min) (*see* **Note 5**).
7. Catalysis solution 2: 80 μL solution of 78.3 mM $MgCl_2$, 833 μM guanosine, 1 mg/mL proteinase K, 50 mM Na-MOPS, pH 7.0 (*see* **Note 6**).
8. Stop solution: 100 mM EDTA, pH 8.0, 0.04 % bromophenol blue (BPB), 0.04 % xylene cyanol, 72 % (v/v) formamide, pH 8.0.

2.3 Denaturing Polyacrylamide Gel Electrophoresis (PAGE)

1. Running gel buffer (10× TBE): 890 mM Tris-base, 890 mM boric acid, 20 mM EDTA, pH 8.3.
2. Denaturing polyacrylamide mix: 20 % acrylamide/bisacrylamide (29:1), 7 M urea, 1× TBE buffer (*see* **Note** 7).
3. Two pairs of glass plates 20 cm × 20 cm × 4 mm with 0.35 mm thick spacers for each pair of plates and a comb that gives 5 μL wells with at least 20 wells per comb.
4. Two gel boxes for running gel electrophoresis and a power supply capable of generating 16 W per gel.
5. Whatman 3MM filter paper.
6. Gel dryer.

2.4 Autoradiography Equipment and Analysis Tools

1. Storm 820 phosphorimager (Amersham Pharmacia Biotech, Inc.—Molecular Dynamics Div.) or comparable equipment.
2. Phosphorimager exposure cassette and screen (Amersham Pharmacia Biotech, Inc.—Molecular Dynamics Div.) or comparable product.
3. ImageQuant TL v7.2 software (Molecular Dynamics) or comparable software: Used for densitometry analysis to determine fraction product produced over time.

3 Methods

We present typical reaction volumes for this assay in Fig. 3b and in the protocol, but reaction volumes can vary on an experiment by experiment basis (*see* **Note 8**). A minimum of six time points should be collected from each folding and cleavage reaction (but *see* Fig. 2 legend). Typical time points for cleavage reactions are 2 μL, such that the volume of solution in the catalysis stage must be at least 12 μL (typically 20 μL). Similarly, the folding stage requires enough volume to accommodate folding stage time points, with each folding time point requiring 6 μL.

3.1 Labeling 5′ Termini of Oligonucleotides with ^{32}P

1. Prepare the kinase reaction solution by mixing 1 μL of 50 μM oligonucleotide substrate, 1 μL of 5× PNK buffer, 1 μL of [γ-^{32}P]-ATP (150 mCi/mL), and 1 μL of T4 PNK (10 U/μL). Note that this protocol produces a 4 μL reaction with a final buffer concentration of 1.25×.
2. Incubate the 4 μL reaction at 37 °C for 1 h in a water bath.
3. Prepare the 20 % non-denaturing gel for isolation of the 5′ labeled substrate.
4. Add 6 μL of stop solution to halt the labeling reaction.
5. Load 5 μL each into two adjacent wells on the gel and run the gel at 12 W for 60 min to ensure complete separation of substrate from unreacted [γ-^{32}P]-ATP.
6. Separate the gel plates to expose the gel and cover the gel face with a layer of plastic wrap. Carefully press X-ray film sealed in envelope lightly to the plastic wrap for about 10–15 s while marking the gel and film for alignment by poking at least three holes with a needle. Develop the film in the film processor.
7. Align the exposed film with the gel and cut out the gel band of the labeled substrate using a flame-sterilized razor blade.
8. Elute the substrate from the gel piece by submerging the gel piece in ~150 μL of TE buffer overnight at 4 °C. The next morning, pellet the gel fragments by centrifugation using the microcentrifuge and transfer the supernatant containing the eluted oligonucleotide into a new tube. The specific activity of the labeled oligonucleotide is conveniently measured using Cerenkov detection by transferring 1 μL to a scintillation vial and detecting radioactive decay in the scintillation counter without adding scintillant. The efficiency of detection for ^{32}P disintegrations is roughly 30 %. The labeled oligonucleotide should be stored at −20 °C and may be used for several weeks before it decays to background levels ($t_{1/2} \approx 2$ weeks).

3.2 Rate of Folding to the Native State of Unfolded Tetrahymena Ribozyme

1. Aliquot 4 μL of stop solution into thirty-six tubes for six cleavage reactions with six time points each.
2. Heat folding solution 1 (without Mg^{2+}) at 95 °C for 2 min and then place it on ice for 2 min. This step improves reproducibility by ensuring that the ribozyme molecules begin folding from the same ensemble of unfolded states in all experiments (*see* **Note 1**).
3. Aliquot 12 μL of catalysis solution 1 into six tubes. These will become the catalysis reactions when aliquots of the folding reaction are added to them in **step 5** below.
4. Place the folding solution at the desired temperature for folding (typically 25 or 37 °C), allow at least one min for the temperature to equilibrate, and then add 8 μL of 50 mM $Mg(CH_3COO)_2$ (also equilibrated at the same temperature) to initiate folding ($t_1 = 0$) (*see* **Note 9**).
5. After 15 s, remove 6 μL of folding solution 1 and add it to one of the tubes prepared above in **step 3** (containing 12 μL of catalysis solution 1) at 25 °C to quench folding (*see* **Note 10**).
6. Incubate the quenched folding reaction from **step 5** for 5 min to allow formation of the native and long-lived misfolded conformations. Note that this protocol only measures the folding progress from the misfolded conformation to the native conformation. To follow the initial phase of folding of this ribozyme, which gives a small fraction of native ribozyme, the folding quench and substrate cleavage reaction are performed the same way but at 0 °C [36].
7. Repeat **steps 5** and **6** for all six values of t_1 to probe the fraction of native ribozyme at each time point as the RNA folds to the native state. Common times of t_1 are 15 s, 1 min, 10 min, 30 min, 90 min, and 180 min (*see* **Note 10**).
8. Approximately 5 min after quenching folding in **step 5**, add 2 μL of excess substrate (twofold excess over the ribozyme concentration) mixed with trace labeled substrate (~20,000 dpm) to initiate the cleavage reaction ($t_2 = 0$). After 15 s, remove 2 μL of the cleavage reaction and add it to the previously aliquoted 4 μL of stop solution (**step 1**) for the 15 s time point (*see* **Note 11**).
9. Repeat **step 8** for all six values of t_2 to probe the fraction of product cleaved at each time point as the substrate is converted into product. Common times of t_2 are 15 s, 30 s, 1 min, 2 min, 4 min, and 10 min (*see* **Note 12**).
10. Prepare the 20 % denaturing polyacrylamide gel by mixing 300 μL of 10 % APS and 30 μL of TEMED with 30 mL of

denaturing polyacrylamide mix. Pour about 15 mL of polymerizing gel into the pre-assembled glass plates and set a comb in the gel with at least twenty 5 μL wells.

11. To facilitate quantitation of individual lanes, load alternate wells (wells 1, 3, 5, etc.) with 3 μL of quenched cleavage reaction from **step 9**. Run the gel for about 20 min at 16 W per gel to allow sample to enter the gel.
12. Once this time has elapsed, add 3 μL of the remaining time points into the remaining unused wells staggered in the same way (wells 2, 4, 6, etc.). Run the gel for 45 min at 16 W per gel to separate the shorter cleaved product from the longer substrate.
13. Separate the plates by leaving the gel attached to one side and cover the gel with plastic wrap before lifting it from the remaining plate. Place the gel on Whatman 3MM filter paper and dry it using the gel dryer. Expose the gel overnight to a phosphor screen and quantify it using a phosphorimager and ImageQuant or comparable software to determine the fraction of product at each time point for each substrate cleavage reaction.
14. Fit the fraction of material present as product against t_2 to an equation that includes an exponential "burst" term and a linear phase for slower subsequent turnovers [22]. Plot the burst amplitude of each cleavage time course against the folding time t_1 and fit the curve to a single exponential equation to determine the rate constant for native folding.

3.3 Rate of CYT-19-Mediated Refolding of Misfolded *Tetrahymena* Ribozyme

1. As in Subheading 3.2, aliquot 4 μL of stop solution into 36 tubes for six reactions with six time points each.
2. Initiate pre-folding of the ribozyme at 25 °C by adding 4 μL of 50 mM $Mg(CH_3COO)_2$ to the pre-folding solution and allow the RNA to fold for approximately 10 min.
3. While waiting, aliquot 12 μL of catalysis solution 2 to six tubes. These will become the catalysis reactions when aliquots of the folding reaction are added to them in **step 5** below. Immediately prior to **step 4** (15 s–1 min), complete the preparation of folding solution 2 by adding 2 μL of 5 μM CYT-19 (*see* **Note 5**).
4. When pre-folding of the ribozyme is complete, remove 6 μL of pre-folding solution and mix it with 34 μL of folding solution 2 at 25 °C to start the CYT-19-mediated folding reaction (*see* **Note 13**).
5. Repeat **steps 5–14** from Subheading 3.2 using the CYT-19-mediated folding reaction generated in **step 4** and catalysis solution 2 to determine the rate of native state formation.

4 Notes

1. To start from a homogeneously unfolded population, the ribozyme mixed with MOPS buffer may first be heated at 95 °C for 2 min and then incubated on ice for 2 min. When all solutions and tubes are prepared for the time course, Mg^{2+} is added to initiate folding. For the *Tetrahymena* ribozyme, this 95 °C step was found to give no detectable effect on folding (R.R., unpublished) and is now routinely omitted. However, when testing a new ribozyme it is necessary to compare the folding properties with and without this step.
2. Because aliquots from the folding reaction contain magnesium and are diluted during the experiment, concentrations of magnesium and guanosine must be adjusted to achieve final desired concentrations.
3. Guanosine has a relatively low solubility. To maintain solubility of the stock solution of guanosine, heat it at 50 °C until it is fully dissolved before adding it to catalysis solutions.
4. The negatively charged triphosphate moiety of ATP binds magnesium ions. To keep the concentration of magnesium unchanged for RNA folding reactions, a stock solution of equimolar amount of magnesium and ATP is necessary.
5. To prevent loss of activity as a result of extended and premature interaction of protein with folding solution components, add CYT-19 immediately before starting the reaction (15 s–1 min).
6. With the solution conditions described and standard concentrations of CYT-19 (<1 μM), proteinase K is not necessary because the high $MgCl_2$ concentration in stage 2 is sufficient to block activity of CYT-19. However, proteinase K addition may be necessary with different proteins or higher CYT-19 concentrations.
7. 7 M urea is used to ensure complete RNA denaturation but is near the solubility limit. As such, the 1 L solution should be periodically mixed over 5–10 min at 37 °C to keep the urea in solution.
8. To determine the necessary reaction volumes, first determine how much catalysis reaction solution is needed in each cleavage reaction. Six time points of 2 μL each implies that at least 12 μL of volume is required in the catalysis stage (reaction volumes of 20 μL are typically used). Next, the volume needed in the folding reaction should be calculated. Aliquots of 6 μL of the folding reaction are typically added to 12 μL of folding quench solution (and 2 μL of substrate solution) for each folding time point

(Fig. 3b). Thus, generating data at six different folding times will require at least 36 μL in the folding reaction.

9. Mixing the solutions well and rapidly is crucial. After each addition be sure to quickly flick the bottom of the tube gently or pipet several times in rapid succession to mix the solution.
10. Once the folding reaction starts, the fraction of native ribozyme needs to be determined at several different times during folding (i.e., several values of t_1). Keep in mind while performing cleavage reactions from the first few folding time points that additional aliquots of the folding reaction may need to be removed, while these cleavage reactions are still in progress (**step 5** in Subheading 3.2.).
11. Excess substrate relative to the ribozyme should be used to ensure that burst amplitudes are stoichiometrically correlated with the fraction of native ribozyme, i.e., twofold excess substrate over the ribozyme results in a burst amplitude of 50 % under conditions such that 100 % of the ribozyme is present in the native state. However, if the substrate binds with comparable rates to both misfolded and native RNA and remains bound on a time scale that is longer than the first turnover of substrate cleavage by the native ribozyme, then trace amounts of labeled substrate may be used. This alternative protocol can increase reproducibility of the burst amplitudes because the burst size is independent of the ratio of substrate to ribozyme in this regime.
12. Once the time course for folding and cleavage reactions are complete, quenched time points can be stored at 4 °C for up to several days before running gel electrophoresis if more time is needed.
13. After pre-folding for 10 min at 25 °C, misfolded *Tetrahymena* ribozyme is stable with a half-life on the order of hours. Because of this long lifetime, the folding reaction does not have to be initiated immediately at 10 min as long as the actual pre-folding time is recorded.

References

1. Herschlag D (1995) RNA chaperones and the RNA folding problem. J Biol Chem 270: 20871–20874
2. Russell R (2008) RNA misfolding and the action of chaperones. Front Biosci 13:1–20
3. Treiber DK, Williamson JR (1999) Exposing the kinetic traps in RNA folding. Curr Opin Struct Biol 9:339–345
4. Jarmoskaite I, Russell R (2011) DEAD-box proteins as RNA helicases and chaperones. Wiley Interdiscip Rev RNA 2:135–152
5. Linder P, Jankowsky E (2011) From unwinding to clamping - the DEAD box RNA helicase family. Nat Rev Mol Cell Biol 12: 505–516
6. Mohr S, Stryker JM, Lambowitz AM (2002) A DEAD-box protein functions as an ATP-dependent RNA chaperone in group I intron splicing. Cell 109:769–779
7. Huang HR, Rowe CE, Mohr S et al (2005) The splicing of yeast mitochondrial group I and group II introns requires a DEAD-box

protein with RNA chaperone function. Proc Natl Acad Sci USA 102:163–168
8. Tijerina P, Bhaskaran H, Russell R (2006) Nonspecific binding to structured RNA and preferential unwinding of an exposed helix by the CYT-19 protein, a DEAD-box RNA chaperone. Proc Natl Acad Sci USA 103: 16698–16703
9. Solem A, Zingler N, Pyle AM (2006) A DEAD protein that activates intron self-splicing without unwinding RNA. Mol Cell 24:611–617
10. Halls C, Mohr S, Del Campo M et al (2007) Involvement of DEAD-box proteins in group I and group II intron splicing. Biochemical characterization of Mss116p, ATP hydrolysis-dependent and -independent mechanisms, and general RNA chaperone activity. J Mol Biol 365:835–855
11. Mohr S, Matsuura M, Perlman PS et al (2006) A DEAD-box protein alone promotes group II intron splicing and reverse splicing by acting as an RNA chaperone. Proc Natl Acad Sci USA 103:3569–3574
12. Bhaskaran H, Russell R (2007) Kinetic redistribution of native and misfolded RNAs by a DEAD-box chaperone. Nature 449:1014–1018
13. Potratz JP, Del Campo M, Wolf RZ et al (2011) ATP-dependent roles of the DEAD-box protein Mss116p in group II intron splicing *in vitro* and *in vivo*. J Mol Biol 411: 661–679
14. Del Campo M, Tijerina P, Bhaskaran H et al (2007) Do DEAD-box proteins promote group II intron splicing without unwinding RNA? Mol Cell 28:159–166
15. Del Campo M, Mohr S, Jiang Y et al (2009) Unwinding by local strand separation is critical for the function of DEAD-box proteins as RNA chaperones. J Mol Biol 389:674–693
16. Chen Y, Potratz JP, Tijerina P et al (2008) DEAD-box proteins can completely separate an RNA duplex using a single ATP. Proc Natl Acad Sci USA 105:20203–20208
17. Sengoku T, Nureki O, Nakamura A et al (2006) Structural basis for RNA unwinding by the DEAD-box protein Drosophila *Vasa*. Cell 125:287–300
18. Del Campo M, Lambowitz AM (2009) Structure of the yeast DEAD box protein Mss116p reveals two wedges that crimp RNA. Mol Cell 35:598–609
19. Yang Q, Del Campo M, Lambowitz AM et al (2007) DEAD-box proteins unwind duplexes by local strand separation. Mol Cell 28: 253–263
20. Russell R, Jarmoskaite I, Lambowitz AM (2012) Toward a molecular understanding of RNA remodeling by DEAD-box proteins. RNA Biol 10:44–55
21. Russell R, Herschlag D (2001) Probing the folding landscape of the *Tetrahymena* ribozyme: commitment to form the native conformation is late in the folding pathway. J Mol Biol 308:839–851
22. Wan Y, Mitchell D 3rd, Russell R (2009) Catalytic activity as a probe of native RNA folding. Methods Enzymol 468:195–218
23. Potratz JP, Russell R (2012) RNA catalysis as a probe for chaperone activity of DEAD-box helicases. Methods Enzymol 511:111–130
24. Russell R, Herschlag D (1999) New pathways in folding of the *Tetrahymena* group I RNA enzyme. J Mol Biol 291:1155–1167
25. Latham JA, Cech TR (1989) Defining the inside and outside of a catalytic RNA molecule. Science 245:276–282
26. Sclavi B, Sullivan M, Chance MR et al (1998) RNA folding at millisecond intervals by synchrotron hydroxyl radical footprinting. Science 279:1940–1943
27. Zarrinkar PP, Williamson JR (1994) Kinetic intermediates in RNA folding. Science 265: 918–924
28. Zhuang X, Bartley LE, Babcock HP et al (2000) A single-molecule study of RNA catalysis and folding. Science 288:2048–2051
29. Russell R, Millett IS, Tate MW et al (2002) Rapid compaction during RNA folding. Proc Natl Acad Sci USA 99:4266–4271
30. Narlikar GJ, Gopalakrishnan V, McConnell TS et al (1995) Use of binding energy by an RNA enzyme for catalysis by positioning and substrate destabilization. Proc Natl Acad Sci USA 92:3668–3672
31. Sinan S, Yuan X, Russell R (2011) The *Azoarcus* group I intron ribozyme misfolds and is accelerated for refolding by ATP-dependent RNA chaperone proteins. J Biol Chem 286:37304–37312
32. Schroeder R, Barta A, Semrad K (2004) Strategies for RNA folding and assembly. Nat Rev Mol Cell Biol 5:908–919
33. Rajkowitsch L, Chen D, Stampfl S et al (2007) RNA chaperones, RNA annealers and RNA helicases. RNA Biol 4:118–130
34. Russell R, Das R, Suh H et al (2006) The paradoxical behavior of a highly structured misfolded intermediate in RNA folding. J Mol Biol 363:531–544
35. Russell R, Herschlag D (1999) Specificity from steric restrictions in the guanosine binding pocket of a group I ribozyme. RNA 5: 158–166
36. Russell R, Tijerina P, Chadee AB et al (2007) Deletion of the P5abc peripheral element accelerates early and late folding steps of the *Tetrahymena* group I ribozyme. Biochemistry 46:4951–4961

Chapter 14

Evaluation of RNA Chaperone Activity In Vivo and In Vitro Using Misfolded Group I Ribozymes

Katharina Semrad

Abstract

Here two methods to measure RNA chaperone activity in vivo and in vitro are described. In both assays folding of a misfolded group I intron RNA into the splicing-competent form in the presence of proteins with RNA chaperone activity is measured. Folding is evaluated indirectly by assessing the difference in splicing activity with or without proteins with RNA chaperone activity. In vitro, we use the *thymidylate synthase* (*td*) group I intron of phage T4 that is split into two halves. As a consequence this split ribozyme is only capable to fold and in turn to splice *in trans* at elevated temperatures. Proteins with RNA chaperone activity enable splicing of the split intron at lower temperatures. This difference in splicing activity is measured to assess the efficacy of the RNA chaperone. In vivo, a mutant variant of the *td* group I intron is trapped in a misfolded conformation, resulting in a reduced splicing activity. Over-expression of proteins with RNA chaperone activity results in an increase in splicing in vivo, as these proteins resolve the misfolded conformation.

Key words RNA chaperone activity, Group I intron, Splicing, Ribozyme, In vivo folding, In vitro folding, *Trans*-splicing

1 Introduction

RNA chaperone activity is the ability of certain proteins to enhance RNA folding in an ATP-independent manner and without actively stabilizing or specific binding to RNA structures [1]. RNA chaperone activity is highly likely to play an important role within cellular processes where RNA folding is required [2]. However, it is challenging to demonstrate RNA chaperone activity in its natural environment due to the fact that the majority of proteins that were shown to possess RNA chaperone activity are involved in multiple processes other than RNA folding. Therefore, a simple knock-out experiment of the protein of interest results in pleiotropic effects. To circumvent this dilemma, RNA chaperone activity is typically measured using nonnative substrates [3].

Christina Waldsich (ed.), *RNA Folding: Methods and Protocols*, Methods in Molecular Biology, vol. 1086, DOI 10.1007/978-1-62703-667-2_14, © Springer Science+Business Media, LLC 2014

Group I introns are large RNA molecules typically sized between 200 nt up to 1,000 nt depending if their open reading frame that encodes for an endonuclease is present or not. Group I introns are capable of catalyzing their excision from flanking exon sequences via two consecutive transesterification reactions. The catalysis of the self-splicing reaction, however, requires correct folding of the ribozyme. In vivo, efficient folding of the group I intron ribozymes is usually guaranteed, in vitro transcribed group I intron ribozymes usually splice significantly slower, suggesting that a subpopulation of ribozymes is detrimentally misfolded [4]. The presence of proteins with RNA chaperone activity significantly increases the population of properly folded RNA and as a consequence splicing occurs (Figs. 1 and 3).

In both assays, the in vitro *trans*-splicing assay and the in vivo folding trap assay, group I intron folding to the native, active state is monitored indirectly by measuring the ability of the ribozyme to splice which is a consequence of productive folding of the ribozyme. In the in vitro *trans*-splicing assay the *td* gene including the ribozyme is split into two halves: H1 (680 nt), consists of 549 nt of exon 1 (E1) and 131 nt of the 5′ part of the intron (I1), while H2 (170 nt) is composed of 147 nt of the 3′ part of the intron (I2; without the internally encoded endonuclease) and 23 nt of exon 2 (E2) [5, 6]. At 55 °C the two parts of the *thymidylate synthase* pre-RNA are able to fold efficiently into the catalytically active fold and splice *in trans*. At lower temperatures (i.e., 37 °C), however, splicing does not occur (Fig. 1). The addition of proteins with RNA chaperone activity can alleviate this effect at low temperatures by resolving misfolded elements and splicing occurs. The in vitro transcribed and internally radioactively labeled RNA transcripts H1 and H2 are folded at different temperatures in the absence and in the presence of proteins with RNA chaperone activity; the difference in catalytic activity of the split ribozyme is then evaluated in the *trans*-splicing assay. The products of the *trans*-splicing assay are finally separated on a standard denaturing PAGE and evaluated using a program like ImageQuant (Fig. 2).

The in vivo folding trap assay is based on measuring the splicing efficiency of the *td* intron in *Escherichia coli* strains [7–9]. These *E. coli* strains contain a multicopy pUC-derived vector that encodes the *thymidylate synthase* gene of phage T4. Two variations of this gene are required for the in vivo folding trap assay: first, the wild-type *td* gene (pTZ18U-*td*Δ*P6-2; td*wt) with a deletion of the open reading frame for the intron-encoded endonuclease to avoid toxicity; second the mutant *td* construct *td*SH1. Compared to the *td*wt, this construct carries two point mutations, an exonic mutation 82 nucleotides upstream of the 5′ splice-site (U-82A) and the intronic mutation C865U. The exonic mutation results in a stop codon that enables nonnative base pairing of exon and intron sequences, thereby abolishing splicing (Fig. 3). The intronic mutation

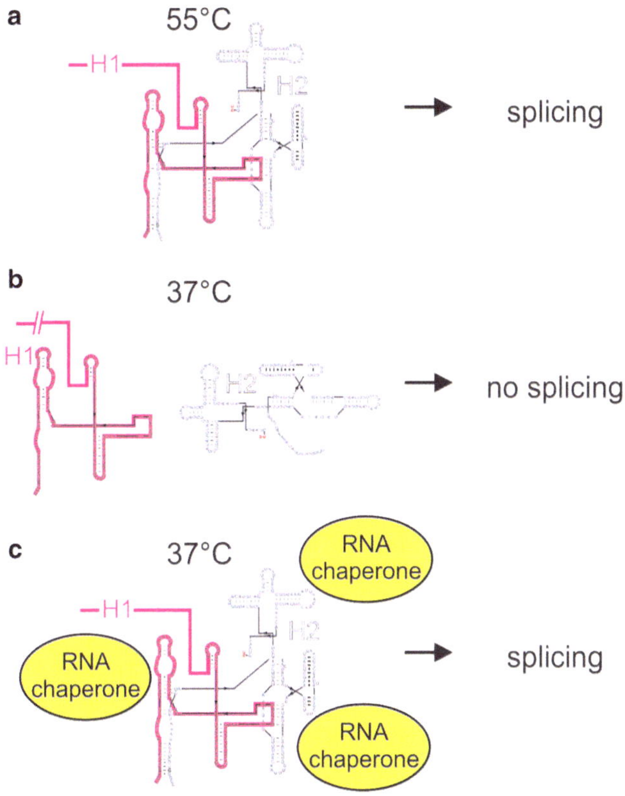

Fig. 1 The principle of the *trans*-splicing assay. (**a**) The two in vitro transcribed RNA constructs H1 (consisting of 549 nt of exon 1 of the *thymidylate synthase* gene of phage T4 and 131 nt of the 5′ part of the intron) and H2 (consisting of 147 nt of the 3′ part of the intron and 23 nt of exon 2) fold into the native, splicing-competent structure at 55 °C and splicing occurs. (**b**) H1 and H2 cannot assemble into the splicing-competent conformation due to misfolding at 37 °C. As a consequence splicing cannot occur. (**c**) At low temperature (37 °C) RNA chaperones are able to facilitate correct folding of H1 and H2 and in turn these proteins promote efficient *trans*-splicing

C865U results in destabilization of the tertiary structure [10]. Both mutations lead to splicing deficiency in the *td*SH1 mutant due to incapability of efficient, correct folding. This defect in folding can be rescued by proteins with RNA chaperone activity (Fig. 3). Thus, in addition to the either wild-type or mutant intron variant (*td*wt, *td*SH1), the *E. coli* cells are co-transformed with a low copy vector: a pSU20 derivative that encodes the protein with (putative) RNA chaperone activity or as empty vector. Both the proteins of interest and the *td* gene are cloned under control of an inducible lac promoter. At early log phase the cells are harvested and total cellular RNA is extracted from the following *E. coli* strains: (1) *td*wt + pSU20-empty, (2) *td*SH1 + pSU20-empty, and

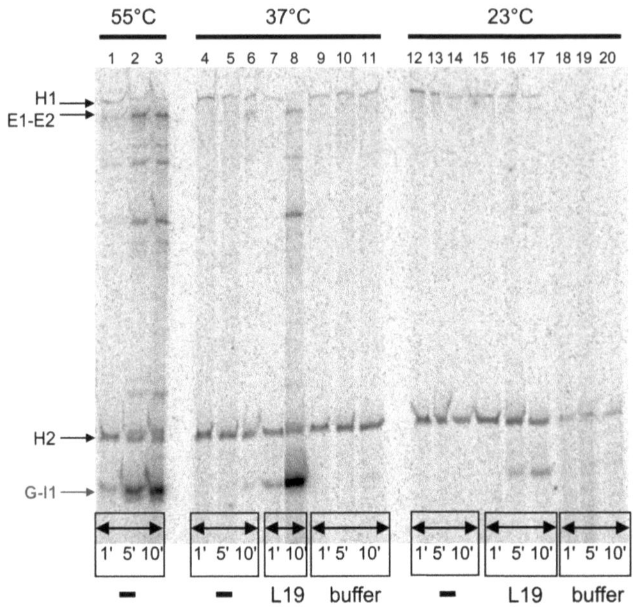

Fig. 2 The *E. coli* ribosomal protein L19 facilitates *trans*-splicing. The first time course shows the positive control, a *trans*-splicing time course at 55 °C in the absence of protein (*lanes 1–3*). Splicing occurs, as indicated by the product of the first step of splicing (guanosine cofactor is covalently linked to the 5′ half of the intron; G-I1). The negative controls show that splicing is not detected at low temperatures in the absence of protein (*lanes 4–6* and *12–14*). However, in the presence of 2 μM L19 splicing occurs efficiently both at 37 and 23 °C, (*lanes 7–8* and *15–17*, respectively), as indicated by the appearance of the G-I1 product. The buffer control reveals that the protein storage buffer alone does not stimulate splicing at low temperatures (*lanes 9–11* and *18–20*). E1-E2 refers to the product of the ligated exons. The product of the first step of splicing G-I1 shows a higher radioactive signal due to the isotope ^{32}P at α- or γ-phosphate of GTP, while H1 and H2 were body-labeled using [α-^{35}S]-UTP during in vitro transcription

(3) *td*SH1 + pSU20-putative-RNA-chaperone (Fig. 3). From our experience, the presence of an RNA chaperone usually does not enhance splicing of wt *td* intron. Therefore, it is not necessary to test a strain containing *td*wt + pSU20-known-RNA-chaperone. The significantly reduced splicing of the *td*SH1, however, can be rescued in the presence of over-expressed proteins with RNA chaperone activity (Figs. 3 and 4b). The total RNA of each extracted *E. coli* culture contains only a small percentage of the *thymidylate synthase* gene, which can be present in three forms (Fig. 4a): the pre-RNA, the spliced mRNA, and a cryptic spliced RNA (the function of the alternative splice product is yet unknown). Using the reverse transcription-based poisoned primer assay the amount of unspliced pre-RNA and spliced mRNA is specifically determined. Most reverse transcriptases tend to fall off their RNA templates

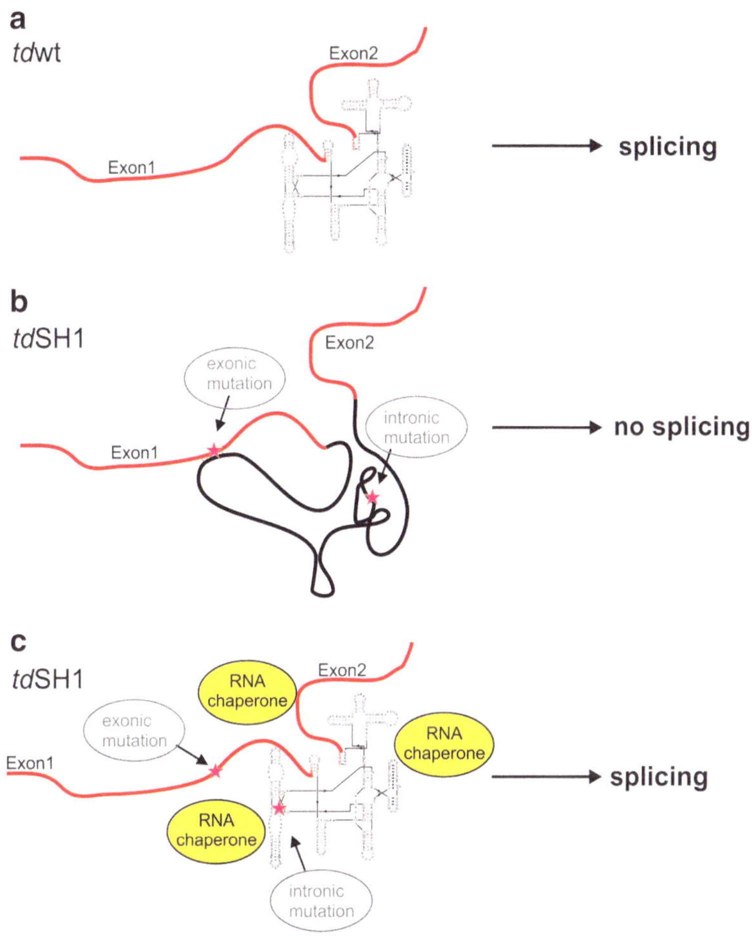

Fig. 3 The principle of the in vivo folding trap assay. (**a**) The wild-type *thymidylate synthase* pre-RNA (*td*wt) splices efficiently in vivo, as the intron folds into the catalytically active structure. (**b**) The mutant *thymidylate synthase* pre-RNA (*td*SH1) misfolds by forming an alternative pairing of exon and intron sequences (due to the exonic mutation U-82A) and is structurally destabilized (due to the intronic mutation C865U). As a consequence, splicing of this mutant pre-RNA is significantly reduced. (**c**) Proteins with RNA chaperone activity rescue group I intron folding and as a result splicing occurs

due to secondary structure elements and G/C-rich sequence stretches in the template and this tendency increases the longer the template RNA. To circumvent this problem and to ensure that the reverse transcription reaction also quantitatively reflects the actual percentages of pre-RNA to mRNA, a modified reverse transcription assay is used (Fig. 4a). In the poisoned primer assay very short sections of the whole template are only reverse transcribed: The radioactively end-labeled oligonucleotide hybridizes to the 5′ end of exon 2, only 2 nucleotides downstream of the 3′ splice-site. The nucleotide mix added to the reverse transcription reaction includes

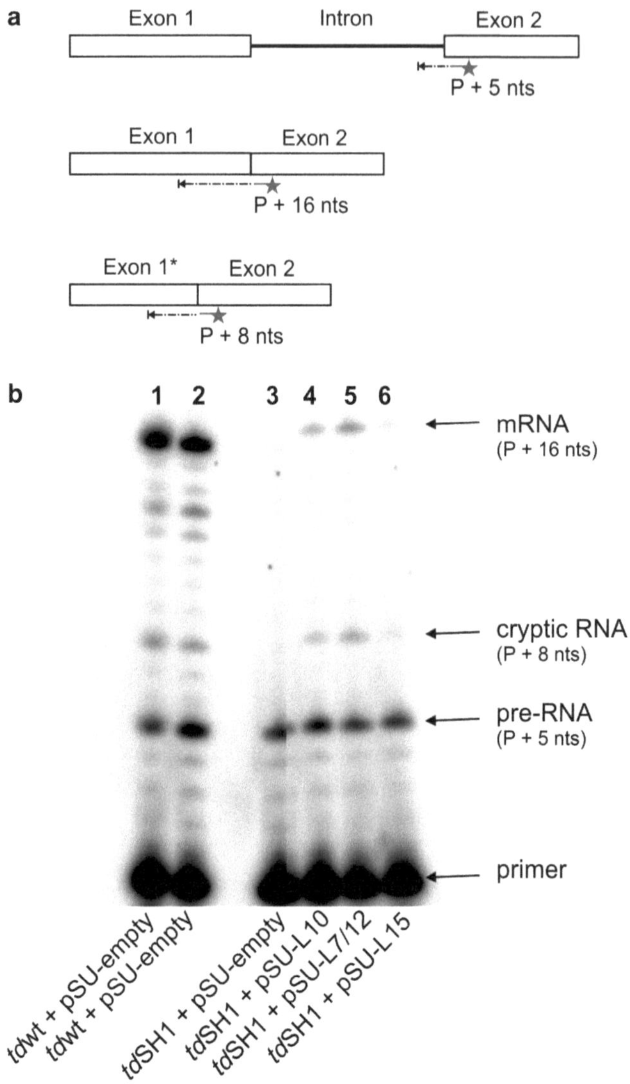

Fig. 4 The RNA chaperone activity of *E. coli* ribosomal proteins L10 and L7/12 monitored using the in vivo folding trap assay. (**a**) Schematic representation of the reverse transcription-based poisoned primer assay. Due to the addition of ddTTP in the reaction the reverse transcriptase stops the extension of the radioactively labeled primer (P) at the first A in the template RNA. In the case of the pre-RNA the first A appears after five nucleotides, resulting in a cDNA fragment of 23 nt (P + 5 nt). In the case of the spliced mRNA the reverse transcript is 34 nt long (P + 16 nt) and for the cryptic spliced mRNA a 26 nt cDNA (P + 8 nt) is received. (**b**) A representative gel of a poisoned primer reaction is shown: the reverse transcription products represent the amount of pre-RNA, of cryptic spliced RNA or of mRNA. *Lanes 1* and *2* show efficient splicing the wild-type *td* construct (*td*wt) in vivo. In contrast, splicing of the mutant *td* construct (*td*SH1) is virtually abolished, as only pre-RNA is detectable (*lane 3*). In the presence of the *E. coli* ribosomal proteins L10, L7/12 splicing of the mutant *td*SH1 is enhanced (*lanes 4* and *5*), while the ribosomal protein L15 does not significantly stimulate *td* intron splicing (*lane 6*)

dATP (2′deoxy-adenine-triphosphate), dCTP, dGTP, and only ddTTP (2′,3′dideoxy-thymidine-triphosphate). This results in a stop of the reverse transcription at the first adenine (A) within the template sequence because ddTMP is incorporated into the nascent cDNA. In the case of the pre-RNA the first A occurs after an elongation of the primer for only 5 nt. In the case of the spliced mRNA the first A occurs after 16 nt and within the cryptic spliced construct the first A is present after 8 nt of extension. These short cDNA fragments are then separated on a standard denaturing PAGE and splicing is analyzed using a software, like Image Quant (Fig. 4b). Upon comparing the splicing activity of the mutant intron *td*SH1 in the absence and presence of the protein of interest to the splicing efficiency of the *td*wt, the potential RNA chaperone activity of that protein can be assessed in vivo (Fig. 4).

2 Materials

Chemically pure water is used to prepare all solutions (except for the bacterial growth medium). DEPC-treated water is not used in our laboratory, because of the toxicity of DEPC (diethylpyrocarbonate, which inactivates any enzyme and RNases in particular). The buffers are usually stored in small aliquots at −20 °C and can be thawed when needed. Multiple cycles of freezing and thawing do not cause any problems. Solutions are sterilized and stored at room temperature unless indicated otherwise.

Materials from Subheadings 2.1 to 2.4 refer to the supplies required for the in vitro *trans*-splicing assay, whereas materials from Subheadings 2.5 to 2.8 refer to those required for the in vivo folding trap assay.

2.1 Linearization of Plasmid Templates

Plasmids encoding H1 and H2 are pTZ18U-*td*Δ1-3 derivatives (Amp^R). H1 is linearized with *Sal*I and H2 is linearized with *Bam*HI, then both are transcribed.

1. Plasmids pTZ18U-*td*Δ1-3-H1 and pTZ18U-*td*Δ1-3-H2.
2. Restriction enzyme *Sal*I (20 U/μL); 10× *Sal*I buffer. Store at −20 °C.
3. Restriction enzyme *Bam*HI (20 U/μL); 10× *Bam*HI buffer. Store at −20 °C.
4. TE-saturated phenol (store at 4 °C) and chloroform.
5. 3 M NaOAc, pH 8.0.
6. 100 % ethanol.
7. 70 % ethanol.

2.2 In Vitro Transcription Using T7 RNA Polymerase

1. 10× transcription buffer: 400 mM Tris–HCl, pH 6.9, 260 mM $MgCl_2$, 30 mM spermidine. Store at −20 °C.
2. 100 mM ATP, 100 mM GTP, 100 mM CTP, 100 mM UTP. Store at −20 °C.
3. [α-^{35}S]-UTP (12.5 µCi/µL, 1,250 Ci/mmol). Store at 4 °C.
4. 1 M DTT. Freeze in aliquots at −20 °C.
5. T7 RNA polymerase (20 U/µL). Store at −20 °C.
6. RNase-free DNase I (1 U/µL). Store at −20 °C.

2.3 Denaturing PAGE

1. 20× TBE: 1,780 mM Tris-base, 1,780 mM boric acid, 40 mM EDTA, pH 8.0.
2. 8 % acrylamide solution (for gel-purification): 7 M urea, 8 % acrylamide/bisacrylamide (19:1), 1× TBE. Store at 4 °C.
3. 5 % acrylamide solution (for the *trans*-splicing assay): 7 M urea, 5 % acrylamide/bisacrylamide (19:1), 1× TBE. Store at 4 °C.
4. 10 % acrylamide solution (for the poisoned primer assay): 7 M urea, 10 % acrylamide/bisacrylamide (19:1), 1× TBE. Store at 4 °C.
5. 10 % (w/v) APS (ammonium persulfate in water). Store at 4 °C.
6. TEMED. Store at 4 °C.
7. Loading buffer: 7 M urea, 0.025 % bromophenol blue (w/v), 0.025 % xylene cyanol (w/v) in 1× TBE.
8. Elution buffer: 0.3 M NaOAc, pH 5.3, 1 mM EDTA, pH 8.0.
9. 100 % ethanol.
10. UV lamp and fluor-coated TLC plate.
11. Whatman 3MM paper and plastic wrap.
12. Gel dryer.
13. Vertical gel electrophoresis equipment, including high voltage power supply, glass plates (20 (W)×22 (L) cm for gel-purification and the *trans*-splicing assay; 20 (W)×42 (L) cm for the poisoned primer assay), combs and spacers (0.4 mm thickness for the *trans*-splicing and poisoned primer assays; 1 mm thickness for gel-purification).
14. Phosphorimager and exposure cassettes (GE Healthcare) or comparable equipment.
15. ImageQuant (GE Healthcare) or comparable software for data analysis.

2.4 Trans-Splicing Assay

1. 10× splicing buffer: 40 mM Tris–HCl, pH 7.6, 30 mM $MgCl_2$, 4 mM spermidine, 40 mM DTT. Store at −20 °C.

2. Protein storage buffer (K. Nierhaus buffer 4): 20 mM Tris–HCl, pH 7.4, 4 mM $MgAc_2$, 400 mM NH_4Cl, 0.2 mM EDTA, 5 mM β-mercaptoethanol (*see* **Note 1**). Store at −20 °C.
3. Stop solution: 40 mM EDTA, pH 8.0, 300 μg/mL tRNA. Store at −20 °C.
4. [α- or γ-^{32}P]-GTP (10 μCi/μL, 3,000 Ci/mmol); dilute 1:10 prior to use to yield a 1 μCi/μL stock. Store at 4 °C.
5. Water-saturated phenol, pH 4.7; store at 4 °C. Chloroform.
6. 3 M NaOAc, pH 8.0.
7. 100 % ethanol.

2.5 Culturing E. Coli Strains

1. *E. coli* strain C600 [F^-, supE44, thi-1, thr-1, leuB6, lacY1, tonA21, thy^-].
2. Plasmid *td*wt: pTZ18U-*tdΔP6-2* is a pUC derivative (high copy plasmid with shortened ColE1 ori, Amp^R, pLac) that contains the *thymidylate synthase* wild-type gene of phage T4 including the group I intron, in which the open reading frame coding for an endonuclease has been deleted.
3. Plasmid *td*SH1: this plasmid has been derived from the *td*wt by introducing two point mutations (U-82A/C865U) that result in reduced splicing activity. Store at −20 °C.
4. Plasmid pSU20-RNA chaperone: a plasmid derived from the pAZ205 (low copy, p15A ori, Cam^R, pLac) vector in which the gene coding for a potential RNA chaperone is cloned into. As a control the empty pSU20 vector is required as well. Store at −20 °C.
5. TBY-E growth medium: 5 g yeast extract, 5 g NaCl, 10 g tryptone per 1 L.
6. Thymine solution (4 mg/mL). Add NaOH to dissolve thymine in water. After sterile filtering store the solution at 4 °C. Add 50 mg thymine to 1 L of TBY-E medium.
7. 1,000× ampicillin stock solution (100 μg/μL). Store at −20 °C.
8. 1,000× chloramphenicol stock solution (30 μg/μL). Dissolve in 70 % ethanol and store at −20 °C.
9. 0.1 M IPTG. Store at −20 °C.
10. TM buffer: 10 mM Tris–HCl, pH 7.0, 10 mM $MgCl_2$.

2.6 RNA Extraction

1. TE: 10 mM Tris–HCl, pH 7.5, 1 mM EDTA, pH 8.0.
2. Buffer A: 150 μL TE, 1.5 μL 1 M DTT, 0.75 μL RNase inhibitor (40 U/μL), 4 μL lysozyme (10 mg/mL), 0.75 μL H_2O. Prepare freshly prior to every RNA extraction. Keep on ice.

3. Buffer B: 4 μL 1 M $MgAc_2$, 3.5 μL RNase-free DNase I (1 U/μL), 0.1 μL RNase inhibitor (40 U/μL), 12.4 μL H_2O. Prepare freshly prior to every RNA extraction. Keep on ice.
4. Buffer C: 10 μL 0.2 M acetic acid, 10 μL 10 % SDS. Prepare in advance or freshly.
5. Water-saturated phenol, pH 4.7; store at 4 °C; chloroform.
6. 3 M NaOAc, pH 5.3.
7. 100 % ethanol.
8. 70 % ethanol.
9. Liquid nitrogen or dry ice–ethanol bath.

2.7 Oligonucleotide Kinase Reaction

1. [γ-^{32}P]-ATP (10 μCi/μL, 3,000 Ci/mmol). Store at 4 °C.
2. T4 polynucleotide kinase (PNK; 10 U/μL) and 10× T4 PNK buffer. Store at −20 °C.
3. DNA oligonucleotide (NBS-2): 5′ GACGCAATATT AAACGGT 3′. Store at −20 °C.
4. 0.5 M EDTA, pH 8.0.
5. 3 M NaOAc, pH 5.3.
6. 100 % ethanol.
7. 70 % ethanol.

2.8 Poisoned Primer Reaction

1. 4.5× hybridization buffer: 225 mM K-HEPES, pH 7.0, 450 mM KCl. Store at −20 °C.
2. 10× extension buffer: 500 mM Tris–HCl, pH 8.0, 50 mM $MgCl_2$, 50 mM DTT. Store at −20 °C.
3. Poisoned mix: 3 μL 10 mM dATP, 3 μL 10 mM dGTP, 3 μL 10 mM dCTP, 25 μL 5 mM ddTTP. Store at −80 °C and thaw an aliquot only 3–4 times.
4. 2× stop solution: 300 mM NaOAc, pH 8.0, 30 mM EDTA, pH 8.0.
5. AMV reverse transcriptase (10 U/μL). Store at −20 °C.

3 Methods

Subheadings 3.1–3.4 describe the *trans*-splicing assay, whereas Subheadings 3.5–3.8 describe the in vivo folding trap assay.

3.1 Linearization of Plasmids Encoding H1 and H2

1. Set up a restriction digest for each plasmid: in case of H1, use 200 μg of plasmid, 1× *Sal*I buffer, 1× BSA, 10 μL *Sal*I (20 U/μL) in a total volume of 500 μL; in case of H2, use 200 μg of plasmid; 1× *Bam*HI buffer, 1× BSA, 10 μL *Bam*HI (20 U/μL) in a final volume of 500 μL.

2. Incubate at 37 °C overnight.
3. Heat-inactivate the *Sal*I enzyme (65 °C, 20 min); perform a standard phenol/chloroform extraction for the digest with *Bam*HI.
4. Precipitate both samples using 1/10 volume of 3 M NaOAc, pH 5.3, and 3× volume of 100 % ethanol. Keep the samples at −20 °C for 30–60 min.
5. Centrifuge the tubes at 18,000 × *g* for 30 min (4 °C). Discard the supernatant, wash the pellet with 70 % ethanol.
6. After drying the pellet, solubilize the DNA in a 20 μL water and measure the DNA concentration.

3.2 In Vitro Transcription Using T7 RNA Polymerase

1. Set up two separate in vitro transcription reaction using either the H1 or the H2 linearized plasmid as template: use 50 μg DNA, 30 μL 10× transcription buffer, 15 μL 100 mM ATP, 15 μL 100 mM GTP, 15 μL 100 mM CTP, 7.5 μL 100 mM UTP, 7.5 μL [α-^{35}S]-UTP (12.5 μCi/μL), 3 μL 1 M DTT, 21 μL T7 RNA polymerase (20 U/μL) and add H_2O to the final volume of 300 μL (*see* **Note 2**).
2. Incubate the reaction at 37 °C for 3 h or until cloudiness occurs (*see* **Note 3**). Mix the reaction from time to time by tapping the tube.
3. Add 10 μL RNase-free DNase I (1 U/μL) and incubate the reaction at 37 °C for another 30 min.

3.3 Gel-Purification

1. Add 50 μL loading buffer to each of the two transcriptions and load them separately on a 8 % denaturing polyacrylamide gel. Run the gel with 25 W.
2. After the gel run, disassemble the gel and cover it with plastic wrap.
3. Detect the RNA by UV shadowing using UV lamp (254 nm) and placing the gel on top of a fluor-coated TLC plate.
4. Cut out the major band of each transcript and put the gel pieces into reaction tubes. Cover the gel slices completely with elution buffer and shake at room temperature overnight.
5. Transfer the elution buffer into a new reaction tube.
6. Precipitate the RNA by adding 3× volume of 100 % ethanol. Keep the tubes at −20 °C for 60 min to overnight.
7. Spin the samples at 18,000 × *g* for 30 min (4 °C). Discard the supernatant and after washing the pellet with 70 % ethanol dry it.
8. Resuspend the RNA pellet in 20 μL of H_2O. Measure the OD_{260} and calculate the RNA concentration.

3.4 Trans-Splicing Assay

For each *trans*-splicing assay a minimum of four reactions should be performed: aside from the sample of interest, i.e., RNA chaperone-mediated *trans*-splicing at 37 °C (a), it is necessary to perform also control *trans*-splicing reactions at 55 °C (positive control; (b)), at 37 °C (negative control; (c)), and the protein storage buffer control at 37 °C (d) to exclude any influence on *trans-splicing* from the protein storage buffer.

1. Prepare a 200 nM solution of each in vitro transcript, H1 and H2.
2. Prepare two reaction tubes for a single reaction. In tube 1 pipette 1.5 μL of the 200 nM H1 and 1.5 μL of the 200 nM H2 RNA transcripts.
3. In tube 2 pipette 1.5 μL of the diluted [α- or γ-^{32}P]-GTP stock (1 μCi/μL), 1.5 μL of 10× splicing buffer and then either 9 μL H_2O (for b, c) or 2 μM protein of interest (for a; *see* **Note 4**) or the same amount of protein storage buffer (for d; *see* **Note 1**). Add H_2O to yield a total volume of 12 μL (i.e., 15 μL including the 3 μL that are present in tube 1).
4. Heat tube 1 at 95 °C for 1 min and cool slowly to 55 °C (for b) or 37 °C (for a, c, d), respectively.
5. When the desired temperature is reached, add the prepared mix of tube 2 to initiate the *trans*-splicing reaction (GTP starts the splicing reaction; *see* **Note 5**).
6. Prior to the start of the *trans*-splicing reaction, prepare tubes for each reaction and each time point that is taken. Add 90 μL stop solution into each tube (in case of 4 reactions and 4 time points per reaction, prepare 16 tubes with 90 μL stop solution).
7. For each time point pipette 3.2 μL of the *trans*-splicing reaction into the 90 μL stop solution.
8. As the protein may affect the quality of the PAGE, perform a standard phenol, pH 4.7, extraction followed by $CHCl_3$ extraction after adding 100 μL TE first to each time point.
9. Precipitate the reactions with 1/10 volume of 3 M NaOAc, pH 8.0, and 3× volume 100 % ethanol (*see* **Note 6**). After incubation at −20 °C for 30–60 min, centrifuge the samples at 18,000 × *g* for 30 min (4 °C). Discard the supernatant and air-dry the pellet.
10. Resuspend the RNA in loading buffer. Load the reactions on a 5 % denaturing polyacrylamide gel (Fig. 2). Run the gel with 30 W to separate the bands.
11. Afterwards disassemble the gel and transfer it onto the Whatman 3MM paper. Cover the gel with plastic wrap and dry it by applying heat and vacuum.
12. Expose the gel in a phosphorimager cassette overnight. Scan the gel and analyze the results using ImageQuant.

3.5 Preparation of E. coli Cultures

1. Transform the *E. coli* strain C600(thy$^-$) with various combinations of the above described plasmids (*see* **Note 7**): (1) C600(thy$^-$) + *td*wt + pSU20, (2) C600(thy$^-$) + *td*SH1 + pSU20, (3) C600(thy$^-$) + *td*SH1 + pSU20-putative RNA chaperone (*see* **Note 8**).
2. Grow overnight cultures of single colonies at 37 °C in 2 mL TBY-E medium containing 100 μg of thymine and add 1× ampicillin and 1× chloramphenicol.
3. Inoculate 100 μL of the overnight culture in 100 mL fresh TBY-E medium containing 5 mg thymine, 1× ampicillin, 1× chloramphenicol, and 1 mM IPTG. Grow the cultures at 37 °C to an OD_{600} of 0.2–0.4 (*see* **Note 9**).
4. Harvest the *E. coli* cultures by centrifuging at 4,500 × *g* for 5 min (4 °C). Discard the supernatant and resuspend the cell pellet in 1 mL ice-cold TM buffer.
5. Transfer the cell suspension into a 1.5 mL tube, spin down at 6,000 × *g* for 1 min. Discard the supernatant and store the cell pellet at −80 °C until all cell pellets are ready for RNA preparation.

3.6 RNA Extraction

1. Resuspend each cell pellet in 157 μL buffer A.
2. Freeze the cell suspension in liquid nitrogen (or dry ice–ethanol bath) and then thaw the samples in a warm water bath (approximately 30 °C). Repeat the freeze–thaw **step 3** times.
3. Add 20 μL buffer B to each sample. Incubate the tubes on ice for 45–60 min.
4. Add 20 μL buffer C and incubate the samples at room temperature for 5 min.
5. Perform a phenol extraction: first add 200 μL phenol, pH 4.7, vortex and spin at 18,000 × *g* for 5 min. Extract the aqueous phase again with 100 μL phenol, pH 4.7, and 100 μL $CHCl_3$, vortex and spin at 18,000 × *g* for 5 min (*see* **Note 10**). Extract the aqueous phase with 200 μL $CHCl_3$, vortex and spin at 18,000 × *g* for 5 min.
6. Precipitate the aqueous phase with 1/10 volume of 3 M NaOAc, pH 5.3, and 3× the volume of 100 % ethanol. Place the tubes at −20 °C for 30–60 min.
7. Centrifuge the precipitated RNA at 18,000 × *g* for 20 min (4 °C) and resuspend the pellet in 20–50 μL H_2O (depending on the size of the RNA pellet).
8. Measure the RNA concentration.

3.7 Primer Kinase Reaction

1. End-label the DNA oligonucleotide NBS-2 (*see* **Note 11**): mix 10 pmol oligo, 10 pmol [γ-^{32}P]-ATP, 1 μL 10× T4 PNK buffer and 1 μL T4 polynucleotide kinase (20 U/μL) in a final volume of 10 μL.

2. Incubate the reaction at 37 °C for 20 min.
3. Heat-inactivate the kinase by incubating the sample at 95 °C for 1 min. Add 1 μL 0.5 M EDTA, pH 8.0, and precipitate with 1/10 volume of 3 M NaOAc, pH 5.3, and 3× the volume of 100 % ethanol. Incubate the tube at −20 °C for at least 30 min.
4. Centrifuge at 18,000 × *g* for 20 min (4 °C). Discard the supernatant, wash the pellet with 70 % ethanol and resuspend the dried pellet in 20–30 μL H_2O.

3.8 Poisoned Primer Reaction

1. Mix 20 μg of the extracted total cellular RNA with 1 μL 4.5× hybridization buffer and 1 μL end-labeled NBS-2 oligonucleotide in a total volume of 4.5 μL.
2. Incubate the tubes at 95 °C for 1 min and cool them to 42 °C.
3. Add the 2.3 μL freshly prepared extension mix. For a single reaction, mix 0.67 μL 10× extension buffer, 0.33 μL poisoned mix, 0.3 μL AMV reverse transcriptase (10 U/μL), 1 μL H_2O.
4. Incubate the samples at 42 °C for 1 h.
5. Add 7 μL 2× stop solution to each tube and precipitate with 60 μL 100 % ethanol.
6. Centrifuge the tubes at 18,000 × *g* for 20 min (4 °C) after precipitating them at −20 °C for least for 30 min. Discard the supernatant and air-dry the pellet.
7. Resuspend the sample in 8 μL loading buffer and separate them on a 10 % denaturing polyacrylamide gel (Fig. 4b).
8. After the gel run (40 W) transfer the gel to the Whatman 3MM paper and cover it with plastic wrap. Dry the gel by applying heat and vacuum.
9. Expose the gel in a phosphorimager cassette overnight. Scan the gel and analyze the results using ImageQuant.

4 Notes

1. This is a storage buffer for ribosomal proteins and is used in this experiment to measure RNA chaperone activity of ribosomal proteins. However, any other protein storage buffer can be used as long as it does not influence the *trans*-splicing assay. To test the influence of the buffer the protein buffer has to be measured in the *trans*-splicing assay as well (Fig. 2).
2. I find that using higher amounts of DNA template in transcription increases the transcription efficiency. However, it has to be ensured that the DNase digest afterwards works efficiently as well to get rid of the excess of DNA.

3. This can already happen after ½ h and then the transcription is usually done.
4. A final concentration of 2 μM protein in the *trans*-splicing assay usually proved to be sufficient for showing RNA chaperone activity. However, it is possible to titrate the protein concentration to demonstrate the amount of protein required to yield maximal RNA chaperone activity.
5. It is usually helpful to take an aliquot of the *trans*-splicing assay prior to adding the starting mix with GTP as a time point 0 min.
6. It proved helpful to use NaOAc, pH 8.0, for this precipitation because the high concentration of EDTA in the stop solution precipitates easily when NaOAc, pH 5.3, was used and interfered with the gel run afterwards.
7. Use either electro-competent or chemically competent *E. coli* cells with high transformation efficiency, because two plasmids have to be transformed simultaneously and the transformation efficiency decreases.
8. The in vivo folding trap assay has its limit because some proteins may be toxic when over-expressed in *E. coli* or may interfere with other cellular processes. It might happen that the protein that should be tested results in no growth of the *E. coli* culture so that the in vivo assay cannot be performed. It might also occur that due to the high toxicity the protein of interest is not expressed and therefore a negative result is received after the poisoned primer assay. It might be necessary to test the expression of the protein prior to performing the in vivo folding trap assay.
9. It is important that the cell cultures are harvested at comparable OD_{600}.
10. If there is a lot of cell material still present, which is shown as a white interphase, an additional phenol–chloroform extraction step may be performed.
11. The DNA oligonucleotide NBS-2 requires gel-purification at least once per year because otherwise the oligonucleotide forms dimers and trimers, thereby interfering with the poisoned primer assay. The oligo may be gel-purified prior or after end-labeling. However, purifying the kinased oligo ensures the quality of the oligo and removes the excess of [γ-^{32}P]-ATP.

Acknowledgements

I thank Wayne Laing and Gwendolyn Semrad for their support. This work is dedicated to the memory of Christina Semrad.

References

1. Herschlag D (1995) RNA chaperones and the RNA folding problem. J Biol Chem 270:20871–20874
2. Schroeder R, Barta A, Semrad K (2004) Strategies for RNA folding and assembly. Nat Rev Mol Cell Biol 5:908–919
3. Rajkowitsch L, Semrad K, Mayer O et al (2005) Assays for the RNA chaperone activity of proteins. Biochem Soc Trans 33:450–456
4. Uhlenbeck OC (1995) Keeping RNA happy. RNA 1:4–6
5. Semrad K, Green R, Schroeder R (2004) RNA chaperone activity of large ribosomal subunit proteins from *Escherichia coli.* RNA 10:1855–1860
6. Ameres SL, Shcherbakov D, Nikonova E et al (2007) RNA chaperone activity of L1 ribosomal proteins: phylogenetic conservation and splicing inhibition. Nucleic Acids Res 35:3752–3763
7. Clodi E, Semrad K, Schroeder R (1999) Assaying RNA chaperone activity *in vivo* using a novel RNA folding trap. EMBO J 18:3776–3782
8. Semrad K, Schroeder R (1998) A ribosomal function is necessary for efficient splicing of the T4 phage thymidylate synthase intron *in vivo.* Genes Dev 12:1327–1337
9. Prenninger S, Schroeder R, Semrad K (2006) Assaying RNA chaperone activity *in vivo* in bacteria using a ribozyme folding trap. Nat Protoc 1:1273–1277
10. Waldsich C, Masquida B, Westhof E et al (2002) Monitoring intermediate folding states of the *td* group I intron *in vivo.* EMBO J 21:5281–5291

Chapter 15

RNA Conformational Changes Analyzed by Comparative Gel Electrophoresis

Sébastien H. Eschbach and Daniel A. Lafontaine

Abstract

The study of biologically relevant native RNA structures is important to understand their cellular function(s). Native gel electrophoresis provides information about such native structures in solution as a function of experimental conditions. The application of native gel electrophoresis in a comparative manner allows to obtain precise information on relative angles subtended between given pair of stems in an RNA molecule. By adapting this approach, it is possible to obtain very specific structural information such as the amplitude of dihedral angles and helical rotation. As an example, we will describe how native gel electrophoresis can be used to study the folding of the *S*-adenosylmethionine (SAM) sensing riboswitch.

Key words RNA conformational changes, Riboswitch, Folding, Native structures, Dihedral angle, Helical rotation

1 Introduction

In structurally complex nucleic acid molecules, secondary structures often contain helical branchpoints, or junctions, that connect helical stems through single-strand segments [1–3]. Such junctions are particularly prevalent in RNA molecules where they form higher-order structures that are important for the adoption of the native state, as exemplified in the ribosomal RNA structures [4, 5]. Three-, four-, and five-way helical junctions are widespread structural elements that have been shown to be crucial for the folding and activity of various RNA molecules, such as riboswitches and ribozymes [1, 6]. In riboswitches, helical junctions are used to create particularly complex ligand binding site exhibiting very high affinity and selectivity toward a diversity of cellular metabolites [6]. However, despite their importance in biological systems and efforts that have been done to understand their roles [2, 7, 8], we only begin to understand the underlying folding processes and to predict their structural and functional influence in host mRNA molecules.

Christina Waldsich (ed.), *RNA Folding: Methods and Protocols*, Methods in Molecular Biology, vol. 1086,
DOI 10.1007/978-1-62703-667-2_15, © Springer Science+Business Media, LLC 2014

Native gel electrophoresis enables to assess the global structure of molecules in solution in various experimental conditions, which is possible due to the fact that the electrophoretic mobility of nucleic acids in polyacrylamide gels is very sensitive to the global shape of the migrating molecule [9]. It has been suggested that nucleic acids migrate according to the "reptation model" in which molecules move through the gel matrix in a "tube" under the influence of the electric field [10]. The rate of migration results from an equilibrium occurring between the electric field and the retardation due to friction with the gel matrix. The migration of helical DNA molecules in a matrix "tube" is modeled as follows [1, 11]:

$$\mu = \frac{Q}{\xi}\left\langle \frac{h_x^2}{L^2} \right\rangle \tag{1}$$

where μ is the migration rate, Q is the effective charge, ξ is the frictional coefficient, h_x is the end-to-end distance in the direction of the electric field, and L is the contour length. The equation predicts that any bending or kinking of a DNA helix causing the end-to-end distance to be reduced would result in a slower rate of gel migration. In the case of helical junctions, we extend defined pairs of helices irradiating from the junction in order to obtain information about the end-to-end distance and thus the angle subtended between elongated helices. Therefore, by comparing different pairs of elongated helices, it is possible to deduce the global arrangement of the junction. By employing such an approach in a comparative manner, the comparative gel electrophoresis (CGE) technique has been very successful to deduce the folding pathway for several DNA and RNA molecules [7], such as the Holliday DNA junction [12] that was recently confirmed by crystallography [13, 14]. Over the years, the CGE technique has been adapted to study very specific structural aspects, such as helical rotation [15] and dihedral angle formation [16], that are very hard to characterize using conventional biochemical or biophysical approaches.

We recently employed the CGE technique to decipher the folding pathway of the *S*-adenosylmethionine (SAM) riboswitch aptamer [15]. The SAM-I aptamer is organized around a four-way helical junction composed of four stems (P1-P4) that are joined through unpaired residues (Fig. 1a). It was previously shown that the aptamer relies on the formation of a pseudoknot structure involving the P2 loop to achieve a productive interaction with the ligand [17, 18]. Therefore, it is not possible to extend the P2 helix as required for CGE assays as this would prevent the formation of the pseudoknot interaction and ligand binding. As a result, only helices P1, P3, and P4 can be extended to allow the obtention of

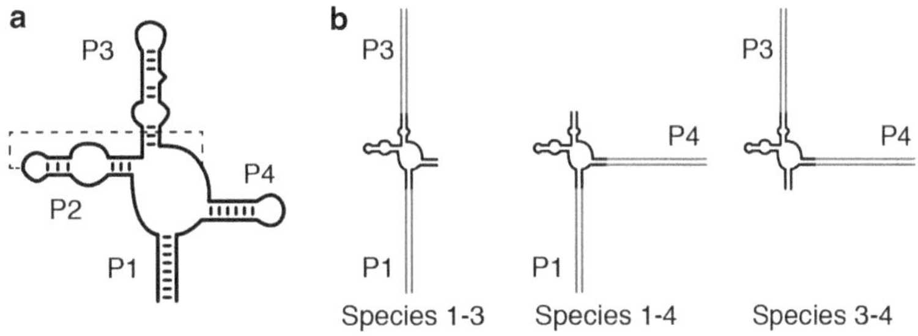

Fig. 1 The SAM-I riboswitch aptamer. (**a**) Schematic representation of the SAM-I aptamer. The aptamer is organized around a four-way junction connecting helices P1, P2, P3, and P4. The P2 stem-loop is involved in the formation of a pseudoknot interaction (*dashed line*) and therefore cannot be extended for CGE analysis. (**b**) Aptamer junctions used for CGE analysis. Given that only P1, P3, and P4 can be extended, this gives rise to the generation of three different species: 1–3, 1–4, and 3–4. Each species is named according to the pair of extended helices, e.g., species 1–3 has helices P1 and P3 extended. Extensions are performed by adding random sequences to obtain a total of 40 bp per helix

three different aptamer species that can be monitored using CGE assays (Fig. 1b). In this chapter, we describe how CGE analysis can be used to study the folding of the aptamer as a function of experimental conditions, and how the obtained patterns can be interpreted to deduce the aptamer folding pathway.

2 Materials

Because RNases are prevalent and very hard to inactivate, it is important to employ RNase-free solutions and to use RNase-free tips (*see* **Note 1**).

2.1 In Vitro Transcription of RNA Strands and End-Labeling

1. DNA template consisting of a T7 RNA polymerase (RNAP) promoter fused to the sequence to be transcribed. Whenever possible, a GCG starting sequence should be used for transcripts to reduce misincorporation during transcription [19].
2. 10× hybridization buffer: 10 mM Tris–HCl, pH 8.0, 10 mM $MgCl_2$, 25 mM NaCl.
3. 10× transcription buffer: 40 mM Tris–HCl, pH 7.9, 20 mM $MgCl_2$, 2 mM spermidine, 0.01 % (v/v) Triton X-100. Store at −20 °C.
4. T7 RNA polymerase (T7 RNAP; homemade or commercially available).
5. 3 M NaOAc (pH not adjusted).
6. 100 % ethanol.

7. Shrimp alkaline phosphatase (1 U/μL) and 1× manufacturer's reaction buffer. Store at −20 °C.
8. T4 polynucleotide kinase (10 U/μL) and 1× manufacturer's reaction buffer. Store at −20 °C.
9. [γ-^{32}P]-ATP (10 μCi/μL).

2.2 Polyacrylamide Gel Electrophoresis

1. 10× TBE: 900 mM Tris-base, 900 mM borate, 20 mM EDTA.
2. Denaturing acrylamide solution: 20 % acrylamide/bisacrylamide solution (19:1), 8 M urea and 1× TBE. The solution is neurotoxic when not polymerized. Keep the solution in the dark at room temperature.
3. Denaturing dilution buffer: 8 M urea in 1× TBE. Store at room temperature (*see* **Note 2**).
4. Native acrylamide solution: 20 % acrylamide/bisacrylamide solution (29:1) and 1× TBE. The same recommendations mentioned in point 2 of this section should be followed here.
5. Solution of 10 % (w/v) ammonium persulfate in water. Store in the dark at room temperature for 2–3 weeks.
6. TEMED (*N*,*N*,*N*′,*N*′-tetramethylethane-1,2-diamine).
7. 50 % (v/v) glycerol.
8. 25 mM NaCl.
9. Denaturant loading buffer: 95 % (v/v) formamide, 50 mM EDTA, pH 8.0, 0.05 % (w/v) bromophenol blue, 0.05 % (w/v) xylene cyanol.

3 Methods

Except for the polymerase, ensure that all reagents have been equilibrated at room temperature before mixing together.

3.1 In Vitro Transcription of RNA Strands

1. Preparation of templates: For templates smaller than 100 nucleotides, a DNA oligonucleotide corresponding to the T7 RNAP promoter and an oligonucleotide corresponding to the reverse and complementary sequence of the promoter fused to the desired transcript are incubated together during a slow cooling procedure (heat at 75 °C and cool slowly at 25 °C) in 1× hybridization buffer. In cases where longer templates are needed, a recursive PCR amplification procedure should be employed using appropriate oligonucleotides.
2. In a total volume of 100 μL, add 2.5 mM rNTPs, 1× transcription buffer, 10 mM DTT, 100 pmol DNA template, and 50 U of T7 RNAP.
3. Gently mix the sample and quick spin in a microcentrifuge.

4. Incubate the sample 3 h at 37 °C (*see* **Notes 3** and **4**).
5. Precipitate the sample by adding 3 volumes of 100 % ethanol and 0.1 volume of 3 M NaOAc. Place the sample at −80 °C for 30 min or at −20 °C overnight.
6. Centrifuge at 13,000 × *g* in a microcentrifuge for 30 min at 4 °C.
7. Discard the supernatant and quick spin to remove any traces of ethanol on Eppendorf walls. Air-dry the pellet for 5 min and resuspend in 75 % (v/v) formamide for PAGE purification (*see* **Notes 5** and **6**).

3.2 Denaturing Polyacrylamide Gel Electrophoresis

1. Prepare a standard gel (25 cm × 16.5 cm) with a 1 mm thick comb and spacers. For most RNA molecules ranging from 60 to 120 nt, we consider that the use of a 10 % acrylamide gel should allow enough resolution to purify RNA molecules. The following protocol assumes the preparation and handling of a 10 % acrylamide gel.
2. Prepare a 10 % gel by mixing 15 mL of denaturing acrylamide solution with 15 mL denaturing dilution buffer. Add 200 μL 10 % (w/v) ammonium persulfate and 20 μL TEMED in a consecutive manner to the mixture. Shake well after each addition and pour the gel in the mounted plates. Insert the comb and let the gel to polymerize for at least 1 h.
3. Remove the comb and carefully wash each well using a syringe filled with 1× TBE. Pre-run the gel for at least 30 min at 18 W using 1× TBE as running buffer.
4. Stop the power supply. Before loading samples, wash again the wells. Heat the samples to 95 °C for 1 min and load the samples on the gel. Run the gel at 18 W to allow RNA molecules to migrate at least 2/3 of the gel.
5. Stop the power supply, remove glass plates, and put the acrylamide gel on a TLC plate to visualize RNA bands by UV shadowing (254 nm).
6. Using clean blades, cut the bands of interest with a maximum size of 1 cm × 1 cm. At this point, acrylamide bands can either be frozen or immediately used for elution.

3.3 RNA Elution

Although various techniques are available to extract RNA molecules from acrylamide, we have found the crush and soak approach to reproducibly yield good quality RNA.

1. Using a clean blade, cut acrylamide bands into smaller pieces and introduce them in a 1.5 mL Eppendorf tube.
2. Add 500 μL water into each tube and leave at 4 °C overnight.
3. Transfer the liquid into a fresh 1.5 mL tube. Add 1 mL 100 % ethanol and 50 μL 3 M NaOAc. Leave at −20 °C overnight or at −80 °C for 30 min.

4. Centrifuge samples at 13,000 × *g* in a microcentrifuge for 30 min at 4 °C.
5. Discard the supernatant and quick spin to remove any traces of ethanol on Eppendorf walls. Air-dry the pellets for 5 min and resuspend them in 50 μL water for UV quantification.

3.4 End-Labeling of RNA Strands

This protocol assumes a standard dephosphorylation step using a shrimp alkaline phosphatase that can be heat-denatured; this does not require phenol–chloroform purification prior to labeling. According to CGE constructs (Fig. 1b), each RNA junction is comprised of three transcripts. Thus, to allow detection of each junction species, it is necessary to label at least one transcript per junction. We find that radio-labeling the shortest transcripts gives a high and reliable efficiency of end-labeling.

1. Dephosphorylation step: in a total volume of 20 μL, incubate 20 pmol of RNA, 1× manufacturer's reaction buffer, 5 U of shrimp alkaline phosphatase, and water. Follow the manufacturer's protocol to ensure efficient dephosphorylation.
2. End-labeling step: in a total volume of 10 μL, mix 5 pmol (5 μL) of dephosphorylated RNA, 10 U of T4 polynucleotide kinase, 1× manufacturer's reaction buffer, and 10 μCi [γ-^{32}P]-ATP. Follow the manufacturer's protocol and stop the reaction by adding 10 μL denaturing loading buffer. Purify by gel electrophoresis (*see* Subheading 3.2). To detect the RNA of interest, stop the power supply, remove glass plates, and expose the gel to a X-ray film or phosphorimager screen. After exposition, cut the band into small pieces and elute as indicated above (Subheading 3.3). Resuspend into a volume smaller than 50 μL to increase the counts per minute (cpm) per microliter.

3.5 Production and Isolation of RNA Junctions

1. Mix together 1 μM of each unlabeled transcripts with at least 10,000 cpm of the end-labeled RNA in the presence of 10 mM Tris–HCl, pH 7.9 and 50 mM KCl.
2. Heat at 75 °C and cool slowly to 25 °C.
3. Quick spin and add 0.5 volume of 50 % glycerol.
4. Prepare a native 10 % gel as in **step 2** in Subheading 3.2. However, use 15 mL of native acrylamide solution and 15 mL of 1× TBE instead of the denaturing mixtures.
5. Mount the gel in a cold room.
6. Carefully wash each well using a syringe filled with 1× TBE. Load the samples and run the gel at 100 V to allow RNA molecules to migrate at least 2/3 of the gel. The duration of migration will vary depending on the size and conformation of the molecule. In the case of the SAM-I construct, a migration duration of at least 15 h is needed.

7. Stop the power supply, remove glass plates, and expose to a X-ray film or phosphorimager screen.
8. Using clean blades, cut the RNA junctions. At this point, acrylamide bands should be used immediately for elution. Follow the Subheading 3.3, but be sure to work in a cold room using cold solutions. After precipitating the RNA, resuspend in 50 μL of cold 1× TBE. RNA junctions can be kept at 4 °C for several days (*see* **Note 7**).

3.6 CGE Analysis

Comparative gel electrophoresis provides information about the relative migration rate of each RNA junction. In the case of the SAM riboswitch, three end-to-end distances are monitored using three different RNA junctions, species 1–3, 1–4, and 3–4 (Fig. 1b).

1. Prepare a 10 % native gel by following the same procedure as described in Subheading 3.5. However, as a way to study the conformation of the junctions in the presence of NaCl, add 25 mM NaCl in the gel and in the running buffer.
2. Prepare a 10 μL solution containing at least 1,000 cpm of the desired junction and 25 mM NaCl. Add 5 μL 50 % glycerol.
3. Carefully wash each well using a syringe filled with 1× TBE. Load the three RNA junctions in different wells and run the gel at 100 V to allow RNA molecules to migrate at least 2/3 of the gel. If the running buffer contains $MgCl_2$ or NaCl, a recirculating system needs to be put in place to exchange the running buffer between both gel tanks.
4. Stop the power supply, remove glass plates, and expose to a X-ray film or phosphorimager screen.
5. Migrations can be performed in various conditions. For example, given that it is expected that Mg^{2+} ions and SAM are important for folding the aptamer, we also monitor CGE constructs in the presence of 25 mM NaCl + 10 mM Mg^{2+} and 25 mM NaCl + 10 mM Mg^{2+} + 5 μM SAM.
6. In the presence of 25 mM NaCl, a migration pattern similar to the one presented in Fig. 2a is obtained. This pattern suggests that the junction adopts a "Y" shape structure in which the smallest angle is defined by P1 and P4 stems, and in which both P1–P3 and P3–P4 subtended angles are similar (*see* Fig. 2d, unfolded state).
7. In the presence of 25 mM NaCl + 10 mM Mg^{2+}, RNA junctions exhibit very similar migration rates, suggesting that all interhelical angles are very similar. The structural change induced by the binding of Mg^{2+} ions is consistent with the close juxtapositions of stems P1 and P3 (Fig. 2d), as observed in the crystal structure [20]. Although the deduction of this intermediate state is not obvious from the migration pattern, it agrees with all of our biophysical and biochemical data [15].

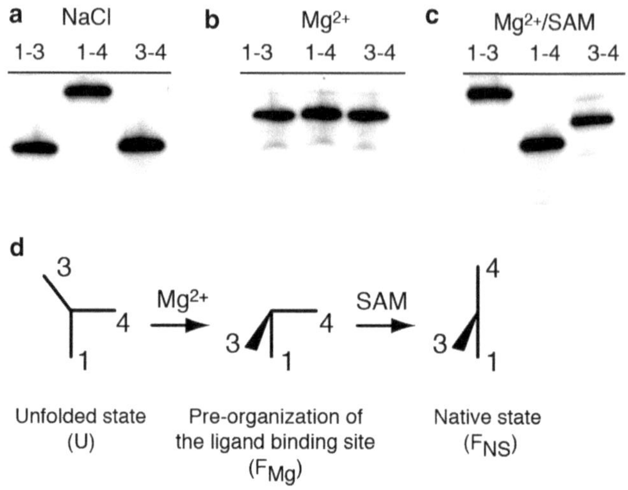

Fig. 2 Comparative gel electrophoresis. CGE data were obtained in the presence of various experimental conditions. (**a**) In 25 mM NaCl, the obtained migration pattern is consistent with the aptamer adopting a "Y" structure, corresponding to an extended and unfolded structure. (**b**) The addition of 10 mM Mg^{2+} ions yields a different migration pattern in which co-migration is observed for all species. (**c**) Further addition of 5 μM SAM induces an additional migration pattern where slower and faster species correspond to 1–3 and 1–4, respectively. (**d**) Schematic representation of the global structures of the SAM aptamer in the unfolded (U), magnesium-induced (Fmg) and SAM-induced (FNS) states. These data agree well with the crystal structure showing that helices P1 and P3 are in close juxtaposition to each other and that P1 and P4 are coaxially stacked [20]

8. When the junction is analyzed in the presence of 25 mM NaCl, 10 mM $MgCl_2$, and 5 μM SAM, a pattern similar to that presented in Fig. 2c is obtained. In this case, the slowest migrating junction corresponds to the species 1–3, agreeing with the P1–P3 close juxtaposition. Moreover, the largest interhelical angle occurs between P1 and P4 stems, indicating the SAM-induced P1–P4 helical stacking (Fig. 2d), consistent with our data [15] and the crystal structure [20].

4 Notes

1. To avoid RNase contamination or RNA degradation, use gloves at anytime even when handling pipette tips so that they do not come into contact with bare hands. Be sure to always keep RNA on ice especially because the presence of a small concentration of ions could accelerate the scission of single-stranded RNA regions. RNase-free solutions should be used. Autoclaved distilled or DEPC-treated water can be used to prepare solutions.
2. Since the dissolution of the urea is endothermic, it is preferable to heat up a bit the solution to help the dissolution of the

mixture. Be sure to cover the solution if there is acrylamide because the fumes are neurotoxic.

3. We find that the presence of a white precipitate during the incubation is generally a good indication that transcription was efficient.
4. The transcription might need optimization depending of the template used. Changing the amount of RNAP, the rNTPs concentration or the incubation time could help to get better transcription yields.
5. It is generally easier to resuspend the pellet after heating at 95 °C.
6. It is advisable to wash the pellet with 70 % ethanol before air-drying. We find that this added step help in the obtention of high-quality CGE data.
7. The RNA junction purification step can be omitted if the detected complex folds into a single conformer. If it is the case, mix around 5,000 cpm of the labeled transcript with 1 μM of other RNA strands. Complete to 10 μL with the required buffer (depending of the monitored conditions). After the slow cool process, add 5 μL of 50 % glycerol. Load on a native gel as discussed in Subheading 3.6.

Acknowledgements

This work was supported by the National Sciences and Engineering Research Council of Canada (NSERC). DAL is a Canadian Institutes of Health Research (CIHR) New Investigator Scholar.

References

1. Lilley DM (1998) Folding of branched RNA species. Biopolymers 48:101–112
2. Lescoute A, Westhof E (2006) Topology of three-way junctions in folded RNAs. RNA 12:83–93
3. Westhof E, Fritsch V (2000) RNA folding: beyond Watson-Crick pairs. Structure 8: R55–R65
4. Ban N, Nissen P, Hansen J et al (2000) The complete atomic structure of the large ribosomal subunit at 2.4 A resolution. Science 289:905–920
5. Wimberly BT, Brodersen DE, Clemons WM Jr et al (2000) Structure of the 30S ribosomal subunit. Nature 407:327–339
6. Serganov A, Patel DJ (2012) Metabolite recognition principles and molecular mechanisms underlying riboswitch function. Annu Rev Biophys 41:343–370
7. Lilley DM (2009) The structure and folding of branched RNA analyzed by fluorescence resonance energy transfer. Methods Enzymol 469:159–187
8. Lilley DM (2004) Analysis of global conformational transitions in ribozymes. Methods Mol Biol 252:77–108
9. Lilley DM (2000) Structures of helical junctions in nucleic acids. Q Rev Biophys 33: 109–159
10. Lilley DM (2000) Analysis of global conformation of branched RNA species using electrophoresis and fluorescence. Methods Enzymol 317:368–393
11. Lumpkin OJ (1982) Mobility of DNA in gel electrophoresis. Biopolymers 21:2315–2316
12. Duckett DR, Murchie AI, Diekmann S et al (1988) The structure of the Holliday junction, and its resolution. Cell 55:79–89

13. Ortiz-Lombardia M, Gonzalez A, Eritja R et al (1999) Crystal structure of a DNA Holliday junction. Nat Struct Biol 6:913–917
14. Eichman BF, Vargason JM, Mooers BH et al (2000) The Holliday junction in an inverted repeat DNA sequence: sequence effects on the structure of four-way junctions. Proc Natl Acad Sci USA 97:3971–3976
15. Heppell B, Blouin S, Dussault AM et al (2011) Molecular insights into the ligand-controlled organization of the SAM-I riboswitch. Nat Chem Biol 7:384–392
16. Lafontaine DA, Norman DG, Lilley DM (2002) The global structure of the VS ribozyme. EMBO J 21:2461–2471
17. McDaniel BA, Grundy FJ, Henkin TM (2005) A tertiary structural element in S box leader RNAs is required for S-adenosylmethionine-directed transcription termination. Mol Microbiol 57:1008–1021
18. Heppell B, Lafontaine DA (2008) Folding of the SAM aptamer is determined by the formation of a K-turn-dependent pseudoknot. Biochemistry 47:1490–1499
19. Pleiss JA, Derrick ML, Uhlenbeck OC (1998) T7 RNA polymerase produces 5' end heterogeneity during *in vitro* transcription from certain templates. RNA 4:1313–1317
20. Montange RK, Batey RT (2006) Structure of the S-adenosylmethionine riboswitch regulatory mRNA element. Nature 441:1172–1175

Chapter 16

Detecting RNA Tertiary Folding by Sedimentation Velocity Analytical Ultracentrifugation

Somdeb Mitra

Abstract

Analytical Ultracentrifugation (AUC) is a highly sensitive technique for detecting global conformational features of biological molecules and molecular interactions in solution. When operated in a sedimentation velocity (SV) recording mode, it reports precisely on the hydrodynamic properties of a molecule, including its sedimentation and diffusion coefficients, which can be used to calculate its hydrated radius, as well as, to estimate its global shape. This chapter describes the application of SV-AUC to the detection of global conformational changes accompanying equilibrium counterion induced tertiary folding of structured RNA molecules. A brief theoretical background is provided at the beginning, aimed at familiarizing the readers with the operational principle of the technique; then, a detailed set of instructions is provided on how to design, conduct, and analyze the data from an equilibrium RNA folding experiment, using SV-AUC.

Key words RNA folding, Counterion, Analytical ultracentrifugation, Sedimentation velocity, Hydrodynamic measurements

1 Introduction

Large noncoding functional RNA molecules often fold into intricate three-dimensional structures to execute their biological functions [1, 2]. Structured RNA molecules also undergo dynamic changes that are quintessential to their cellular functions, ranging from biological catalysis to regulation of gene expression [3]. A property of RNA molecules, like all other bio-polymers, which allows such folding and conformational dynamics to be monitored in solution, is their hydrodynamic radius, also known as Stokes radius (R_H). In low ionic strength solutions, electrostatic repulsion between the negatively charged phosphate groups on the phosphodiester backbone of RNA molecules results in rigid extended structures [4–6]. Upon addition of positively charged counterions (e.g., mono or multivalent metal ions, polyamines, etc.), RNA molecules adopt secondary and tertiary folds from the extended

Christina Waldsich (ed.), *RNA Folding: Methods and Protocols*, Methods in Molecular Biology, vol. 1086,
DOI 10.1007/978-1-62703-667-2_16,

© Springer Science+Business Media, LLC 2014

chain-like states [7, 8]. Counterions induce RNA folding by exerting predominantly two effects: first, screening of electrostatic repulsion to generate a relaxed ensemble of secondary structure modules which can approach each other to form tertiary contacts [5, 9], and, second, site specific binding of ions that stabilizes the tertiary structure by neutralizing pockets of high negative charge densities [10, 11]. Both of these effects lead to compactly folded native states that have smaller R_H than the unfolded state [12, 13]. RNA can also undergo large scale conformational changes in response to binding or release of biological molecules [14]. Thus, both folding and conformational switching can change the hydrodynamic properties of an RNA. One highly sensitive solution technique that can precisely detect small changes in R_H, relies on measurement of the velocity at which an RNA molecule sediments to the bottom, when spun in a closed compartment and at high angular velocities in an analytical ultracentrifuge (AUC) (Fig. 1a) [12]. A detailed theoretical description of all the hydrodynamic variables governing the bulk-transport of solutes under centrifugal force is beyond the scope of this article (c.f. [15–17]).

As a homogeneous solution of RNA is subjected to centrifugation in a closed chamber, the solute particles move towards the bottom of the chamber, thereby, increasing concentrations of the solute particles towards the bottom and creating a solute-depleted region at the top (Fig. 1b). Therefore, a solvent–solution boundary is generated (Fig. 1c) that moves radially outwards from the center of rotation, at a rate that is governed by two opposing physical forces: *sedimentation* and *diffusion*. The property of a solute particle that determines the sedimentation force it feels, under a given set of solution conditions, is known as its sedimentation coefficient (s). It is formally defined as the ratio of the linear velocity (u) to the angular acceleration ($\omega^2 x$) of the solute particle subjected to centrifugation ($s = u/\omega^2 x$), where ω is the angular velocity and x is the radial distance of the particle from the center of rotation. The linear velocity, u, of the particle can be expressed as: $u = (M_b/f)\omega^2 x$, where M_b is its effective buoyant mass and f is the frictional coefficient that determines the force of viscous drag, which it feels, in a direction opposite to that of its movement. Effective buoyant mass is calculated as $M_b = m\left(1-\bar{v}\rho\right)g$, where m is the molecular mass, $\bar{v}$ is the partial specific volume, ρ is the solvent density, and g is the acceleration due to gravity. In a dilute solution of viscosity η containing non-interacting, perfectly spherical solute particles of radius R_H, the frictional coefficient of these particles, f_s, is given by the Stokes law, $f_s = 6\pi\eta R_H$. Therefore, the expression for sedimentation coefficient can be rewritten as, $s = \left(m\left(1-\bar{v}\rho\right)g \,/\, 6\pi\eta R_H\right)$, depicting that s depends both on the mass and the size of the particle [16, 18].

At the solvent–solution boundary, the force of diffusion tends to reestablish solution-homogeneity in the solute-depleted region.

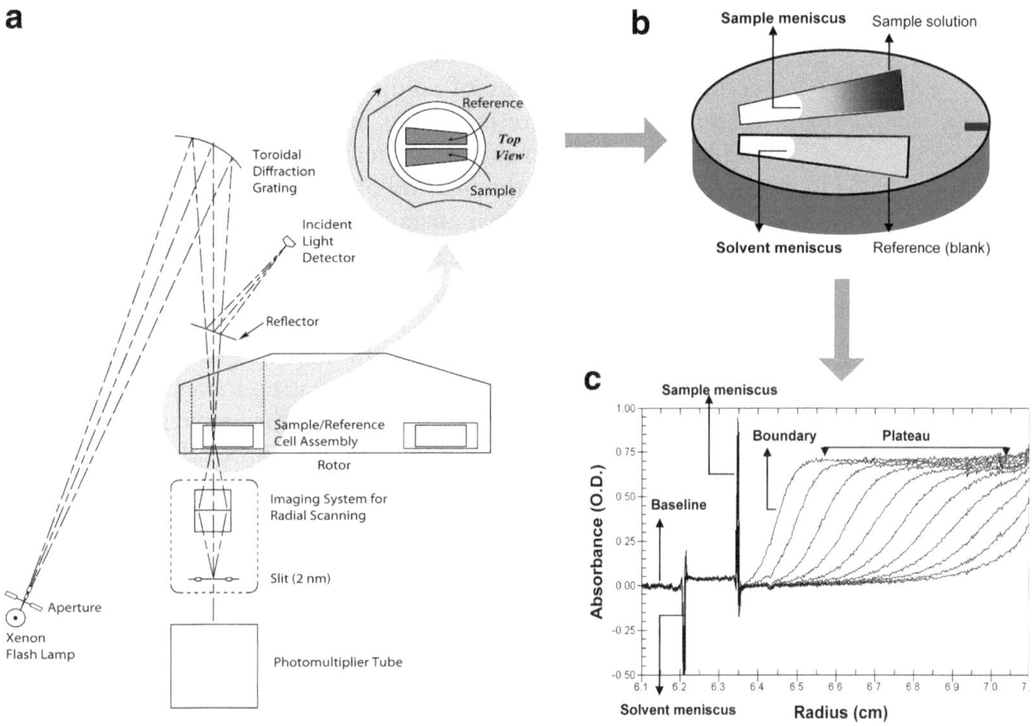

Fig. 1 Analytical ultracentrifugation: from the optical system to data output. (**a**) Overview of the optical system and rotor arrangement in the Beckman Optima analytical ultracentrifuge. One cell holder is zoomed in, to show the positioning of the cells within the rotor. *Courtesy of Beckman Coulter, Inc.* [Adapted from Greg Ralston; *Introduction to Analytical Ultracentrifugation*; http://www.beckman.com/literature/Bioresearch/361847.pdf]. (**b**) Schematic diagram of a centerpiece inside an assembled cell. The slightly greater height of the solvent column compared to the sample column helps in separation of the solvent and the sample menisci in the optical scans. (**c**) Examples of optical scans obtained from an AUC run, showing progressive movement of the boundary in a radial direction towards the bottom of the cell. 11 scans are shown, spaced at regular time intervals. The scans are recorded as absorbance (O.D.) versus radius (cm). The first set of (*downwards*) spikes represent the air–solvent meniscus while the second set of (*upwards*) spikes represent the air–sample solution meniscus [The entire figure and figure legend has been reprinted from: Methods in Enzymology, Volume 469, *Using analytical ultracentrifugation (AUC) to measure global conformational changes accompanying equilibrium tertiary folding of RNA molecules*, Somdeb Mitra, Pages 209–236, Copyright 2009 [12], with permission from Elsevier]

Diffusion is determined by yet another physical property of the solute, its diffusion coefficient (D). For a perfectly spherical particle, D is measured as: $D=(kT/f_s)$, where k is Boltzmann constant and T is the absolute temperature. Therefore, under a given solution condition, D is predominantly dependent on the size of the molecule.

AUC experiments can be conducted either as a sedimentation equilibrium (SE) run [19] or as a sedimentation velocity (SV) run [20–22]. In an SE experiment, the sample solution is centrifuged at lower angular velocity (~8,050 × g to 18,000 × g) for several hours (up to days) so that sedimentation and diffusion forces balance each other, resulting in the formation of an equilibrium concentration gradient of solutes. Under these conditions, the distribution of solute molecules at each point in the gradient is predominantly determined by their mass and not by their shapes, thereby making this the method of choice for studying stoichiometries and equilibrium constants of reversible macromolecular association reactions. For studying reactions like RNA folding, aimed at measuring the changes in global dimensions of the molecules, SV runs are conducted [23–27]. In an SV experiment, molecules are subjected to high angular velocities (typically 50,310–289,786 × g) so that sedimentation forces dominate over diffusion and the solvent–solution boundary moves all the way to the bottom of the compartment by the end of the run, typically in 3–5 h. Under conditions of SV, the size and shape of the molecule play a crucial role in determining the rate at which a molecule moves through the solvent; a compact and/or spherical molecule moves faster than an extended and/or cylindrical one, of the same mass. Therefore, an RNA, when folded into its tertiary structure with a smaller R_H, would sediment faster than when it exists in its unfolded form with a larger R_H [23].

In a typical SV experiment, solutions of RNA molecules are loaded into chemically inert chambers with transparent quartz windows (optical cells). These cells are then placed in a rotor and centrifuged at high angular velocities. UV absorption scans (260 nm) of the solutions are taken at regular intervals by shining light from a UV lamp, filtered through a monochromator, on to the optical cells and detecting the transmitted light by a photomultiplier tube (Fig. 1a). Since RNA absorbs light of wavelength 260 nm, after subtracting the background from a blank solution, the solvent–solution boundary is readily detected (Fig. 1c). The radial positions of the boundary, reporting on the radial concentration gradient ($\partial c/\partial x$), are recorded as a function of time and stored digitally. The sedimentation and diffusion coefficients (s and D) are then obtained by fitting this experimental data to approximate analytical solutions of the Lamm equation, which describes the time-dependent evolution of the radial concentration gradient ($\partial c/\partial t$), and is formulated as:

$$\frac{\partial c}{\partial t} = D\left[\frac{\partial^2 c}{\partial x^2} + \frac{1}{x}\frac{\partial c}{\partial x}\right] - s\omega^2\left[x\frac{\partial c}{\partial x} + 2c\right],$$

where c is the predetermined concentration of the RNA [16, 22, 28].

The premise of an SV experiment to detect cation induced RNA folding is to determine the hydrodynamic parameters, s and D, for an RNA molecule and, using those, to calculate its f and R_H in a given solution condition. In a series of reaction samples, starting from conditions in which the RNA is predominantly in an extended unfolded state (e.g., low ionic strength or absence of cations; Fig. 2d, e), the solution conditions are systematically varied (increase of ionic strength or addition of cations) so that the RNA folds into a compact structure (Fig. 2d–g). Each of these samples, when analyzed by AUC, operated in an SV mode, yields the *s* and *D* of the RNA, and thus, its *f* and R_H in that particular solution condition. Hence, a series of AUC runs can report on progressive changes in R_H and the transition midpoints of equilibrium RNA tertiary folding induced by mono or multivalent cations [23] (Fig. 2d). Alternatively, transition midpoints of equilibrium unfolding reactions can also be measured by adding denaturing agents like urea or metal ion chelating agents like ethylenediaminetetraacetic acid (EDTA).

Although calculation of Stokes radius is based on the premise that molecules assume roughly spherical shapes in solution, in reality, most macromolecular shapes are better represented by ellipsoids of revolution [15]. These could be either flat disc-shaped oblate ellipsoids or elongated rod-like prolate ellipsoids (Fig. 2e). Unlike a perfect sphere of radius R_H, whose frictional coefficient f_s can be described by the Stokes law, the frictional coefficients (*f*) for different types of ellipsoids are dependent on their respective axial ratios, i.e., the ratios of the long semi-axis (*a*) to the short semi-axis (*b*). For an RNA molecule resembling an ellipsoid in solution, the ratio of the observed frictional coefficient to the frictional coefficient of a perfect sphere of the same volume (f/f_s), is termed as the shape factor, *F*. The observed frictional coefficient, *f*, is calculated from the experimentally derived hydrodynamic properties *s* and *D*. As an RNA folds, it assumes a more compact globular shape, resembling a sphere. Hence, a decrease in the ratio, f/f_s, as a function of the folding reagent concentration, also reports on folding transition midpoints (Fig. 2e–g) [13, 24].

Here I briefly discuss, from a practical standpoint, the most essential aspects that would help an investigator to design, execute, analyze and interpret an AUC based experiment aimed at detecting counterion-induced RNA folding under equilibrium conditions.

2 Materials

An important consideration for conducting a successful RNA folding experiment using AUC is to make sure that the buffers and all the equipment are free of ribonucleases that degrade RNA molecules. Most of the ribonuclease-free buffer components are purchased from Ambion® and the final buffer stock solutions are filtered through 0.2 μm filters (Pall Corporation®). Nuclease-free water of high purity is either purchased from Ambion® or ultrapure water (resistivity 18.2 MΩ.cm at 25 °C, TOC < 10 ppb) is obtained from a water purification device, like a Milli-Q® water purifier (Millipore).

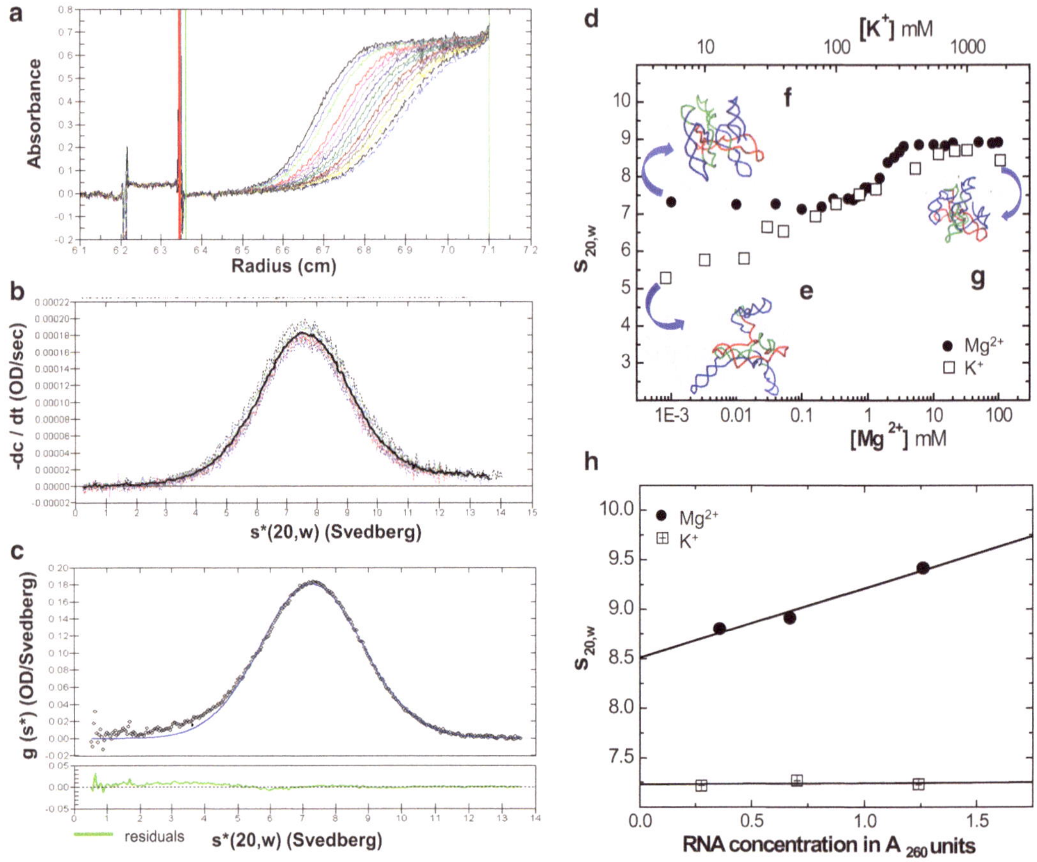

Fig. 2 Analyzing counterion-induced folding of the *Twort* group I intron ribozyme by SV-AUC. (**a–c**) Steps in the analysis of SV-AUC data, using DCDT+, collected on the *Twort* group I intron ribozyme [27, 42], in CEK buffer at 25 °C and at 98,608 × *g*. (**a**) Only 16 raw scans are depicted for clarity. The *red line* indicates the sample meniscus, while the *two green lines* indicate the radial range of data scans used in the analysis. (**b**) The $-dc/dt$ versus $s^*_{20,w}$ curves for the same scans, derived by transformation of the X-axis. (**c**) The $-dc/dt$ versus $s^*_{20,w}$ distribution is converted to the $g(s^*)$ versus $s^*_{20,w}$ distribution and fit for a single species using a Gaussian function. The slight deviation from the fit at the left end does not improve with two Gaussians. The *green line* at the bottom shows the residuals of the fit. (**a–c**) have been reprinted from: Methods in Enzymology, Volume 469, *Using analytical ultracentrifugation (AUC) to measure global conformational changes accompanying equilibrium tertiary folding of RNA molecules*, Somdeb Mitra, pages 209–236, Copyright 2009 [12], with permission from Elsevier. (**d**) SV-AUC derived $s_{20,w}$ values for Mg^{2+}-mediated (*filled black circles*) and K^+-mediated (*open squares*) equilibrium folding of the *Twort* ribozyme in CEK and CE buffer, respectively. The lower and upper abscissa represent Mg^{2+} or K^+ concentrations, respectively, while the ordinate represents $s_{20,w}$ in Svedberg units ($S = 10^{-13}$ s); increasing $s_{20,w}$ depicts that the molecules assume a more compact, faster sedimenting conformation with increasing ionic strength. The insets in *panel d* depict representative models of the *Twort* ribozyme at three different ionic strengths, consistent with the measured hydrodynamic properties: (**e**) CE buffer, extended structures ($s_{20,w} = 4.47$ S, $R_H = 72.5$ Å), (**f**) CEK buffer, partly compact structures ($s_{20,w} = 7.26$ S, $R_H = 44.8$ Å), (**g**) CEK buffer with 80 mM Mg^{2+} or CE buffer with 2 M K^+, tertiary folded structure ($s_{20,w} = 8.91$ S, $R_H = 36.5$ Å). The blue region around each model schematically depicts its global shape; ellipsoid structures become progressively spherical ($f/f_s = 2.8$, 1.76, 1.43 for (**e–g**), respectively). The data presented in (**d**) have been replotted from [27]. The structural models and the hydrodynamic measurements have been extracted from [43]. (**h**) Test for concentration dependent aggregation of the *Twort* ribozyme, either in the folded state (CEK with 10 mM Mg^{2+}; *filled black circles*) or in the unfolded state (CEK; *open squares*). The upward slope of the $s_{20,w}$ values of the folded RNA, with increasing RNA concentration (measured in 260 nm absorbance units), indicates formation of higher molecular weight species, suggestive of aggregation. The constant $s_{20,w}$ values of the unfolded ensemble, with increasing RNA concentration, suggests that the aggregation is specific to the folded state of the ribozyme

2.1 RNA Transcription

1. A bacterial plasmid which is constructed by cloning the DNA sequence corresponding to the RNA of interest, downstream of a T7 promoter sequence. Changing the first nucleotide after the T7 promoter (+1) to G greatly improves transcription yield. Usually, the 3′ side of the RNA sequence is flanked by an HδV ribozyme sequence to generate homogeneous 3′ ends (*see* **Note 1**).
2. Polymerase chain reaction (PCR) machine (thermal cycler) and forward and reverse primers for amplification of the linear transcription template from the plasmid. The reverse primers are often designed to carry 2′ O-methylation modification at the first two positions (*see* **Note 1**).
3. MEGAscript® T7 kit (Ambion®), or comparable product, for in vitro transcription of RNA.
4. Denaturing gel solution containing 5 % acrylamide/bisacrylamide (19:1), 7 M urea in 1× Tris-borate EDTA (TBE) buffer. Store at 4 °C.
5. 1× TBE buffer: 89 mM Tris, 89 mM boric acid, 2 mM EDTA.
6. 10 % (w/v) ammonium persulfate (Sigma-Aldrich®). Store at 4 °C.
7. N,N,N′,N′-tetramethylethylenediamine (TEMED > 97 %; Fisher Scientific).
8. Vertical gel running apparatus (Thermo Scientific, Owl Dual-Gel Vertical Electrophoresis Systems), glass plates (10 × 10 cm), combs and spacers (1.5 mm thick), and a high voltage programmable power supply.
9. Hand-held UV torch and fluor-coated TLC plate (Ambion®) for UV shadowing.
10. Elution buffer: 10 mM sodium cacodylate, 0.1 mM EDTA, 500 mM KCl (or NaCl), pH 7.3.
11. 100 % and 70 % ethanol.
12. Refrigerated tabletop centrifuge and SpeedVac for drying RNA pellets.
13. 10–50 μM RNA stock solutions. Store at −20 °C.

2.2 Buffers for Folding Reactions

1. 1× CE buffer: 10 mM sodium cacodylate trihydrate ($(CH_3)2AsO_2Na.3H_2O$; Sigma-Aldrich®), 0.1 mM EDTA, pH 7.3. A 10× stock solution is made and stored at 4 °C. The CE buffer is the starting point for experiments in which RNA folding is induced by monovalent ions, like sodium or potassium [23, 29] (*see* **Note 2**). It is also the buffer in which the RNA samples are routinely dissolved and stored.
2. 1× CEK buffer: 10 mM sodium cacodylate, 0.1 mM EDTA, 100 mM potassium chloride, pH 7.3. This is the buffer routinely used while conducting Mg^{2+} (or other divalent ion)

induced RNA folding experiments [7, 27, 30] (*see* **Note 3**). A 10× stock solution is made and stored at 4 °C.

3. Ribonuclease-free solutions of 1 M $MgCl_2$, 0.5 M EDTA, pH 8.0, 2 M KCl and 5 M NaCl.

2.3 Analytical Ultracentrifugation

1. Beckman Proteome Lab™ XL-I analytical ultracentrifuge (formerly known as Optima XL-I; Beckman Coulter) (*see* **Note 4**).
2. An-60 Ti rotor.
3. AUC optical cells (Beckman Coulter): aluminum-epon charcoal centerpieces with sector shaped cavities of 1.2 cm optical path length (Fig. 1b), centerpiece gaskets, quartz windows, window liners and gaskets, upper and lower window holders, cell housing, screw ring, red polyethylene gasket sheet for sealing sample loading ports, brass screws for closing loading ports. It is strongly recommended to have at least six working sets of optical cells so that at least two analytical runs can be performed on a given day.
4. An opaque metal reference cell (Beckman Coulter).
5. Round gel-loading tips (200 μL).
6. Torque stand, torque wrench and screw drivers.
7. RNaseZap (Ambion®) solution, or comparable product, for cleaning the optical cells.
8. Kimwipe® (Kimberly-Clark®) extra low-lint 2-ply paper wipers for delicate surfaces.

2.4 Software for Calculating Hydrodynamic Properties and for Data Analysis

1. SEDNTERP (http://www.jphilo.mailway.com/download.htm).
2. DCDT+ [31].
3. Origin® (Origin Labs®).

3 Methods

3.1 RNA Sample Preparation

1. Perform a standard PCR reaction to generate a linear DNA template from the plasmid containing the RNA sequence, with appropriate primers such that the RNA sequence is flanked on its 5′ side by a T7 promoter sequence and on its 3′ side by a HδV ribozyme sequence.
2. Transcribe the RNA in vitro, using the MEGAscript® T7 Kit (Ambion®) at 37 °C for 2 h.
3. Add an additional 20 mM $MgCl_2$ to the reaction and incubate the transcript at 60 °C for 15 min. This ensures maximum self-cleavage of the HδV ribozyme, resulting in RNA samples with a homogeneous 3′ end.

4. Run the reaction products on a 5 % denaturing polyacrylamide gel. After completion of the run, take the gel off the glass plates, place it on a fluor-coated TLC plate and illuminate it with short wavelength UV to visualize the RNA and the cleaved ribozyme bands.
5. Cut out the RNA band using clean razor blades. Elute the RNA by crushing the gel piece by passing it through a 5 mL syringe, soaking the crushed pieces in the elution buffer in a 15 mL Falcon tube and rocking it overnight in a cold room.
6. Give the solution a quick spin so that the gel pieces settle to the bottom and remove the supernatant carefully.
7. Precipitate the gel-eluted RNA from the supernatant by adding 100 % ethanol (3× volume of the supernatant), incubating at −20 °C for 1 h and spinning down the samples in a refrigerated tabletop centrifuge at ~20,000 × *g* for 1 h at 4 °C.
8. Wash the pellet twice with 70 % ethanol, dry it in a SpeedVac, and dissolve it in 1× CE buffer to a concentration of 10–50 μM.

3.2 Setting Up an Equilibrium Folding Reaction

1. Set up the folding reaction in a total volume of 300 μL with a final RNA concentration of 0.1–1.0 O.D. (A_{260}), given a path-length of 1.2 cm for the sector-shaped AUC optical cells (*see* **Notes 5** and **6**).
2. Prepare the folding buffer in a 1.5 mL microfuge tube. For each monovalent or divalent ion titration point, add the appropriate amount of the corresponding ion from its stock solution (e.g., 2 M KCl, 5 M NaCl, or 1 M $MgCl_2$) to reach the desired concentration of the ion in the sample (usually 0–1.5 M for monovalent ions and 0–100 mM for divalent ions).
3. Subsequently, add 30 μL of the 10× CE or CEK buffer stock solutions to the reaction (*see* **Notes 2** and **3**) and adjust the volume to $(300 - x)$ μL with nuclease-free water, where, x is the amount of RNA that needs to be added from a separate solution to reach a final RNA concentration of ~0.1–1.0 O.D. in the reaction sample.
4. For each titration point, prepare separately a blank solution of 350 μL volume in the corresponding buffer with the same concentration of the counterion but without the RNA. Set up three folding reaction samples (along with their corresponding blanks) for each SV run (*see* **Note 7**).
5. Dilute the stock solution of the RNA in a separate 1.5 mL tube, with 1× CE or CEK buffer, to a volume slightly greater than $3x$ μL, where x is the amount of this solution to be added to each of the three folding reaction points (prepared as described above). The slight excess accounts for the loss in volume by evaporation during the subsequent heating step.

6. Heat the RNA solution, diluted in 1× CE (or CEK) buffer, to 95 °C on a heating block, to denature its secondary structure. Cool it slowly to 50 °C over ~15 min to allow re-formation of native secondary structure (*see* **Note 8**). At this point, incubate the tubes containing the folding buffer (without RNA) at 50 °C as well.
7. Briefly centrifuge the RNA solution in a tabletop centrifuge to spin down the condensates on the lid and keep it back at 50 °C. Then, pipette out x μL of the RNA solution into each of the three reaction tubes containing the folding buffer, which are being concurrently incubated at 50 °C. The reaction volume reaches 300 μL at this point. Mix the solutions by pipetting up and down 3–4 times and continue incubation at 50 °C for 30 min (*see* **Note 9**). Treat the three corresponding blank solutions of 350 μL volume identically, except for the addition of the RNA.
8. Move the reaction tubes from 50 °C to another heat block that is set to the temperature at which the equilibrium tertiary structure is monitored (usually, 25 °C or 37 °C), also referred to as the "folding temperature". Allow the RNA molecules to reach their equilibrium structure at the folding temperature and at the given ion concentration for 1 h. At the end of 1 h, the folding reaction is usually complete (*see* **Note 10**) and the samples are ready to be loaded into the optical cells.

3.3 Optical Cell Assembly and Loading Samples into the Cells

1. Clean all the cell components rigorously using RNaseZap and non-abrasive low-lint Kimwipes (to prevent scratching of the quartz windows and the delicate center pieces), wash with ultrapure water and wipe dry with low-lint Kimwipes. It is highly recommended to dedicate a set of optical cells for RNA experiments only. This helps to minimize chances of ribonuclease contamination of the cells from previously used DNA/protein samples which may not have been prepared optimally to minimize RNase contamination.
2. Follow the instructions for assembling the optical cells as detailed in the Beckman Proteome Lab™ XL-A/XL-I AUC manual. Briefly, first place the two quartz windows inside their corresponding holders containing the window liners and gaskets; then place the centerpiece between the two windows and finally, place the whole assembly inside the metal cell housing (*see* **Note 11**).
3. Close the unit with the screw ring at the top, first tightening it by hand and then further tightening to 120 psi using a torque wrench. The final torqueing prevents sample leakage during AUC runs.
4. Place each optical cell horizontally on a bench with the screw ring facing the experimenter and carefully load the folding

reaction sample (300 μL) into the right sector-shaped cavity (Fig. 1b) through the hole in the top, using a round gel-loading tip. Insert the tip as far down into the cavity as possible and release the solution slowly to avoid formation of air-bubbles. Similarly, load the blank solution (350 μL) into the left cavity (Fig. 1b) (*see* **Note 12**).

5. Repeat **step 4** for loading the two other optical cells, each with a different folding reaction sample and the corresponding blank solution. Mark the outer surface of the metal housing of each cell with a unique number and make a note of the contents of each cell.
6. Seal the loading holes of the centerpieces with small circular pieces of gaskets, cut out from the red polyethylene sheet that comes with the centrifuge accessories. Close the ports by threading in the brass screws; exercise care to ensure that the screws are snug but not over tightened.
7. Finally, blow dust particles off the quartz windows of the loaded and sealed optical cells using compressed air.

3.4 Loading Optical Cells into the Rotor

1. Switch on the AUC machine and adjust the temperature settings to the experimental temperature, which is usually 25 °C (*see* **Note 13**).
2. Place the rotor inside the centrifuge and close the sliding door to let the rotor equilibrate to the experimental temperature for at least 1 h. Start the equilibration before assembling the optical cells.
3. Weigh the loaded and sealed optical cells and set aside the two cells that match most closely with each other. Then adjust the weight of the opaque reference cell by placing small weight-pieces in it, to match it to the weight of the optical cell that deviates the most from the other two. The weights of a pair of cells should not differ by more than 0.5 g.
4. Take the temperature equilibrated rotor out of the centrifuge and place the reference cell in the slot marked "4". Place the cell counter balanced against the reference cell opposite to it at position "2" and the other two cells at positions "1" and "3", respectively. Place the cells such that the screw ring sides face upward and the loading ports face toward the center of the rotor (*see* **Note 14**).
5. Lifting the rotor up carefully, align the cells using the tool supplied by Beckman, such that the marking on the side of a cell opposite to the screw ring aligns perfectly with the marking on the periphery of the corresponding slot, on the lower surface of the rotor.
6. Place the rotor back inside the centrifuge, attach the monochromator firmly to its holder inside the centrifuge chamber

(*see* **Note 15**), close the sliding door and turn on the vacuum pump. Allow the pressure to come down to at least 100 microns before starting preliminary runs.

3.5 Setting Up a Sedimentation Velocity Run and Data Acquisition

1. The computer attached to the AUC instrument controls centrifugation runs and automated optical scan acquisitions, with the help of user-input commands via the graphical user interface (GUI) of the "Beckman XL-x" software. In the Beckman XL-x GUI, open a "New File" window from the "File" section and enter the following parameters in their corresponding boxes: "Rotor"→4 hole, "Speed"→724×*g* (for an initial diagnostic run) and "Temperature"→25 °C. This set of parameters can be entered automatically in all the four boxes (the first three for the optical cells and the fourth one for the reference cell) by selecting the option "All settings identical to cell 1". Make sure to change the names of the samples in the "Comment" section of each box to ascribe a set of scans to the sample contained in the corresponding cell.
2. Select "Velocity" as the scanning option from the box designated "Cell#" and "Absorbance" as the type of optics used during scanning. Specify the radial position range to be scanned by entering 5.9 cm for "Rmin" and 7.1 cm for "Rmax" (*see* **Note 16**). Since the samples being analyzed are RNA, select the scanning wavelength as 260 nm in the box designated "W1".
3. Perform a test scan, using the option "wavelength scan", to ensure that the xenon lamp inside the AUC produces light of sufficient intensity in the UV region. If the light intensity is insufficient, clean the lamp following instructions provided in the Proteome Lab™ XL-A/XL-I AUC manual.
4. Perform a preliminary scan to check for sample leakage. Once the pressure reaches ~30 microns, open the "Options" window and initiate spinning of the rotor by selecting "Start single scan". Once the rotor speed goes up to 724×*g* (as set in **step 1**), perform about 5–6 single scans by clicking on "Start single scan" each time. A progressive drop in the 260 nm absorption peak or an increase in the chamber pressure upon initiating the spinning of the rotor indicates sample leakage. In such a case, stop the run, take out the rotor, remove the optical cells, unscrew the brass screws on the loading ports of the leaking cell and reseal the ports with a fresh pair of polyethylene gaskets pieces.
5. In the absence of leakage, increase the speed of the rotor to that required for SV analysis (*see* **Note 17**) by changing the value in the "Speed" box to the desired speed and selecting "Start single scan". Once the rotor reaches the desired speed, allow it to spin for ~10 min and then perform an absorption scan by selecting "Start single scan".

6. If there are no indications of leakage at the high rotor speed, set up the actual SV run by first specifying the "Number of scans" as 60 in the "Method" window (usually 60 scans give more than enough radial positions of the moving solvent–solution boundary to calculate the rate of sedimentation), and then checking the "No delay calibration" box in the "Options" window. It is also recommended to check the box "Stop XL after last scan", which automatically stops the rotor after the last absorption scan. Save these settings by selecting "Save file as" under the "File" menu and initiate the experimental run by selecting the "Start method scan" in the "Options" window. Usually, a typical SV run takes about 3 h for completion (*see* **Note 7**).
7. After the SV run is complete and the rotor stops spinning, turn off the vacuum pump and let the pressure come back to normal. Open the door, unscrew the monochromator and remove it from its holder and finally, take the rotor out. Remove the cells from the rotor, disassemble the parts and clean them thoroughly with RNaseZap and ultrapure water before assembling them for loading the next set of samples (*see* **Note 11**).

3.6 Analyzing SV Data to Obtain Hydrodynamic Properties

The software DCDT+ [28, 31] is widely used to analyze the information embedded in the absorption scans. Briefly, in the DCDT+ analysis, time-derivative of concentration at fixed radial positions, $(\partial c/\partial t)_r$, is calculated by subtracting pairs of radial concentration scans spaced at close and uniform time intervals, $[(c_2 - c_1)/(t_2 - t_1)]$ [21, 22]. Plotting $(\partial c/\partial t)_r$ as a function of r generates a Gaussian shaped distribution. Since the apparent sedimentation coefficient s^*, defined as the sedimentation coefficient of a molecule in the complete absence of diffusion, is directly related to r, the $(\partial c/\partial t)_r$ versus r curves can be replotted as $(\partial c/\partial t)_r$ versus s^*. Moreover, since the time-derivatives of radial concentrations can be converted to the apparent sedimentation coefficient distribution function, $g(s^*)$, the $(\partial c/\partial t)_r$ versus s^* curve can be further replotted as $g(s^*)$ versus s^* curves. For a homogeneous, non-interacting species of molecules possessing a unique global conformation, this Gaussian plot has a single peak centered at the mean weight-averaged sedimentation coefficient, s, of the molecule. The presence of more than one peak can indicate the presence of either multiple conformations or different aggregated states of RNA molecules. The standard deviation of the Gaussian fit to the $g(s^*)$ distribution can be additionally used to calculate the diffusion coefficient, D, of the molecule. Finally, s and D, are used to calculate the shape factor (F) and the Stokes radius (R_H) of the molecule, which report on its global conformation under a given set of solution conditions. The steps involved in data analysis are outlined as follows:

1. Estimate the viscosity (η) and density (ρ) of the solution by using the software SEDNTERP and the standard reagents listed in the same. These values are required for the subsequent steps in data analysis (*see* **Note 18**). Assume the partial specific volume $(\bar{v})$ and the hydration of the RNA molecule to be 0.53 cm^3/g and 0.59, respectively.
2. The absorbance scans, stored as digital data in dated folders in the hard drive of the computer controlling the AUC, are exported and uploaded to a computer in which the DCDT+ software is preinstalled. Within each dated folder, the data from each optical cell is automatically stored in a separate sub-folder. To analyze the scans from a given optical cell, initiate the software DCDT+ and open the corresponding sub-folder; the digital absorbance scans are automatically loaded into the software.
3. The set of scans are displayed as absorbance versus radial position (Fig. 2a). Specify the sample meniscus position, the start point of the scans and the end point, by clicking the cursor at the appropriate positions over the displayed scans. Select "Next" to let the software calculate the $(\partial c/\partial t)_r$ values, as described above, and convert the A_{260} versus r scans to the $-dc/dt$ versus s^* distribution plots (Fig. 2b).
4. Generate an ensemble of good Gaussian distributions by selecting a subset of the plots (and hence, data scans) either manually or automatically (by selecting the "Auto adjust" option). Convert the apparent sedimentation coefficient to the standard apparent sedimentation coefficient, $s^*_{20,w}$ (with reference to a standard temperature, 20 °C, and a standard solvent, water; *see* **Note 19**).
5. Choose the default range of $s^*_{20,w}$ selected by the software to generate a good $g(s^*)$ versus $s^*_{20,w}$ distribution. This distribution should ideally produce a single peak centered around the weight averaged $s^*_{20,w}$, which, can be shown mathematically, to correspond to the actual standard sedimentation coefficient, $s_{20,w}$, of the molecule (Fig. 2c) (*see* **Note 20**).
6. In the next step, fit the $g(s^*)$ versus $s^*_{20,w}$ distribution to a Gaussian function. As for the number of species comprising the distribution, use a single or multiple species fitting option based on the number of peaks observed in the $g(s^*)$ versus $s^*_{20,w}$ distribution. The most common fitting method involves first selecting the option, "s/D" for molecular weight (according to the Svedberg equation, $M = (s / D) \times (RT / 1 - \bar{v}\rho)$, where M and $\bar{v}$ are the molecular weight and the partial specific volume of the RNA molecule and ρ is the solvent density) and then, "fit D for each species". The fit, thus produced, provides the following information: $s_{20,w}$ from the maxima of

the fit, $D_{20,w}$ from the standard deviation (*see* **Note 20**) and C_o (initial RNA concentration) from the total area under the Gaussian curve (Fig. 2c).

7. Repeat **steps 1–6** with the data from all the analytical cells from the series of AUC runs, corresponding to the different folding conditions in order to obtain the equilibrium $s_{20,w}$ and the $D_{20,w}$ of the RNA under each cation concentration used for folding.
8. Plot the $s_{20,w}$ values as a function of cation concentration to demonstrate that the RNA molecules assume an increasingly compact shape (higher $s_{20,w}$) with increasing cation concentration (Fig. 2d). Also, plot the $s_{20,w}/D_{20,w}$ value, at each cation concentration, to check for counterion-dependent RNA aggregation (*see* **Note 21**).
9. Enter the C_o, $s_{20,w}$ and $D_{20,w}$ values in SEDNTERP to calculate the Stokes radius (R_H) of the RNA at each cation concentration and the axial ratio (a/b) of the ellipsoid of revolution that best represents it at the given cation concentration.
10. Calculate the "shape factor" (F) using the R_H and a/b in SEDNTERP. Either R_H or F, when plotted against the corresponding cation concentration, should show a systematic decrease with increase in cation concentration; decrease in R_H indicates that the RNA molecules become progressively more compact, while the decrease in F indicates adoption of a more spherical shape (Fig. 2d–g).
11. Both, R_H and F, report on tertiary folding transition of an RNA induced by a counterion; the midpoint of the transition and the cooperativity of the folding reaction, can be obtained by phenomenological fits (*see* **Note 22**) of the plots of R_H (or F) versus counterion concentration to the Hill equation: $Y = K_d^{n_H} [M]^{n_H} / 1 + K_d^{n_H} [M]^{n_H}$; where, Y denotes the signal quantitated (R_H or F), $[M]$ is the counterion concentration, K_d is the midpoint of the equilibrium folding transition, and n_H is the Hill coefficient that indicates the cooperativity of the folding transition. Fit the data using the "nonlinear least squares fitting" routine implemented in standard software like Origin and report the best-fit values at 65 % confidence intervals (*see* **Note 23**).

4 Notes

1. Generation of RNA molecules with heterogeneous 3′ ends, due to addition of one or two extra nucleotides at the end by the T7 RNA polymerase, is a common problem during in vitro transcription. To avoid this, transcription templates can be

generated to contain a Hepatitis δ virus (HδV) ribozyme sequence, immediately downstream to the RNA sequence of interest. Addition of the HδV ribozyme sequence can be either implemented at the stage of PCR generation of the DNA template by using a reverse primer bearing the reverse complementary sequence of the HδV ribozyme, or, by inserting the DNA sequence corresponding to the RNA of interest into the *Eco*RI-*Ngo*MIV sites of the vector pRAV12 [32]. Following transcription of the entire template, the HδV ribozyme autocatalytically and precisely cleaves itself out of the nascent RNA transcript, thereby, producing the RNA of interest with a homogeneous 3′ end. The self-cleavage reaction of the HδV ribozyme is stimulated by the addition of an excess amount of Mg^{2+} after transcription, which favors folding and stabilization of the HδV ribozyme. An alternate way to reduce 3′ heterogeneity of the transcribed RNA is to use modified reverse primers during PCR amplification of the DNA template such that the last two nucleotides of the template are methylated at 2′-O position, which prevents addition of extra nucleotides at the 3′ end by the T7 RNA polymerase.

2. If the goal of the experiment is to measure the midpoint and cooperativity of an RNA tertiary folding transition induced by monovalent ions, the starting buffer of choice is CE buffer, which has approximately 8 mM of free Na^+ ions. Although the amount of free Na^+ ions from 10 mM Na-cacodylate is negligible (~8 mM) compared to the amount required for either RNA backbone charge neutralization or tertiary folding, to be technically accurate while conducting monovalent ion titrations with other counterions (e.g., potassium), one might choose to use the corresponding salt of cacodylic acid for making the experimental buffer.
3. If Mg^{2+} (or other multivalent cation)-induced tertiary folding is being studied, it is recommended to use buffers like CEK, which have at least 100 mM monovalent ions. The 100 mM K^+ ions in the CEK buffer neutralizes the negatively charged phosphodiester backbone of RNA molecules, thereby, reducing intermolecular electrostatic repulsion that can generate artifacts during measurements of hydrodynamic properties. Furthermore, it also screens intramolecular repulsive forces between the backbones of the helices, thereby, allowing for greater conformational flexibility (Fig. 2e vs. f) [5]. Finally, it helps avoid the idiosyncratic Mg^{2+}-mediated folding behavior observed for some large structured RNA molecules, like the self-splicing group I intron ribozymes (*Tetrahymena* [33] and *Azoarcus* [34]), under conditions of very low ionic strength, most likely due to nonnative docking of secondary structure modules in a rigid ensemble.

4. One can either use a Proteome Lab™ XL-A instrument in which the sample detection is based only on absorption optics or a Proteome Lab™ XL-I instrument which can detect both optical absorption and Rayleigh interference. The interference method, based on detecting differences in refractive index between the sample and the reference solution, is particularly useful in certain cases when the sample does not absorb significantly in the UV range. It also provides higher signal-to-noise ratio and covers a greater sample concentration range. The XL-A/XL-I instruments can now also be equipped with a fluorescence detection system (AU-FDS manufactured by Aviv Biomedicals), which provides the convenience of studying a broader range of samples, including nucleic acids and proteins, labeled with fluorescent dyes.
5. It is important to optimize the RNA concentration before initiating a folding experiment because very low RNA concentrations (<0.1 O.D.) produce low signal to noise ratio while at higher RNA concentrations (≥1 O.D.) the detector saturates and does not yield accurate data. While calculating RNA concentrations, it is extremely important to remember that the optical path length of the sector-shaped cavities of the AUC centerpieces is 1.2 cm and not 1 cm. For example, for an RNA molecule of about 240 nucleotides with molecular weight (MW) of approximately 75 kDa, the 0.1–1.0 A_{260} range in a 1.2 cm optical path length spans a concentration range of approximately 30–300 nM.
6. Since RNA samples can aggregate at high concentrations and deviate from ideal non-associating solution behavior, preliminary SV runs should be performed, before setting up an experiment, to test for concentration dependent self-assembly (Fig. 2h). In order to check for concentration dependent aggregation a series of dilutions of the RNA sample are analyzed, ranging between 0.08 OD and 1.2 OD. Two concentration series are determined. The first in 1× CEK (or CE) buffer while the second in CEK buffer with 10 mM Mg^{2+} (or CE buffer with 2 M NaCl or KCl). These series correspond to the unfolded and folded states of the RNA molecule being analyzed respectively. Absorption data are collected, analyzed and $s_{20,w}$ values of both the unfolded and the folded samples are plotted as a function of RNA concentration (in absorption units) (Fig. 2h). If $s_{20,w}$ remains constant with increasing RNA concentration, the sample is considered to behave as an ideal solution of monodisperse particles. A significant downward slope, usually observed in the absence of Mg^{2+}, indicates nonideal behavior at high concentrations due to electrostatic repulsion between negatively charged RNA backbones [24]. An upward slope reflects RNA aggregation

with increasing concentration (Fig. 2h). Samples which aggregate may also show multiple components in the sedimentation boundaries.

7. With a four slot An-60 Ti rotor and the two-chambered centerpieces, only three reaction samples (along with their corresponding blanks) can be analyzed in an SV run. Therefore, conducting a full series of monovalent or divalent metal ion titration with the RNA of interest, involves multiple SV-AUC runs. Since, conducting an equilibrium cation induced RNA tertiary folding reaction takes about 3 h and loading the samples in to the cells and conducting an SV run with three samples at a time takes up to another 3–4 h, typically, on a given day, six titration points can be set up and analyzed. For effective time utilization, after initiating the first SV-AUC run, three more folding reactions are set up and the second set of three optical cells are washed and assembled during the incubation phases of the second set of folding reactions.
8. Due to the possibility of formation of stable alternate secondary structures by mispairing of bases, RNA molecules are highly susceptible to nonnative secondary structure formation. Heating the initial ensemble melts all the secondary structures and slow cooling allows formation of native secondary structure. In a few instances, there might be highly stable nonnative secondary structures in the unfolded ensemble [35]. It is, therefore, highly recommended to perform a secondary structure mapping of the RNA, using methods like chemical footprinting (e.g., SHAPE [36] or DMS [37]), prior to conducting SV-AUC studies, to ensure the presence of correctly base paired secondary structures in the initial state ensemble, under the particular experimental conditions. If nonnative base pairing is detected (compared to that observed in atomic resolution structures, when available), the solution conditions might need to be optimized to achieve native base pairing in the unfolded state ensemble.
9. RNA molecules can also get kinetically trapped into misfolded intermediates during tertiary folding at physiological or lower temperatures, following addition of counterions. A short incubation at higher temperatures (e.g., 50 °C) can provide additional thermal energy that helps avoid formation of such kinetically trapped states during folding [38]. It should be noted though that the presence of divalent metal ions, like Mg^{2+}, can cause significant hydrolysis of RNA molecules at higher temperatures. Therefore, it is recommended to not prolong the 50 °C incubation.
10. Most structured RNA molecules fold into their native tertiary structure in 1 h under common experimental conditions. However, the presence of extremely long-lived misfolded

intermediates, lasting for several hours, has also been reported [39]. Therefore, it is recommended to employ alternate strategies, like tertiary structure mapping by hydroxyl radical footprinting [30] or functional assays [40], to monitor formation of biologically active native structures, before performing AUC experiments, to determine the appropriate equilibration time required for an RNA to fold into its native tertiary structure.

11. Sometimes it might be hard to push the assembly into the cell housing. It is recommended to not apply too much force which can damage the windows and the centerpiece; instead, all the cell components, including the housing, are placed inside a refrigerator at 4 °C for about 15 min, taken out, reassembled and then gently pushed into the housing; it should go in easily. Similarly, it is often hard to push the assembly out of the housing after an AUC run, in which case, the cells are again placed at 4 °C for about 15 min to allow the assembly to be gently pushed out.
12. Each centerpiece has two sector shaped cavities with 1.2 cm optical path length, one for the sample solution and one for the blank solution (Fig. 1b). Since the blank solution is prepared in slight excess of the sample solution (350 μL vs. 300 μL in our experiments), after loading, the solution-air interfaces (meniscus) of the two solutions are at slightly different heights. This helps in distinguishing the sample meniscus from the solvent meniscus (Fig. 1c).
13. Although, the XL-A/XL-I instruments can be operated within a temperature range of 0–40 °C, practical difficulties like too much pump oil vapor at higher temperatures and unstable temperature maintenance at lower temperatures, restrict the effective operational temperature range. We prefer to conduct AUC runs at 25 °C, close to the ambient room temperature. Accordingly, our equilibrium RNA folding experiments are conducted at 25 °C. If it is absolutely essential to conduct the folding experiments at more physiological temperatures, like 37 °C, it is advisable to replace the usual oil diffusion pump of the AUC with a turbomolecular pump to avoid excessive oil vapors.
14. Make a note as to which sample (based on the numbering outside the optical cell) is placed at which position in the rotor. The position number "4" in the 4-hole An-60 Ti rotor is fixed for the opaque cell, which the instrument uses as a reference point for detecting the positions of the experimental cells. It is a common mistake to put the opaque cell in another slot. If this error occurs, it will be obvious during the first set of diagnostic scans at 724 × *g*, as no boundary position will be observed in the digital images from the slot carrying the reference cell; instead a jagged line will be displayed. If this is observed, the

run should be immediately stopped, the rotor taken out and the position of the reference cell and the wrongly placed analytical cell switched.

15. Utmost care should be taken while handling the monochromator, which is one of the most delicate and expensive, detachable parts of the XL-A/XL-I instrument. While attaching the monochromator to its holder, the users should keep in mind that it sits in the slot at an acute angle. After securing the device in its holder the base-ring should be tightened optimally, without destroying the threading, and before closing the chamber door the monochromator should be thoroughly checked to ensure it is not wobbling. Serious damages to the instrument can occur if the monochromator gets dislodged during an AUC run.
16. The "Rmin" and the "Rmax" options specify the radial position range along the optical cells that the instrument would scan, during an AUC run, to monitor the solvent–solution boundary movement. It is generally recommended to start collecting scans between 5.9 cm and 7.1 cm and if necessary, reduce the range according to the number and quality of the scans at the beginning and at the end.
17. Since the rate of sedimentation of a molecule is directly proportional to its size, the choice of rotor speed in a SV experiment depends inversely on the size of the RNA molecule being studied. Users should choose an appropriate velocity so that at least 50–60 absorbance scans can be collected before the solution–solvent boundary hits the bottom of the cell. For a molecule of about 400 nucleotides (MW ~130 kDa) velocities around 30,000 × *g* should be chosen, whereas for a molecule of about 240 nucleotides (MW ~75 kDa), higher velocities, around 95,000 × *g*, are more appropriate. It is strongly recommended to conduct AUC runs at different velocities, prior to initiating folding studies on a new RNA molecule, to find the appropriate velocity range. Once it is determined, all the experiments with a given RNA molecule should be performed at the same velocity to allow comparison between data sets.
18. Since RNA folding is mediated by titrating in increasing concentrations of the counterion, it is important to recalculate the buffer density and viscosity of each sample by entering the altered ionic strength in SEDNTERP.
19. Both, sedimentation and diffusion coefficients, s and D, are dependent on the solvent viscosity and temperature. Therefore, these parameters will differ for a given RNA molecule based on the experimental conditions in which its folding is analyzed. Hence, s and D, measured under a given experimental condition, are transformed to standard values with reference to a

standard temperature (20 °C) and a standard solvent (water) and reported as $s_{20,w}$ and $D_{20,w}$. This standardization is especially important in counterion-mediated RNA folding studies, as it reports on the changes in the intrinsic properties of an RNA molecule, and therefore, facilitates comparison between folding behaviors studied in different solution conditions.

20. The weight averaged sedimentation coefficient, s_w can be expressed as, $s_w = \int_0^{s_p^*} s^*.g\left(s^*\right).ds^* / \int g\left(s^*\right).ds^*$, where, $g(s^*)$ is the apparent sedimentation coefficient distribution function, expressed as: $g(s^*) = (\partial c/\partial t)_{corr}(1/C_0)[\omega^2 t^2/\ln(r_m/r)](r/r_m)^2$. The quantity, $(\partial c/\partial t)_{corr}$ is the corrected value of $(\partial c/\partial t)_r$ taking into account contributions from the plateau region, C_o is the initial loading concentration and r_m is the fixed position of the sample meniscus [41]. The maxima of the $g(s^*)$ versus s^* curve, denotes the sedimentation coefficient of the molecule, s. Appearance of more than one peak in this distribution would indicate the existence of distinct conformational states with different hydrodynamic properties or self-association of the RNA molecules or both. Users are advised to carefully check the distribution to rule out the presence of multiple peaks. The diffusion coefficients can be obtained from the standard deviation (σ) of the fits of the $g(s^*)$ distribution to a Gaussian function, according to the relation: $D = (\sigma r_m \omega^2 t)^2/2t$, where r_m is the fixed position of the sample meniscus, ω is the angular velocity and t is the time [41].

21. Since, s/D is proportional to the molecular weight, the plot of $s_{20,w}/D_{20,w}$ versus counterion concentration should produce a straight line parallel to the abscissa, provided that the molecules do not aggregate at high ionic strengths. It is not uncommon to observe deviation from this behavior in samples containing very high concentrations of Mg^{2+}.

22. Cooperative ligand binding to a molecule is quantitatively described by the Hill equation, $Y = K_d^{n_H}\left[M\right]^{n_H} / 1 + K_d^{n_H}\left[M\right]^{n_H}$, in which K_d represents the equilibrium dissociation constant, $[M]$ is the ligand concentration and n_H is the Hill coefficient that depicts the cooperativity of the process. The fitting of equilibrium RNA folding data points to the Hill equation is purely phenomenological and the fitted curves do not represent counterion binding to RNA; instead, the fitted curve at a given temperature, also called the folding isotherm, simply represents a transition between two states, mediated by the counterion. For the purpose of analyzing counterion-mediated RNA folding isotherms, K_d and n_H represent, respectively, the midpoint and the cooperativity of the folding transition, whereas $[M]$ denotes the concentration of the counterion.

23. In some cases where the RNA folding reaction is not a simple two-state transition, the folding isotherms may not be adequately fit by a single-site binding model. In such cases, the data points should be fit to modified Hill equations that represent multiple-site cooperative ligand binding. For example, an isotherm that is best described by two transitions can be phenomenologically fit to a two-site binding Hill equation:

$$\Upsilon = f\left\{\frac{K_{d1}^{\;n_{H1}}[M]^{n_{H}1}}{1+K_{d1}^{\;n_{H1}}[M]^{n_{H1}}}\right\}+(1-f)\left\{\frac{K_{d2}^{\;n_{H2}}[M]^{n_{H2}}}{1+K_{d2}^{\;n_{H2}}[M]^{n_{H2}}}\right\};$$

where Υ denotes the signal quantitated, f represents the fraction of the first transition, K_{d1}, K_{d2} and n_{H1}, n_{H2} are the midpoints and cooperativities of the two transitions, respectively, and $[M]$ is the counterion concentration.

Acknowledgements

The author wishes to express his gratitude to Dr. Ruben L. Gonzalez, Jr. for hosting him as a postdoctoral research scientist during the preparation of this manuscript, to Dr. Daniel MacDougall, Dr. Kelvin Caban, and Mr. Colin Kinz-Thompson for critically reading and commenting on the manuscript, to Dr. Michael Brenowitz for training him on AUC, and to Dr. Alain Laederach and Dr. Chunxia Chen for kindly providing the structural models of the *Twort* ribozyme. This work was supported partly by a Susan G. Komen for the Cure Postdoctoral Fellowship (PDF12231199) to the author and partly by a Research Project Grant (R01-GM 084288) from the National Institute of General Medicine and Sciences of the National Institute of Health to Dr. Ruben L. Gonzalez, Jr.

References

1. Gesteland RF, Cech TR, Atkins JF (2006) The RNA World, 3rd edn. CSHL Press, Cold Spring Harbor
2. Fresco JR (1998) RNA structure and Function. Cold Spring Harbor Monograph Archive, vol 35, CSHL Press, Cold Spring Harbor, ISBN 978-087969-941-3
3. Dethoff EA, Chugh J, Mustoe AM et al (2012) Functional complexity and regulation through RNA dynamics. Nature 482:322–330
4. Bai Y, Das R, Millett I et al (2005) Probing counterion modulated repulsion and attraction between nucleic acid duplexes in solution. Proc Natl Acad Sci USA 102:1035–1040
5. Takamoto K, Das R, He Q et al (2004) Principles of RNA compaction: insights from the equilibrium folding pathway of the P4-P6 RNA domain in monovalent cations. J Mol Biol 343:1195–1206
6. Russell R, Millett IS, Tate MW et al (2002) Rapid compaction during RNA folding. Proc Natl Acad Sci USA 99:4266–4271
7. Kwok LW, Shcherbakova I, Lamb JS et al (2006) Concordant exploration of the kinetics of RNA folding from global and local perspectives. J Mol Biol 355:282–293
8. Cole P, Crothers D (1972) Conformational changes of transfer ribonucleic acid. Relaxation

kinetics of the early melting transition of methionine transfer ribonucleic acid (*Escherichia coli*). Biochemistry 11:4368–4374
9. Das R, Kwok LW, Millett IS et al (2003) The fastest global events in RNA folding: electrostatic relaxation and tertiary collapse of the *Tetrahymena* ribozyme. J Mol Biol 332:311–319
10. Grilley D, Soto A, Draper D (2006) Mg^{2+}–RNA interaction free energies and their relationship to the folding of RNA tertiary structures. Proc Natl Acad Sci USA 103:14003–14008
11. Draper D, Grilley D, Soto A (2005) Ions and RNA folding. Annu Rev Biophys Biomol Struct 34:221–243
12. Mitra S (2009) Using analytical ultracentrifugation (AUC) to measure global conformational changes accompanying equilibrium tertiary folding of RNA molecules. Methods Enzymol 469:209–236
13. Kieft JS, Costantino DA, Filbin ME et al (2007) Structural methods for studying IRES function. Methods Enzymol 430:333–371
14. Brautigam CA, Wakeman CA, Winkler WC (2009) Methods for analysis of ligand-induced RNA conformational changes. Methods Mol Biol 540:77–95
15. Cantor CR, Schimmel PR (1980) Ultracentrifugation. In: Bartlett AC (ed) Biophysical chemistry, part II: techniques for the study of biological structure and function, vol II. W.H. Freeman and Company, San Francisco
16. Fujita H (1975) Foundations of ultracentrifugal analysis. Wiley, New York
17. Scott DJ, Schuck P (2005) A brief introduction to the analytical ultracentrifugation of proteins for beginners. Analytical ultracentrifugation: techniques and methods. Royal Society of Chemistry, Cambridge, UK
18. Tanford C (1961) Physical chemistry of macromolecules. Wiley, New York
19. Laue TM (1995) Sedimentation equilibrium as thermodynamic tool. Methods Enzymol 259:427–452
20. Laue TM, Stafford WF 3rd (1999) Modern applications of analytical ultracentrifugation. Annu Rev Biophys Biomol Struct 28:75–100
21. Stafford WF 3rd (1992) Boundary analysis in sedimentation transport experiments: a procedure for obtaining sedimentation coefficient distributions using the time derivative of the concentration profile. Anal Biochem 203:295–301
22. Stafford WF 3rd (1994) Boundary analysis in sedimentation velocity experiments. Methods Enzymol 240:478–501
23. Takamoto K, He Q, Morris S et al (2002) Monovalent cations mediate formation of native tertiary structure of the *Tetrahymena thermophila* ribozyme. Nat Struct Biol 9:928–933
24. Costantino D, Kieft JS (2005) A preformed compact ribosome-binding domain in the cricket paralysis-like virus IRES RNAs. RNA 11:332–343
25. Dann CE 3rd, Wakeman CA, Sieling CL et al (2007) Structure and mechanism of a metal-sensing regulatory RNA. Cell 130:878–892
26. Agalarov SC, Sogorin EA, Shirokikh NE et al (2011) Insight into the structural organization of the omega leader of TMV RNA: the role of various regions of the sequence in the formation of a compact structure of the omega RNA. Biochem Biophys Res Commun 404:250–253
27. Mitra S, Laederach A, Golden BL et al (2011) RNA molecules with conserved catalytic cores but variable peripheries fold along unique energetically optimized pathways. RNA 17:1589–1603
28. Philo JS (2000) A method for directly fitting the time derivative of sedimentation velocity data and an alternative algorithm for calculating sedimentation coefficient distribution functions. Anal Biochem 279:151–163
29. Shcherbakova I, Gupta S, Chance M et al (2004) Monovalent ion-mediated folding of the *Tetrahymena thermophila* ribozyme. J Mol Biol 342:1431–1442
30. Shcherbakova I, Mitra S (2009) Hydroxyl-radical footprinting to probe equilibrium changes in RNA tertiary structure. Methods Enzymol 468:31–46
31. Philo JS (2006) Improved methods for fitting sedimentation coefficient distributions derived by time-derivative techniques. Anal Biochem 354:238–246
32. Kieft JS, Batey RT (2004) A general method for rapid and nondenaturing purification of RNAs. RNA 10:988–995
33. Sclavi B, Sullivan M, Chance MR et al (1998) RNA folding at millisecond intervals by synchrotron hydroxyl radical footprinting. Science 279:1940–1943
34. Chauhan S, Woodson SA (2008) Tertiary interactions determine the accuracy of RNA folding. J Am Chem Soc 130:1296–1303
35. Pan J, Woodson SA (1998) Folding intermediates of a self-splicing RNA: mispairing of the catalytic core. J Mol Biol 280:597–609
36. Wilkinson KA, Merino EJ, Weeks KM (2006) Selective 2′-hydroxyl acylation analyzed by primer extension (SHAPE): quantitative RNA

structure analysis at single nucleotide resolution. Nat Protoc 1:1610–1616

37. Tijerina P, Mohr S, Russell R (2007) DMS footprinting of structured RNAs and RNA-protein complexes. Nat Protoc 2:2608–2623
38. Herschlag D, Cech TR (1990) Catalysis of RNA cleavage by the *Tetrahymena thermophila* ribozyme. 1. Kinetic description of the reaction of an RNA substrate complementary to the active site. Biochemistry 29: 10159–10171
39. Russell R, Das R, Suh H et al (2006) The paradoxical behavior of a highly structured misfolded intermediate in RNA folding. J Mol Biol 363:531–544
40. Wan Y, Mitchell D 3rd, Russell R (2009) Catalytic activity as a probe of native RNA folding. Methods Enzymol 468:195–218
41. Stafford WF (1997) Sedimentation velocity spins a new weave for an old fabric. Curr Opin Biotechnol 8:14–24
42. Golden BL, Kim H, Chase E (2005) Crystal structure of a phage *Twort* group I ribozyme-product complex. Nat Struct Mol Biol 12:82–89
43. Chen C, Mitra S, Jonikas M et al (2013) Understanding the role of three-dimensional topology in determining the folding intermediates of group I introns. Biophys J 104:1326–1337

Chapter 17

RNA Folding Dynamics Using Laser-Assisted Single-Molecule Refolding

Bishnu Paudel and David Rueda

Abstract

RNA folding pathways can be complex and even include kinetic traps or misfolded intermediates that can be slow to resolve. Characterizing these pathways is critical to understanding how RNA molecules acquire their biological function. We have previously developed a novel approach to help characterize such misfolded intermediates. Laser-assisted single-molecule refolding (LASR) is a powerful technique that combines temperature-jump (T-jump) kinetics with single-molecule detection. In a typical LASR experiment, the temperature is rapidly increased and conformational dynamics are characterized, in real-time, at the single-molecule level using single-molecule fluorescence resonance energy transfer (smFRET). Here, we provide detailed protocols for performing LASR experiments including sample preparation, temperature calibration, and data analysis.

Key words RNA folding, Single-molecule FRET, Kinetics, Thermodynamics, Temperature jump, LASR

1 Introduction

RNA molecules play numerous, important biological roles in gene regulation and expression, RNA processing, viral replication, and protein synthesis [1, 2]. In general, they must fold into a native tertiary structure to perform their biological function (Fig. 1a) [3–7]. Studying RNA folding can be challenging due to their rugged potential energy folding landscapes and local minima (Fig. 1b) [8–10]. These folding potential energy surfaces may entrap RNA molecules in deep energy wells leading to misfolded intermediates or kinetic traps [11–14]. Characterization of these kinetic traps is critical to obtain a complete picture of the RNA folding pathway. To describe the energy folding surface, it is necessary to measure the enthalpy, entropy, and free energy of each intermediate along the folding pathway [3, 15, 16]. To determine enthalpies and entropies of tertiary structure formation, it is key to measure the temperature-dependent equilibrium constants and folding rate

Christina Waldsich (ed.), *RNA Folding: Methods and Protocols*, Methods in Molecular Biology, vol. 1086,
DOI 10.1007/978-1-62703-667-2_17, © Springer Science+Business Media, LLC 2014

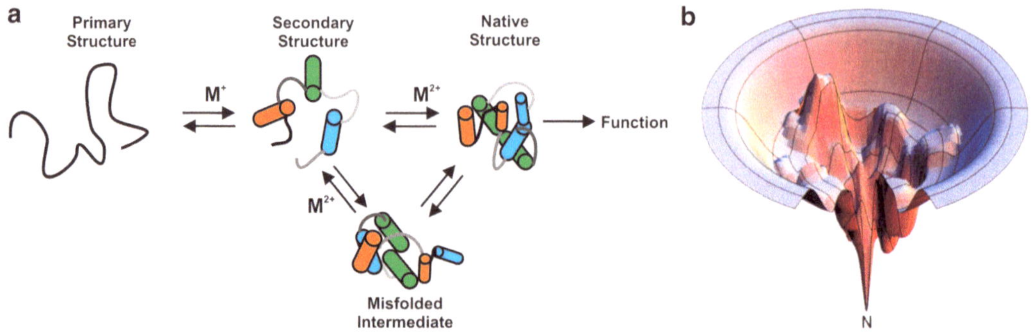

Fig. 1 RNA folding pathways. (**a**) RNA folds hierarchically: A linear chain of nucleic acids (primary structure) folds into its secondary structure in the presence of monovalent metal ions (M^+). The active tertiary (native) structure forms in the presence of divalent metal ions (M^{2+}) to acquire its biological function. Misfolded intermediates may be formed in the search of the native state. (**b**) RNA folding potential energy surface. The folding funnel leads towards the lowest energy state, the native state. But the energy landscape can be rugged and full of local minima that slow down folding and even trap the RNA in a kinetic trap. Adapted from reference [5] and [8] with permission

constants [17]. This can be achieved with traditional methods, such as isothermal titration calorimetry (ITC), differential scanning calorimetry, gradient gel electrophoresis, and time-resolved FRET methods [18–23]. In recent years, IR lasers have also been applied to induce conformational changes of biomolecules [24–28]. In these cases, the IR laser is used to increase the sample temperature very rapidly, which disrupts the chemical equilibrium. However, all these methods yield ensemble-averaged results, which may conceal key intermediates on the pathway.

To overcome the averaging problem, we have developed a new technology that combines temperature-jump kinetics with single-molecule fluorescence resonance energy transfer (smFRET) [29]. Laser-assisted single-molecule refolding (LASR) is a very useful approach to characterize trapped intermediates kinetically and thermodynamically at the single-molecule level. During a LASR experiment, infrared radiation near the water OH stretch absorption region (~1,400 nm) is shed onto a surface-immobilized RNA molecule of interest for a short period of time (typically 1 s). Consequently, LASR causes a sudden, transient, and reversible temperature increase of the solvent. Heat diffuses into solution within a few microseconds (<100 μs), a time scale that is much faster than the folding time of many large biomolecules (ms–s) [30]. Therefore, this rapid increase in thermal energy in solution allows kinetically trapped molecules to resample alternative conformations and refold into the native state [24–26, 28]. LASR is a useful method to study the refolding of biomolecules. One of the great advantages of LASR is its ability to determine transition state parameters, such as energy barriers [29]. Measuring such parameters can be challenging but important to understand how RNA molecules fold and acquire their biological function. Here, we present a detailed protocol for using LASR, including temperature calibration, to measure the melting temperature, thermodynamics, and kinetics of RNA and DNA molecules.

To understand the LASR technique, it is important to have background knowledge on total internal reflection microscopy (TIRFM). TIRFM is a powerful approach to study biomolecules. It is equally applicable to cell-imaging studies and in detecting single molecules [31–33]. For a detailed description on TIRFM, we direct the reader to a vast number of reviews on this topic [34–38]. Here, we only present a brief summary on TIRFM. The main advantage of TIRFM is to virtually eliminate all sources of background by reducing the excitation volume to a thin (~200 nm) sheet at the interface between the microscope slide and the solution. Owing to its low background signal, TIRFM enables the detection of fluorescence signals from single molecules on microscope slides. TIRFM also enables the observation of a wide field, increasing data throughput [34, 39, 40].

To achieve TIRFM (Fig. 2), the excitation light beam must reach the microscope slide at an angle more acute than the critical angle (θ_c). At this angle, the excitation light beam is fully reflected and does not penetrate into solution. The critical angle can be determined using Snell's law,

$$\theta_c = \sin^{-1}\left(\frac{n_w}{n_q}\right) \tag{1}$$

where n_w and n_q are the refractive indices of the solution and the microscope quartz slide ($n_q > n_w$). As a result, an evanescent wave penetrates into the solution. The intensity of the evanescent wave decays exponentially with depth, and therefore, it can only excite fluorophores immediately at the microscope slide (<200 nm), reducing background from other sources.

As a proof of principle, we have successfully applied LASR to measure melting temperatures of DNA duplexes (with an accuracy of ±1 °C) and RNA tertiary structure elements, as well as the transition state energy barriers of kinetically trapped RNAs [29]. To track the conformational changes before and after the T-jump, nucleic acids labeled with a common FRET pair (Cy3 and Cy5, as donor and acceptor, respectively) are typically used [34]. These labels have been used in numerous studies and are thought to only minimally affect the nucleic acid secondary and tertiary structures [41–43].

In our initial experiments, we chose a small RNA hairpin (Table 1) as a model system because it had been previously characterized and its folding pathway includes a transiently stable intermediate that is an effective kinetic trap [44]. In the specific

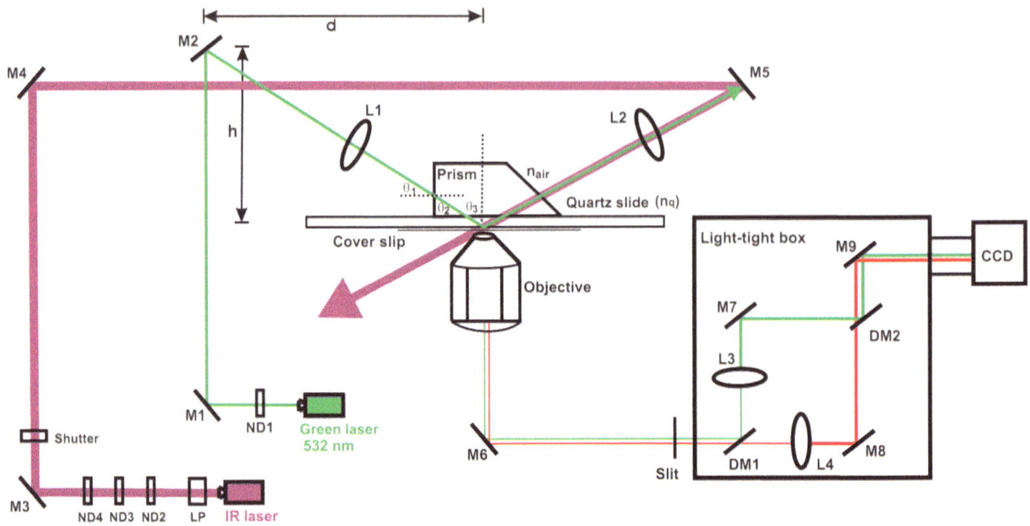

Fig. 2 Schematic diagram of LASR. A *green* laser (*green line*, 532 nm) is used to excite the FRET donor. A neutral density filter (ND1) adjusts the laser intensity. A series of aluminum mirrors (M1 and M2) are used to align the laser beam onto the sample at total internal reflection angle (θ_3). A lens (L1) focuses the beam on the sample. The angle of incidence of the beam can be aligned by adjusting the height (*h*) and the distance (*d*) of M2 to the sample. The IR radiation of a supercontinuum laser (*red line*) is filtered using long-pass filter (LP) and its intensity adjusted through series of neutral density filters (ND2-4). Gold mirrors (M3-5) are used to direct the IR beam onto the sample from the opposite end of the prism. The exposure time of IR laser is controlled using computer-driven shutter. The donor and acceptor fluorescence is collected using an achromatic objective lens. A mirror (M6) directs the emitted fluorescence towards a light tight box that splits the donor and acceptor fluorescence. An adjustable slit decreases the width of the image as indicated in the text. Inside the light tight box, a dichroic mirror (DM1) reflects the donor emission and transmits the acceptor to image both intensities separately on different halves of the CCD camera. A second dichroic mirror (DM2) redirects the separated beams onto the camera, with the help of mirrors M7-9. Each image is focused on the camera using independent lenses (L3-4). Each frame is recorded by the CCD camera in a computer for further analysis. Adapted from ref. [49]

case of the RNA hairpins used here, their complementary loops enable them to form a kissing complex through loop–loop base pairing (characterized by a ~0.45 FRET value). Their complementary stem sequences also enable them to eventually form an extended duplex (characterized by a ~1.0 FRET value) [29]. Although the duplex is the thermodynamically most stable conformation, it is only reached very slowly at room temperature, which makes the kissing complex a good model system to study RNA kinetic traps.

Future applications of LASR may include the characterization of the activation barriers involved in RNA folding memory effects (such as the hairpin ribozyme or riboswitches) [12, 13, 45], and kinetic traps in the folding reactions of large, multidomain RNAs (such as the group II intron ribozyme) [16, 46]. Currently, a limitation of LASR is that it cannot directly measure the kinetics of

Table 1
DNA and RNA oligonucleotide sequences

Name	Sequence[a]
dsDNA1	5′ Cy3 ATC TAT AAA AAT ATA GAT dT(Cy5)TTT 3′ B
	3′ TAG ATA TTT TTA TAT CTA 5′
dsDNA2	5′ Cy3 ATC ATC TCT CTC TAA GAT GAT dT(Cy5)TTT 3′ B
	3′ TAG TAG AGA GAG ATT CTA CTA 5′
dsDNA3	5′ Cy3 ATT GCG ATA GAG AGA GAT CGC AAT dT(Cy5) TTT 3′ B
	3′ TAA CGC TAT CTC TCT CTA GCG TTA 5′
HP1	5′ B AUA ACA AGG GGA AAU GCC UUG U 3′ Cy3
HP2	5′ Cy5 ACG AGG CAU UUC CCC UUG U 3′

[a]B = Biotin. Cy3 and Cy5 are linked to the nucleic acid through 6-carbon flexible linkers

interconversion during the T-jump, because of the necessity to close the camera shutter. Thus, only the initial and final conformations can be observed. However, this limitation has been solved in other similar approaches such as nanobathtubs, developed in the Nesbitt laboratory [47].

2 Materials

2.1 Synthetic RNA Substrate Purification and Labeling

1. Synthetic RNA or DNA oligonucleotides (Table 1, *see* **Note 1**). Store at −20 °C.
2. Dimethyl sulfoxide.
3. Triethylamine trihydrofluoride (Store in a cool, dry place).
4. *N*,*N*-dimethylformamide.
5. 1-butanol.
6. 80 and 100 % (v/v) ethanol.
7. 14 mL Falcon centrifuge tube.
8. High performance liquid chromatography (HPLC) system with C_8-reverse phase column.
9. HPLC stationary phase A: 100 mM triethylammonium acetate, pH 7.0.
10. HPLC mobile phase B: 100 % acetonitrile.
11. Cy5-succinimidyl ester.
12. Glucose oxidase and catalase. Store at −20 °C.

13. Standard buffer: 50 mM Tris–HCl, pH 7.5, and 10 mM $MgCl_2$.
14. T50 buffer: 50 mM Tris–HCl, pH 7.5, 50 mM NaCl.
15. Acidic piranha solution: 100 mL of 40 % H_2SO_4 and 50 mL of 30 % hydrogen peroxide.
16. Basic piranha solution: 100 mL ddH_2O, 20 mL 30 % hydrogen peroxide and 20 mL 30 % ammonium hydroxide.
17. Streptavidin.
18. Oxygen scavenger system (OSS) solution: dissolve 50 μL (by powder volume) of glucose oxidase in 100 μL of T50 buffer. Store at 4 °C. Mix 2 μL of this solution with 196 μL of standard buffer and 2 μL of 2-mercaptoethanol.

2.2 LASR Setup

1. Broadband supercontinuum laser (SC450) for IR radiation generation (*see* **Note 2**).
2. Green laser for FRET excitation (532 nm, 50 mW, Crystals Laser GCL-532-L).
3. A quartz, Pellin-Broca prism (CV1, Melles-Griot).
4. Three Au-mirrors for IR laser and six Al-mirrors for green laser (Newport).
5. Lenses (L, *f* = 100 mm, UV-fused silica, Thorlabs or Newport).
6. Inverted microscope frame (IX-71, Olympus).
7. Objective (20×, Olympus).
8. Uplans APO Objective (60×, oil immersion, numerical aperture = 1.2, WD = 0.28, W/CC = 0.13–0.21, Olympus, *see* **Note 3**).
9. Quartz microscope slides.
10. Cover slips.
11. Immersion oil (1.51 index of refraction).
12. High quantum yield, back illuminated, amplified CCD camera ($Ixon^+$, DV-897E, Andor, *see* **Note 4**).
13. Neutral density filters (0.04–2.5 OD; Newport).
14. Computer-driven shutter (Uniblitz electronics).
15. Two dichroic mirrors (635DCXT, Chroma).
16. A drill (Dremel 300-N) and diamond drill bits (1.0 mm diameter).
17. Long-pass filter (RG665; Schott).

2.3 Temperature Calibration

1. Microfabricated gold wire (Fig. 2).
2. Temperature-controlled oven.
3. Multimeter (Agilent) or comparable equipment.

4. Glass substrate in which gold wire is fabricated to make sensor plate.
5. Polydimethylsiloxane (PDMS).

3 Methods

3.1 Assembly of Microchannels onto Microscope Slides

1. Boil slides in double-distilled water (ddH_2O) for 30–40 min. If necessary, remove the cover slip and any debris from previous applications using razor blade (*see* **Note 5**).
2. Clean boiled slides using a 1:1 mixture of powder detergent and ddH_2O. Scrub slides vigorously with the fine mixture. Rinse slides thoroughly with ddH_2O to remove any remaining soap.
3. Scrub and rinse slides twice with ethanol and distilled water on both sides for 1–2 min each.
4. Boil slides for 30 min in either acidic piranha or basic piranha solution to regenerate the slide surface.
5. Rinse thoroughly with ddH_2O, and flame slide with a Bunsen burner to pyrolyse organic residues on the surface.
6. Immediately apply double-sided sticky tape to create a straight, 4 mm wide channel on either side of the holes. Apply a second layer of tape on the top of first one to make the channel depth approximately 100 μm.
7. Deposit a cover slip on the top of the sticky layer with caution so that both holes are covered on either side. Gently press the cover slip on the sticky tape to ensure it is well stuck.
8. Remove any tape excess outside the cover slip using a razor blade.
9. Apply epoxy glues to seal the gap between the cover slip and the tape.
10. The channel can be filled using a trimmed 200 μL pipette tip.

3.2 Sample Immobilization and Preparation

1. Heat-anneal DNA or RNA sample to be immobilized (Table 1) in standard buffer (1 μM in 10 μL final volume, *see* **Notes 6** and 7).
2. In two serial dilutions prepare a 200 μL solution of the RNA or DNA in standard buffer (final concentration 25–50 pM, *see* **Note 8**).
3. Inject 60 μL biotin-linked bovine serum albumin (B-BSA, 1 mg/mL) in T50 buffer into the quartz slide microchannel. Incubate for 10 min at room temperature.
4. Wash excess B-BSA by injecting 100 μL T50 buffer in microchannel.

5. Inject 100 μL streptavidin (0.2 mg/mL) in T50 buffer. Incubate for 10 min.
6. Flush excess streptavidin twice with 100 μL standard buffer.
7. Inject 100 μL of the 25–50 pM sample solution and incubate for 10 min (*see* **Note 9**).
8. Wash excess sample with 100 μL standard buffer solution.
9. Inject 100 μL of OSS with the second RNA (HP2, Table 1, final concentration can be variable, but 1 nM is typical) (*see* **Note 10**).

3.3 LASR Setup

1. Build a microchannel using a quartz microscope slide and a cover slip where the sample of interest will be surface immobilized, as previously described [29, 36, 38] (*see* Subheadings 3.1 and 3.2 and *see* **Note 1**).
2. Position slide on inverted microscope slide-holder with cover slip facing objective to collect the fluorescent signals from the sample (Fig. 2).
3. Place quartz prism onto the quartz slide to bring the laser at the critical angle on the slide (Fig. 2). Fill the air gap at the prism–slide interface with refractive index-matched immersion oil to avoid effects on the laser beam propagation.
4. Locate a mirror (M2) at a distance d from the prism to help direct the laser beam onto the sample. Adjust mirror height (h) to reach the critical angle. To reach the critical angle, the maximum height of the mirror should be

$$h_{\max} = d\sqrt{\frac{n_q^2 - n_w^2}{n_{air}^2 - \left(n_q^2 - n_w^2\right)}} \qquad (2)$$

where n_{air} is the refractive index of air.

5. Focus green laser onto the sample using a lens (L1, focal length $f_1 = 100$ mm, Fig. 2). To align the green laser beam accurately, install L1 on a three-dimensional micrometer-accuracy translation stage (*see* **Note 11**).
6. Install remaining mirrors and lenses as shown in Fig. 2.
7. Use L1 and the low magnification objective (20×) to bring the laser beam near the center of the objective axis on the quartz slide.
8. Switch to the high magnification objective (60×) for the final alignment (*see* **Note 3**).
9. Use the supercontinuum laser to generate the IR light that will cause the T-jump. This laser can generate white-light radiation (450–1,800 nm, 2 W, *see* **Note 2**).
10. Filter IR radiation through a long-pass filter (LP, Fig. 2) and direct the resulting beam to the sample through mirrors as

shown in Fig. 2. The filter used here allows a small amount of red light (≥665 nm) to propagate with the IR beam, which can be very useful for beam alignment purposes. Gold mirrors are recommended for the IR laser to minimize losses at reflective surfaces.

11. Focus IR laser beam onto the sample using a lens (L2, 100 mm focal length, Fig. 2) to increase its temperature rapidly. The IR laser angle of incidence should be more acute than the critical angle to avoid total internal reflection. It is important to make the green laser and IR laser overlap at the same position. The waist of the IR laser at the incident spot should be kept wider than the green laser to maintain uniform IR laser intensity in the probing region and therefore, a constant jump temperature.
12. Adjust height of the IR laser using a mirror (M5) to achieve the desire angle of incidence.
13. Use a combination of neutral density filters (ND, Fig. 2) to attenuate the IR laser power and to reach the desired temperature.

3.4 Separation of Colors for FRET

1. Carefully position a variable width slit at the image plane of the microscope to avoid interference effects on the image edge (Fig. 2).
2. Adjust microscope image size to match half of the CCD chip size using variable width slit. Adjust slit width while monitoring the image on the CCD camera.
3. Separate donor and acceptor signals into two paths with a first dichroic mirror (DM1, Fig. 2) that reflects the green fluorescence and transmitting the red fluorescence (*see* **Note 12**).
4. Focus donor and acceptor signal onto the CCD chip through lenses (L3-4, Fig. 2).
5. Recombine both signals in a side-by-side fashion with a second dichroic mirror (DM2, Fig. 2).
6. Align the donor and acceptor signals separately by mirrors M7 and M8, respectively (Fig. 2).
7. Align donor and acceptor images onto different halves of the CCD chip using DM2 and M9 (*see* **Notes 12** and **13**).

3.5 Temperature Calibration

1. Micro-fabricate a narrow (140 μm) gold wire sensor onto a masked glass substrate (Fig. 3a). Apply a thin layer of polydimethylsiloxane (PDMS) to insulate the gold wire (*see* **Note 14**).
2. Pre-calibrate the temperature dependence of the gold sensor resistibility in a temperature-controlled oven using a multime-

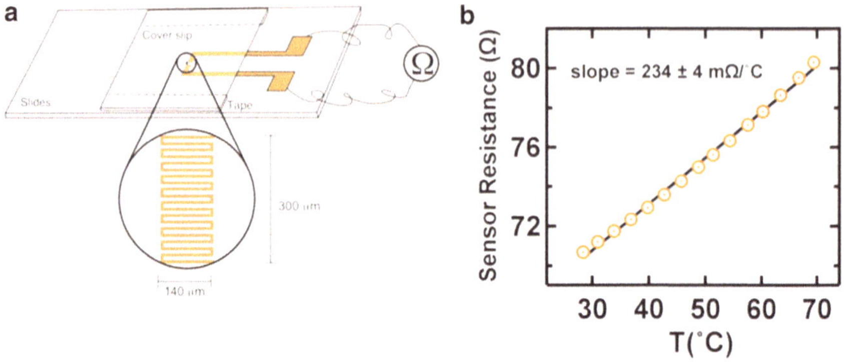

Fig. 3 Temperature calibration. (**a**) Temperature is calibrated using a small gold wire sensor. The gold wire is immobilized on the glass surface, insulated using polydimethylsiloxane (PDMS) and covered by a cover slip to mimic the experimental conditions. (**b**) Gold wire resistance as a function of temperature, measured in a pre-calibrated oven. Resistance increases linearly with temperature. Slope represents the thermal gradient of gold wire. Adapted from ref. [29] with permission

ter to measure the gold electrical resistance. The gold wire resistance increases linearly with temperature (Fig. 3b).

3. Plot the resistance as a function of temperature to obtain the temperature-resistance coefficient of the gold wire (slope, Fig. 3b).
4. Assemble the micro-fabricated gold wire into a microfluidic channel in the same geometry as the sample (Subheading 3.1 and Fig. 3a).
5. Inject standard buffer into the micro-fluidic channel to mimic the final experimental conditions.
6. Position the assembled sensor-slide on the objective, and focused IR beam onto the sensor through the prism as described in Subheading 3.3.
7. The IR laser alignment should be optimized using a micrometer-accurate translation stage by maximizing the observed resistance (Fig. 2). The gold wire sensor resistance is measured directly using the multimeter.
8. Turn IR laser on for 1 s using a computer-driven shutter. Adjust the intensity with neutral density filters to fine-tune the final jump-temperature (*see* **Note 15**).
9. Finally, the obtained resistance is converted to temperature using the pre-calibration curve (Fig. 3b). Temperature calibration should be done before conducting each experiment (*see* **Note 16**).

3.6 Checking the Temperature Calibration

1. Measure the ensemble-averaged UV melting curves of the DNA duplexes as previously described [48] (*see* **Note 17**).
2. Label one of the DNA strands with a 3′ biotin for surface immobilization, a 5′ Cy3 as a FRET donor, and a Cy5 near to 3′ end as a FRET acceptor (Subheading 3.2, Fig. 4a, Table 1).
3. Assemble microscope slides as described in Subheading 3.1 and immobilize each duplex DNA on the slide surface, as described in Subheading 3.2 [35, 36, 38]. The green laser is used to locate the surface-immobilized duplexes and determine their FRET value (Fig. 4c), and therefore their initial state (duplex or hairpin).
4. Turn IR laser on for 1 s to jump the solution temperature rapidly and reversibly (Fig. 4b).
5. Switch back to green laser to determine the state of each molecule after the jump.
6. Repeat these steps at increasing final jump temperatures.

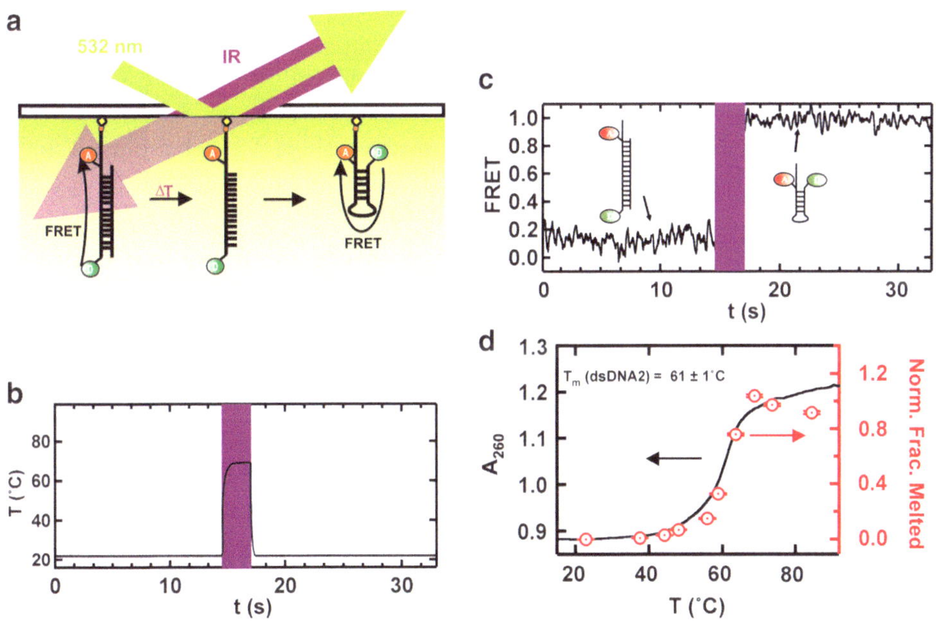

Fig. 4 Single-molecule melting of dsDNA. (**a**) A doubly-labeled dsDNA is surface immobilized onto a microscope slide using streptavidin and biotin as described in the text. DNA is designed (dsDNA2, Table 1) in such a way that it melts upon heating with the IR laser. Consequently, the immobilized strand forms an RNA hairpin that results in high FRET. (**b**) Schematic representation of the temperature change during the temperature jump. *Purple vertical line* represents the duration of heating. (**c**) Representative FRET time trajectory, showing the formation of hairpin structure (high FRET) from dsDNA (low FRET) after heating with IR laser. (**d**) Representative DNA melting curve helps to validate the temperature calibration of LASR. *Black line* is a melting curve obtained from ensemble-averaged UV–vis melting, whereas *red circles* represent LASR data. Both melting curves are within 1 °C accuracy. Adapted from ref. [29], with permission

7. Calculate the fraction of melted molecules (high FRET) among all molecules (low and high FRET) at each jump-temperature to obtain a single-molecule melting curve (Fig. 4d). Details on the procedure to determine melting temperatures and to compare the single-molecule and ensemble data is described in Subheading 3.8 (*see* **Note 18**).

3.7 Sample Imaging

1. Place slide with immobilized sample on microscope slide holder.
2. Align green laser through the TIRFM prism to excite the surface-immobilized molecules as described in **step 5** in Subheading 3.3.
3. Focus objective to detect single molecules.
4. Run acquisition software to record the data. Recording for 30 s before jumping the temperature is typically sufficient to characterize the initial conformational state of the surface-immobilized molecules. Program the computer-controlled IR shutter to open for 1 s after initial imaging to jump temperature. To avoid damaging the CCD chip, the camera shutter should be closed during the T-jump.

3.8 Determining the Refolding Temperature (T_c) of RNA Molecules

1. Calibrate temperature using standard buffer as described in Subheading 3.5.
2. Incubate sample of interest (here RNA hairpins, Table 1) in the microfluidic quartz slide, which is loaded as described in Subheading 3.2 (Fig. 5a).
3. Turn green laser on to excite the donor and record single-molecule movies.
4. After 70 s, turn IR laser on for 1 s to jump the temperature as described in **steps 9–13** in Subheading 3.3 (*see* **Note 19** and Fig. 5b).
5. Molecules whose FRET ratio changes from ~0.45 to 1.0 during the T-jump have interconverted from the kissing hairpin conformation (kinetic trap) to extended duplex conformation (native state). An example trajectory is shown in Fig. 5b (*see* **Note 20**).
6. Count manually the number of interconverting molecules at increasing jump temperatures to measure the refolding temperature.
7. The fraction of dissociated (f_D) or refolded (f_R) molecules is calculated as

$$\begin{cases} f_R = N_R / N_{tot} \\ f_D = N_D / N_{tot} \end{cases} \tag{3}$$

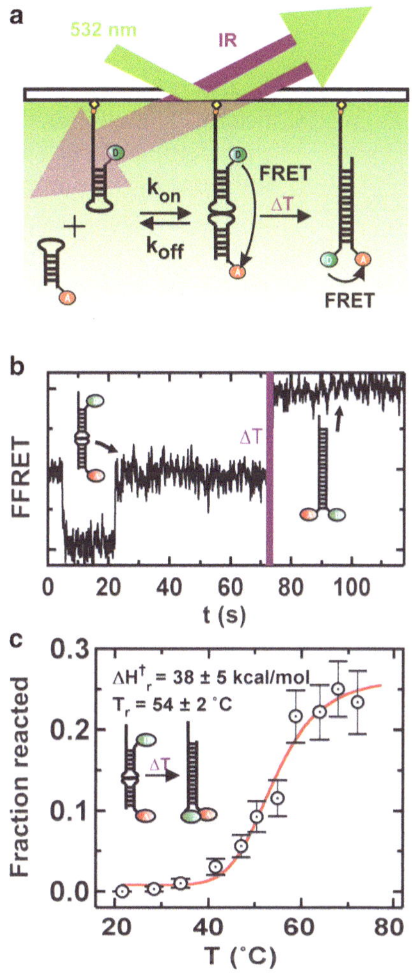

Fig. 5 Single-molecule RNA refolding. (**a**) An RNA hairpin (HP1, Table 1) labeled with a FRET donor is surface immobilized onto the slide using biotin–streptavidin as described in text. A complementary RNA hairpin (HP2, Table 1) labeled with a FRET acceptor approaches HP1 and forms a kissing complex. At room temperature, conversion to the extended duplex (thermodynamically most stable) is hindered by a large energy barrier (unfolding of the hairpins). A temperature jump (ΔT) induced by the IR laser, increases the local temperature helping the hairpins unfold and form the extended duplex. (**b**) Characteristic FRET time trajectory which shows interconversion of the kissing complex (FRET = 0.5) to the extended duplex (FRET = 1.0) after heating with the IR laser. (**c**) Fraction of molecules forming the extended duplex as a function of the jump temperature. The data is fitted to obtain the critical temperature (54 ± 2 °C) and the transition state enthalpy (38 ± 5 kcal/mol) as described in the text. Adapted from ref. [29], with permission

8. Fit calculated fractions to Eq. 4 to determine the melting or refolding critical temperatures (T_c):

$$f(T) = f_0 + (f_{max} - f_0)\frac{T^n}{T_C^n + T^n} \quad (4)$$

where $f(T)$ is the fraction of reacted molecules (melted or refolded) at a temperature T, f_0, and f_{max} are the initial and final fractions reacted, respectively, and n is the cooperativity coefficient. The cooperativity coefficient is related to the slope of the melting curve around the critical temperature and is an indicator of the number of interactions disrupted cooperatively to pass the transition state [29].

3.9 Transition State Analysis

1. Fit the fraction refolded as a function of temperature (Fig. 5c) to Eq. 5 to determine the transition state energy barriers:

$$f(T) = (f_{max} - f_0)\frac{e^{-\frac{\Delta H^{\pm}}{RT}\left(1-\frac{T}{T_C}\right)}}{1 + e^{-\frac{\Delta H^{\pm}}{RT}\left(1-\frac{T}{T_C}\right)}} \quad (5)$$

where T_C is the critical temperature (obtained from the fit to Eq. 4), $\Delta H^{\pm}$ the transition state enthalpy and R the ideal gas constant. This expression is derived from the combination of Eyring's transition state theory and Gibbs-Helmholtz equation (*see* **Note 21**).

2. Calculate the transition state entropy at the critical temperature as (assumes equilibrium at critical temperature between initial state and transition state; *see* **Note 22**):

$$\Delta S^{\pm} = \frac{\Delta H^{\pm}}{T_C} \quad (6)$$

3. Calculate the transition state free energy at temperature T using Eq. 7.

$$\Delta G^{\pm} = \Delta H^{\pm} - T\Delta S^{\pm} \quad (7)$$

4 Notes

1. The accurate measurement of distances from energy transfer efficiencies can be very challenging as several factors, such as overall quantum yields and relative fluorophore orientation can be difficult to determine. However, apparent energy transfer efficiencies are in most cases sufficient to identify conformational states. The apparent energy transfer efficiency is simply calculated as

$$FRET = \frac{I_A}{I_A + I_D} \quad (8)$$

where I_A and I_D are the acceptor and donor intensities, respectively. It is expected that temperature increases result in lower fluorophore emission. To a first approximation, the emission from both fluorophores decreases similarly, such that the overall change in the FRET ratio can be negligible.

2. Alternatively, a fiber coupled 1,445 nm continuous-wave laser can also be used [47].
3. It is important to use a high numerical aperture (NA) objective for single-molecule experiments. Typical objectives for prism-based TIRFM experiments have an NA = 1.2 and 60× magnifying power.
4. A high quantum yield camera capable of detecting very few photons per pixel is recommended. Typically, a cooled, electron multiplied charged coupled device (EM-CCD) camera is used. The CCD chipset is cooled to −80 °C to reduce background noise. The signal to noise (S/N) ratio can be optimized by adjusting the electron multiplying gain value (the highest value does not necessarily yield the highest S/N ratio). The S/N ratio can be further improved by adjusting the frame rate and the excitation intensity. Frame rates of ~30 frames/s and laser powers of ~1 mW typically enable the observation of structural dynamics in the 30 ms timescale or slower for several minutes.
5. In LASR experiments, the sample is immobilized onto a microscope slide with a microfluidic channel. The microfluidic channel is used to flow-in the different reagents for immobilization, the sample, the buffer, and any required cofactors. Our simple, home-built slides are prepared as previously described [38, 49].
6. The nucleic acid must be labeled with a biotin for surface immo bilization and both donor and acceptor fluorophores for imaging (Table 1). Typically, solid-state nucleic acid synthesis enables incorporation of only one fluorophore [50]. To incorporate the second fluorophore, amino-modified nucleic acids can be reacted with chemically activated fluorophores (for example, with succinimidyl ester groups) as previously described [22].
7. If the RNA construct is assembled from multiple strands, we recommend using a two- to four fold excess of the unlabeled or non-biotinylated strands to ensure complex formation with the labeled strand. The buffer can be adjusted depending on the system under investigation.
8. We recommend storing each solution on ice, protected from ambient light.
9. Incubation times and concentrations can be adjusted to optimize the surface coverage of immobilized molecules. It is very important to avoid the injection of bubbles in the channel.
10. Alternative OSSs can be used to extend the lifetime of the fluorophores [51, 52].

11. We recommend lowering the laser intensity for the initial alignment to minimize pre-photobleaching.
12. The separation of emission from donor and acceptor is achieved using dichroic mirrors between the camera and the microscope. The cut-off wavelengths must be matched to the emission spectra of the fluorophores. We recommend 635 nm for the Cy3 and Cy5 FRET pair. A schematic diagram for a home-built, light tight box to separate the signals is shown in Fig. 2.
13. The analysis software locates the donor and acceptor signals from individual molecules and records the intensity of each molecule as a function of time. Because donor and acceptor intensities are simultaneously recorded on the each half of the CCD chip, the signal from each half must be located and matched. A slide with bright fluorescence beads is used just before each experiment to map the two images onto each other. A detailed protocol has been published elsewhere [36, 38].
14. LASR is a technique in which the temperature of the system is increased to study the folding kinetics of biomolecules. A sudden, transient increase in the temperature of the system disrupts the folding equilibrium allowing molecules to overcome activation energy barriers of kinetic traps, which are typically not accessible at room temperature. Therefore, obtaining an accurate and reliable calibration of the temperature is key to interpreting the experiments appropriately. To this end, we use a temperature calibration method based on the resistivity of a small gold wire.
15. The final incident power of the IR laser is estimated to be ~100 mW including energy loss due to reflection and water absorption in air. Exposure time of the IR laser can be controlled using a computer-driven shutter. In our experiments, 1s exposure time is typically used to obtain uniform heating.
16. Alternative temperature calibration approaches consist of measuring the fluorescence lifetime of single fluorophores at increasing temperatures using time correlated single photon counting (TCSPC) [47]. The basic idea is that the excited state lifetime depends on the temperature. The fluorescence lifetime can be determined as a function of laser intensity, thus generating a correlation between laser intensity and final temperature.
17. To check the temperature calibration, we recommend using a sample with known temperature dependence. The melting temperature of DNA duplexes can be readily measured in most UV–vis spectrophotometers and can serve as a convenient model system to check the temperature calibration. The challenge is to determine when a duplex has actually melted. To this end, we

have designed DNA duplexes such that the melted single-stranded DNA forms a hairpin upon cooling (Table 1 and Fig. 4a). By labeling the ends of this DNA strand with Cy3 and Cy5, the duplex form can be identified by a characteristic low FRET signal (long distance between the fluorophore), whereas the single-stranded hairpin yields a high FRET signal (short distance between the fluorophores). We recommend using three different duplexes with melting temperatures varying between 40 and 70 °C. The melting temperature can be adjusted by varying the length and GC content of the duplex (Table 1).

18. The resulting melting curves should be in close agreement with the ensemble UV melting data (within 1 °C accuracy), as shown in Fig. 4d, confirming that the temperature calibration is accurate.
19. A key step is to distinguish trajectories arising from real molecules from background noise accidentally picked up by the analysis software, which can be challenging. Key criteria are continuous fluorescence emission, single step photobleaching and anti-correlated donor and acceptor emission. The fluorescence trajectories from bonafide single molecules are then used to calculate FRET trajectories using Eq. 8. Typical single-molecule FRET time trajectories are shown in Figs. 4c and 5b.
20. Before the T-jump, the FRET values observed randomly switch between 0 and 0.45 indicating formation and dissociation of the kissing complex at normal temperatures, but the high FRET value of the duplex is virtually never observed. Using LASR a fraction of molecules (up to 60 %) interconvert to the extended duplex within the 1 s T-jump, showing that molecules can use the heat energy to overcome the kinetic trap energy barrier [29].
21. For a complete derivation of this Eq. 7, including a discussion on the assumptions please *see* [29].
22. In the example of the kissing hairpins (Fig. 5c), the refolding temperature obtained is 54 ± 1 °C and the activation enthalpy obtained is 38 ± 5 kcal/mol. The activation enthalpy is in close agreement with a value measured independently by the isothermal titration calorimeter [53].

Acknowledgements

This is supported by the National Science Foundation (NSF CAREER award 0747285 to D.R.).

References

1. Burge CB, Tuschl TA, Sharp PA (1999) Splicing of precursors to mRNAs by the spliceosomes. In: Gesteland RF, Cech TR, Atkins JF (eds) The RNA world, 2nd edn. Cold Spring Harbor Press, Cold Spring Harbor, NY, pp 525–560
2. Eddy SR (2001) Non-coding RNA genes and the modern RNA world. Nat Rev Genet 2:919–929
3. Tinoco I Jr, Bustamante C (1999) How RNA folds. J Mol Biol 293:271–281
4. Brion P, Westhof E (1997) Hierarchy and dynamics of RNA folding. Annu Rev Biophys Biomol Struct 26:113–137
5. Pyle AM (2002) Metal ions in the structure and function of RNA. J Biol Inorg Chem 7:679–690
6. Woodson SA (2005) Metal ions and RNA folding: a highly charged topic with a dynamic future. Curr Opin Chem Biol 9:104–109
7. Greenleaf WJ, Frieda KL, Foster DA et al (2008) Direct observation of hierarchical folding in single riboswitch aptamers. Science 319:630–633
8. Dill KA, Chan HS (1997) From levinthal to pathways to funnels. Nat Struct Biol 4:10–19
9. Zhuang X (2005) Single-molecule RNA science. Annu Rev Biophys Biomol Struct 34:399–414
10. Chen SJ (2008) RNA folding: conformational statistics, folding kinetics, and ion electrostatics. Annu Rev Biophys 37:197–214
11. Treiber DK, Williamson JR (2001) Beyond kinetic traps in RNA folding. Curr Opin Struct Biol 11:309–314
12. Zhuang X, Kim H, Pereira MJ et al (2002) Correlating structural dynamics and function in single ribozyme molecules. Science 296:1473–1476
13. Ditzler MA, Rueda D, Mo J et al (2008) A rugged free energy landscape separates multiple functional RNA folds throughout denaturation. Nucleic Acids Res 36:7088–7099
14. Solomatin SV, Greenfeld M, Chu S et al (2010) Multiple native states reveal persistent ruggedness of an RNA folding landscape. Nature 463:681–684
15. Aleman EA, Lamichhane R, Rueda D (2008) Exploring RNA folding one molecule at a time. Curr Opin Chem Biol 12:647–654
16. Karunatilaka KS, Rueda D (2009) Single-molecule fluorescence studies of RNA: a decade's progress. Chem Phys Lett 476:1–10
17. Fiore JL, Kraemer B, Koberling F et al (2009) Enthalpy-driven RNA folding: single-molecule thermodynamics of tetraloop-receptor tertiary interaction. Biochemistry 48:2550–2558
18. Szewczak AA, Podell ER, Bevilacqua PC et al (1998) Thermodynamic stability of the P4-P6 domain RNA tertiary structure measured by temperature gradient gel electrophoresis. Biochemistry 37:11162–11170
19. Nixon PL, Giedroc DP (2000) Energetics of a strongly pH dependent RNA tertiary structure in a frameshifting pseudoknot. J Mol Biol 296:659–671
20. Hammann C, Cooper A, Lilley DM (2001) Thermodynamics of ion-induced RNA folding in the hammerhead ribozyme: an isothermal titration calorimetric study. Biochemistry 40:1423–1429
21. Feig AL (2007) Applications of isothermal titration calorimetry in RNA biochemistry and biophysics. Biopolymers 87:293–301
22. Rueda D, Walter NG (2006) Fluorescent energy transfer readout of an aptazyme-based biosensor. Methods Mol Biol 335:289–310
23. Reymond C, Bisaillon M, Perreault JP (2009) Monitoring of an RNA multistep folding pathway by isothermal titration calorimetry. Biophys J 96:132–140
24. Phillips CM, Mizutani Y, Hochstrasser RM (1995) Ultrafast thermally induced unfolding of RNase A. Proc Natl Acad Sci USA 92:7292–7296
25. Ballew RM, Sabelko J, Gruebele M (1996) Observation of distinct nanosecond and microsecond protein folding events. Nat Struct Biol 3:923–926
26. Proctor DJ, Ma H, Kierzek E et al (2004) Folding thermodynamics and kinetics of YNMG RNA hairpins: specific incorporation of 8-bromoguanosine leads to stabilization by enhancement of the folding rate. Biochemistry 43:14004–14014
27. Dyer RB, Brauns EB (2009) Laser-induced temperature jump infrared measurements of RNA folding. Methods Enzymol 469:353–372
28. Kirschner K, Eigen M, Bittman R et al (1966) The binding of nicotinamide-adenine dinucleotide to yeast d-glyceraldehyde-3-phosphate dehydrogenase: temperature-jump relaxation studies on the mechanism of an allosteric enzyme. Proc Natl Acad Sci USA 56: 1661–1667

29. Zhao R, Marshall M, Aleman EA et al (2010) Laser-assisted single-molecule refolding (LASR). Biophys J 99:1925–1931
30. Thirumalai D, Woodson SA (1996) Kinetics of folding of proteins and RNA. Acc Chem Res 29:433–439
31. Stephens DJ, Allan VJ (2003) Light microscopy techniques for live cell imaging. Science 300:82–86
32. Walter NG, Huang CY, Manzo AJ et al (2008) Do-it-yourself guide: how to use the modern single-molecule toolkit. Nat Methods 5:475–489
33. Mattheyses AL, Simon SM, Rappoport JZ (2010) Imaging with total internal reflection fluorescence microscopy for the cell biologist. J Cell Sci 123:3621–3628
34. Axelrod D (2001) Total internal reflection fluorescence microscopy in cell biology. Traffic 2:764–774
35. Ha T (2001) Single-molecule fluorescence resonance energy transfer. Methods 25:78–86
36. Roy R, Hohng S, Ha T (2008) A practical guide to single-molecule FRET. Nat Methods 5:507–516
37. Pljevaljcic G, Millar DP (2008) Single-molecule fluorescence methods for the analysis of RNA folding and ribonucleoprotein assembly. Methods Enzymol 450:233–252
38. Zhao R, Rueda D (2009) RNA folding dynamics by single-molecule fluorescence resonance energy transfer. Methods 49:112–117
39. Haustein E, Schwille P (2004) Single-molecule spectroscopic methods. Curr Opin Struct Biol 14:531–540
40. Rueda D, Walter NG (2005) Single molecule fluorescence control for nanotechnology. J Nanosci Nanotechnol 5:1990–2000
41. Bassi GS, Murchie AI, Walter F et al (1997) Ion-induced folding of the hammerhead ribozyme: a fluorescence resonance energy transfer study. EMBO J 16:7481–7489
42. Ha T (2001) Single-molecule fluorescence methods for the study of nucleic acids. Curr Opin Struct Biol 11:287–292
43. Zhuang X, Rief M (2003) Single-molecule folding. Curr Opin Struct Biol 13:88–97
44. Gregorian RS Jr, Crothers DM (1995) Determinants of RNA hairpin loop–loop complex stability. J Mol Biol 248:968–984
45. Wood S, Ferre-D'Amare AR, Rueda D (2012) Allosteric tertiary interactions preorganize the c-di-GMP riboswitch and accelerate ligand binding. ACS Chem Biol 7:920–927
46. Steiner M, Karunatilaka KS, Sigel RK et al (2008) Single-molecule studies of group II intron ribozymes. Proc Natl Acad Sci USA 105:13853–13858
47. Holmstrom ED, Nesbitt DJ (2010) Real-time infrared overtone laser control of temperature in picoliter H(2)O samples: "Nanobathtubs" for single molecule microscopy. J Phys Chem Lett 1:2264–2268
48. Santalucia J Jr (2000) The use of spectroscopic techniques in the study of DNA stability. In: Gore GM (ed) Spectrometry and spectrofluorimetry. Oxford University Press, New York, pp 329–354
49. Lamichhane R, Solem A, Black W et al (2010) Single-molecule FRET of protein-nucleic acid and protein–protein complexes: surface passivation and immobilization. Methods 52:192–200
50. Walter NG, Burke JM (2000) Fluorescence assays to study structure, dynamics, and function of RNA and RNA-ligand complexes. Methods Enzymol 317:409–440
51. Crawford DJ, Hoskins AA, Friedman LJ et al (2008) Visualizing the splicing of single pre-mRNA molecules in whole cell extract. RNA 14:170–179
52. Aitken CE, Marshall RA, Puglisi JD (2008) An oxygen scavenging system for improvement of dye stability in single-molecule fluorescence experiments. Biophys J 94:1826–1835
53. Salim N, Lamichhane R, Zhao R et al (2012) Thermodynamic and kinetic analysis of an RNA kissing interaction and its resolution into an extended duplex. Biophys J 102:1097–1107

Chapter 18

RNA Refolding Studied by Light-Coupled NMR Spectroscopy

Harald Schwalbe and Boris Fürtig

Abstract

Conformational transitions (refolding) between long-lived conformational states constitute the time-limiting step during the folding process of large RNAs. As the dynamics of these reactions dominate the regulatory and other functional behavior of RNA molecules, it is of importance to characterize them with high spatial and temporal resolution. Here, we describe a method for the investigation of RNA refolding reactions based on the photolytic generation of preselected conformations in a non-equilibrium state, followed by the observation of the folding trajectory with real-time NMR spectroscopy.

Key words Real-time NMR, NMR laser coupling, RNA refolding, Photo-cleavable protecting groups, Long-lived conformational states

1 Introduction

NMR spectroscopy is the only experimental biophysical technique available to characterize dynamics of bio-macromolecules at atomic resolution. A wide range of experimental parameters is available describing dynamic events at various amplitudes, underlying activation energies and timescales, covering in total more than nine orders of magnitude. With real-time NMR techniques dynamics events can be studied on a timescale of several milliseconds up to several hours, or ranging from $k = 10\ s^{-1}$ to $k = 0.1\ s^{-1}$ [1]. The combination of high-resolution spatial data with kinetic experiments is not achieved by any other biophysical method. Molecular dynamics in this time range include late events in protein folding such as domain reorientations, assembly of macromolecular complexes and large-scale conformational transitions (including RNA refolding).

While RNA secondary structure often forms within milliseconds both in vitro as well as during cellular RNA synthesis [2–4], refolding between formed stable secondary structure elements occurs on a slower timescale. Such refolding reactions, as opposed to folding reactions, are of importance as many ribonucleic acids are able to

Christina Waldsich (ed.), *RNA Folding: Methods and Protocols*, Methods in Molecular Biology, vol. 1086,
DOI 10.1007/978-1-62703-667-2_18, © Springer Science+Business Media, LLC 2014

adopt more than one single three-dimensional structure [5, 6]. For RNA, the different conformations often represent metastable states, since their interconversions are slow. Conformational states of almost equal stability are separated by high energetic barriers. Interestingly, the structural transitions between a single or several different metastable RNA states and the final functional state often constitute the rate-limiting steps on the folding pathway towards a functional RNA fold [7, 8].

In order to monitor these events with NMR experiments two preconditions have to be fulfilled: (1) first the initiation of the reaction should be performed fast as compared to the velocity of the monitored process and preferentially in the NMR spectrometer to avoid long dead times, (2) second the NMR techniques should be conducted in a manner that the temporal resolution is optimal. In order to meet these requirements, we developed a method for the investigation of RNA refolding reactions based on the photolytic generation of preselected conformations in a non-equilibrium state, followed by the observation of the folding trajectory with real-time NMR spectroscopy. RNA sequences that form two interconverting states with different secondary structures in a temperature-dependent equilibrium (termed bistable RNAs [9]) are modified with photo-labile protecting groups. This modification results in the selective destabilization of one of the conformational states and thereby generates a non-equilibrium conformational state, where only the state that remains unpertubed upon addition of the photo-labile group is populated [10, 11]. Using a coupled laser-NMR system [12, 13] release of the parental RNA sequence is achieved directly within the spectrometer. The relaxation of the RNA into its conformational equilibrium is subsequently followed at atomic resolution with the use of NMR experiments [14].

2 Materials

2.1 RNA Constructs

Molecules used in refolding studies can be either designed bistable RNAs for the examination of fundamental aspects of RNA folding behavior or domains of larger RNA molecules that show intrinsic bistable behavior and are responsible for the bistability of the whole RNA molecule [15, 16, 17]. Generally, designed bistable RNAs are of small size (20–40 nt), constitute two stable conformations characterized by (almost) equal thermodynamic stability and (partial) mutual exclusive base pairing pattern (Fig. 1) [18]. The design of these sequences can be conducted in silico using RNA folding prediction programs such as the Vienna Package [19] or Mfold [20]. In contrast, for bistable domains of larger RNAs the exact wild-type sequence is used as the modular architecture of RNAs allows for excision of whole domains without compromising

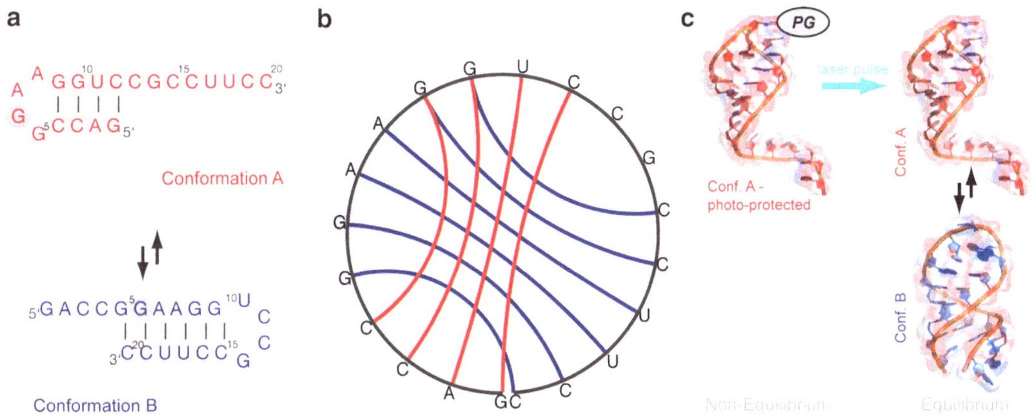

Fig. 1 Photo-protection strategy of bistable RNA system. (**a**) Example of a bistable RNA sequence that can form two distinct secondary structures in equilibrium. (**b**) Mutual exclusive base pairing pattern. The *gray circle* indicates the position of the photo-protecting group that exclusively destabilizes conformation B. (**c**) Comparison of the photo-protected non-equilibrium state that relaxes towards equilibrium after deprotection by a laser pulse

structural integrity of secondary structure elements. The size of these RNAs is limited to ~120 nt as the current inherent size limit for nucleic acid molecules studied by NMR.

All RNAs are investigated in solution under appropriate buffer conditions (*see* **Note 1**). Additionally, the use of small molecular ligands, metal-ions, and proteins as modulating co-solvents is possible [21].

2.2 Choice of Position of Photo-protecting Group and Chemical Synthesis

At first a sequence position has to be identified that exclusively in one of the two bistable conformations forms a strong base pairing interaction. This selected nucleobase is then exchanged to a 1-(2-nitrophenyl)ethyl-substituted (=NPE-substituted) nucleobase, which is designed to impair the formation of a base pair. This photo-cleavable modification leads subsequently to a specific destabilization of the preselected conformation. The bulky and photo-labile (2-nitrophenyl)ethyl group (=NPE group) is expected to disrupt Watson–Crick base pairs and thereby destabilizes associated secondary structures.

As the RNA used in these studies has to be photochemically modified, the molecules have to be synthesized chemically on solid phase. For experimental details concerning chemical synthesis of photo-protected nucleoside building blocks and solid-phase RNA synthesis we refer to following reviews [22–24].

2.3 Laser System

The in situ laser system allowing illumination of NMR samples within the NMR spectrometer consists of a CW-Ar-ion laser (BeamLok 2060 (Spectra Physics) 9.7 W power operating at 340–365 nm) whose beam is conducted to the sample tube via an

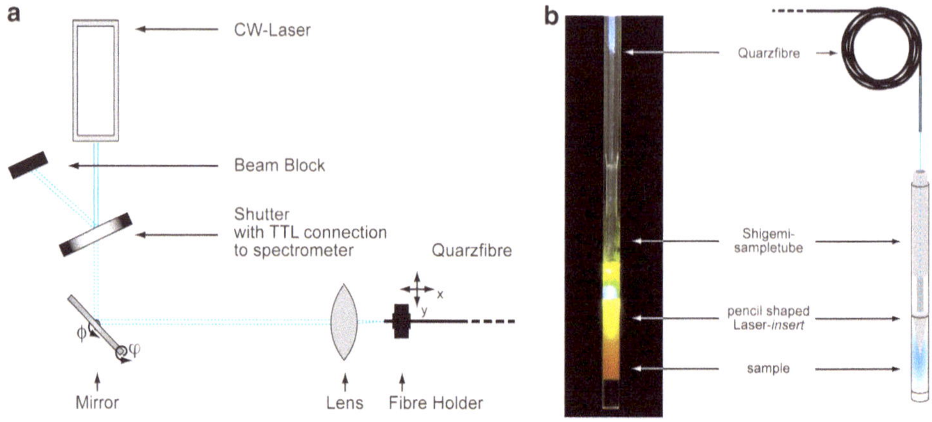

Fig. 2 Laser setup for in situ illumination of NMR samples. (**a**) Laser setup outside the magnet. (**b**) Modified NMR sample tube

optical fiber (LWL-Kabel UV1000/1100 N (Ceram Optec)). A high power resistant lens (lens: U13X (Newport) mounted in a combined lens and fiber holder: F-91-C1-T, FPR1-C1A (Newport)) obtains coupling of the laser beam into the fiber (other optical components used -beamsplitter 200-0 (CVI), -mirror 20010 (CVI)). Irradiation of the complete sample is maintained using pencil shaped Shigemi tips (*see* **Note 2**). A shutter (shutter: uniblitz electronic VS14S2ZM1 (Vincent Associates); shutter-driver: uniblitz VCM-D1 (Vincent Associates)), connected to the spectrometer blocks the laser beam in front of the fiber and can trigger the laser irradiation (Fig. 2).

2.4 NMR System

The NMR system has to consist of a high field liquid state spectrometer equipped with a cryogenic probe with *z*-axis gradients.

3 Methods

3.1 Sample Preparation

1. Carry out all sample preparation steps in a dark room and/or use light protection labware.
2. Prepare sample of bistable RNA after application of a proper folding protocol (*see* **Note 3**) at a final concentration of $c_{RNA} = 0.1$ mM (*see* **Note 4**).
3. Fill a Shigemi NMR tube with 190 μL to maximum 230 μL of RNA solution. Arrange the NMR tube in the spinner turbine such that the tip of the pencil shaped Shigemi plunger resides close to the center of the NMR-coil.

3.2 Laser Coupling

1. Setup the optical components of the laser system according to the scheme of Fig. 2.
2. Check for power losses of all optical components at low power levels of the laser (100 mW). Optimize coupling of laser light into the optical fiber by changing the angle and position of the fiber holder-lens system or mirror. Measure the output power at the end of the fiber that will be connected to the NMR sample tube (*see* **Note 5**). The coupling efficiency should be as high as 75 %, when comparing the power output at the end of the fiber to the power output at the laser head.
3. Connect the shutter driver to the spectrometer via a TTL connection and check if your pulse-code properly addresses the shutter driver.
4. Attach the end of the optical fiber to the NMR sample tube that is already placed in a NMR spinner turbine. The end of the fiber reaches to the very bottom inside of the Shigemi plunger. It is connected to the NMR tube by taping plastic paraffin film around the fiber and the NMR tube at the upper end of the tube.
5. The sample is inserted manually into the NMR spectrometer, to avoid breaking the optical fiber by using the air flow.
6. Increase the output power of the laser to maximum, but block the laser beam in front of the shutter till the real-time experiments is started, in order to avoid damages of the shutter system.

3.3 Real-Time NMR Measurements

1. Set up the NMR experiments using state-of-the-art-technology and conduct all necessary steps (tuning the probe, determination the on-resonance frequency, determination of 1H pulse strength) identical to standard routine of NMR experiments. Except, take special care when optimizing the field homogeneity (shimming) over the pencil shaped sample tube. It is advisable to generate a good starting shim before the RT-NMR measurement on a regular Shigemi tube with similar sample volume. For replicates of the RT-NMR measurement measure the exact position of tube and plunger in relation to the NMR spinner turbine and use these settings always with the same starting shim. Let the sample thermally equilibrate for minimum 20 min before each of the experimental series.
2. Investigate the RNA refolding reaction by an appropriate NMR pulse sequence (*see* **Note 6**) that allows detection of RNA imino protons (*see* **Note 7**).
3. Apply the pulse sequence in the following pseudo-2D or pseudo-3D scheme (Fig. 3): Run the kinetic experiments with as little as possible scans per time increment. Execute first several scans (e.g., ns = 32), followed by irradiation with one laser pulse

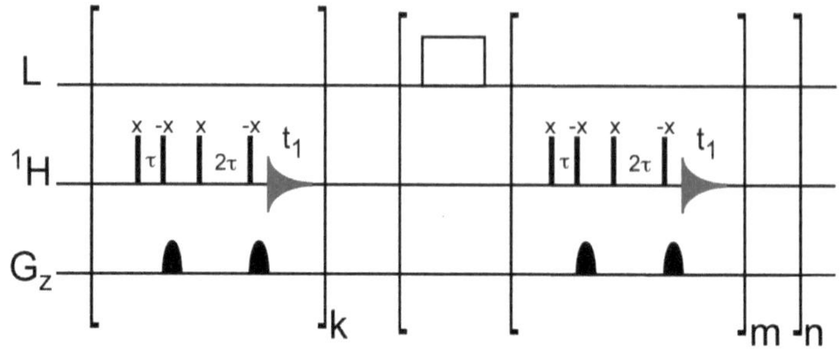

Fig. 3 Exemplary pulse-sequence for a pseudo n-D RT experiment. The scheme highlights the jump-return echo sequence that gives the best S/N ratio for the detection of imino protons in RNA and allows recording single scan kinetic experiments; in a pseudo 2D experiment $n=1$ and k and m are variable, L describes the channel triggering the laser-shutter

of a duration that is short compared to the time frame of the refolding reaction (e.g., 1 s). Then record as many points (e.g., ns = 128) as needed to digitize the kinetic traces properly. Repeat this experimental sequence (laser pulse, number of scans, times points for recording the kinetics) twice (*see* **Note 8**).

4. It is advisable to record identical static control experiments before and after the RT-NMR experiment (e.g., ^{1}H 1D or ^{15}N HSQC) with a high signal-to-noise (S/N) value to characterize the photo-protected non-equilibrium state as well as the fully relaxed equilibrium state (*see* **Note 9**).

3.4 Analysis and Data Treatment

1. Fourier-transform the resulting pseudo 2/3D data-set with appropriate number of data points in the direct dimension after multiplication with a window function (generally a shifted squared sine function is used), submit the resulting spectra to phasing and polynomial baseline correction.
2. Integrate each of the imino proton signals at half-peak height along the time-axis. Export these values (integral of imino peak vs. time) and analyze with appropriate software (e.g., SigmaPlot or Mathematica).
3. For the kinetic analysis employ pairs of disappearing/appearing signals.
4. Determine the residual baseline in the indirect dimension by averaging the intensity values of the first scans (recorded before photolysis) at the position of the appearing signal. Subtract this value from both appearing and disappearing signals.
5. Determine the degree of photolysis after the first laser-pulse from the appearing signals, by comparing the average of the scans at the end of first laser pulse period with the average of scans at the end of the second laser pulse period. Normalize

the signals accordingly and such that each pair of imino protons adds up to unity.

6. Choose an appropriate kinetic mechanism and fit the experimental data accordingly (*see* **Note 10**).

4 Notes

1. Generally good starting buffers are either 25 mM potassium arsenate buffer in H_2O/D_2O 9:1, pH 7.0, or 50 mM potassium phosphate, pH 6.4, with 25 mM potassium chloride in H_2O/D_2O 9:1. But generally all buffers that do not compromise the optical density at the wavelength of the laser are acceptable; pH values slightly below pH 7.0 are advantageous as peaks of RNA-imino protons are not broadened by increased exchange with solvent protons.
2. Although Kuprov et al. report that usage of a tapered fiber allows more efficient and uniform irradiation of the sample, we could not reproduce this better irradiation by a self-made tapered fiber prepared according to their protocol [25].
3. Photo-protected samples are folded similarly to their parental sequences. Dilute RNA (in pure water) is heated for 2–5 min at 95 °C then snap cooled with appropriate volumes of concentrated ice cold buffer to reach the final buffer concentration at a one tenth to one thousand dilution of RNA (*see* **Note 1**). The sample RNA is then concentrated by centrifugal ultrafiltration devices.
4. RNA concentrations higher than 0.1 mM lead to a better S/N ratio in NMR experiments but derogate the yield of the photolytic release. If a sufficient S/N is not achieved at this concentration it is advisable to repeat the experiment at low concentration for several times under identical conditions and sum up the NMR data post acquisition.
5. The output-power at the end of the fiber depends strongly on the condition of the cut surface. It is of advantage to prepare several new cut fiber-ends and determine which one of the ends shows the highest output power; this one should then be used.
6. The advantage of the 1D method is its intrinsic high sensitivity at highest temporal resolution. Normally ^{1}H, ^{19}F and ^{31}P are the nuclei on which the experiments are conducted regarding their high natural abundance (99.9885 %, 100 %, 100 %, respectively) and their high sensitivity (2.79, 2.62, 1.13 relative value of z-component of nuclear magnetic moment in units of nuclear magneton, respectively). Despite its intrinsically high sensitivity the spectral resolution of the 1D method

is limited. It is best applied to spectral regions of high intrinsic resolution as the imino signal region for RNA. Furthermore this limitation can be overcome by using selectively labeled compounds and application of filter experiments. As an example in the case of kinetic studies on RNA it is possible to increase resolution by using either a selective ^{15}N-uridine- or ^{15}N-guanine-labeled RNA which allows varying detection of either the ^{15}N-bound protons, or the ^{14}N-bound protons. In order to facilitate detection of ^{15}N-bound signals as well as ^{14}N-bound signals during a single kinetic experiment, application of selective filter techniques [26] in HMQC-fashion [27] is necessary.

Although 1D methods represent the intrinsically highest temporal resolution they are limited in this respect by the relaxation delay that is needed to start every experiment with a magnetically relaxed system. The parameter that describes the behavior of the system depending on the relaxation delay is the signal-to-noise ratio per experimental time and therefore this has to be optimized. The methodology of choice to tweak this parameter is Ernst-angle excitation; meaning that for fast repetitive pulse experiments, small pulse rotation angles are required to achieve maximum signal amplitude. Even higher dimensional experiments can be utilized in real-time studies exploiting this excitation in conjunction with band selective excitation (SOFAST-experiments [28]).

7. The folding of RNA molecules is monitored primarily on the imino proton resonances. The advantage of that attempt is the highest intrinsic spectral resolution obtained. Comparison of signals per spectral region for a given RNA molecule in non-deuterated solvent indicates that no other spectral region omits a better resolution. Furthermore the imino resonances monitor the base pairing behavior of the nucleotides in the RNA molecule, which are the major determinant of the structural elements and the stability of the whole RNA. Therefore the presence of an imino signal can be directly understood as a long-range restrain that represents a base pair between two distinct nucleotides of the RNA molecule. But since the imino protons are labile with respect to exchange it is important to know if there may be a difference in the exchange rate of the two (or more) states in-between the folding is characterized. If there would be a major difference in the exchange rate, the apparent rates of folding would be influenced by that and could not directly be interpreted as rates of folding.

8. This approach leads always to an NMR data set that is constructed as pseudo 2D or 3D experiments. The dimensions of the experiment are the chemical shift (δ^1H along $\omega 2/\omega 3$) and the real-time (RT along $t1/t1$ and $t2$). The first plane in a

pseudo 3D is recorded before application of the first trigger, whereas all following planes are recorded after a respective trigger event. All $1+n$ planes comprise therefore a single kinetic trace. If a single trigger event is sufficient to transfer the system from non-equilibrium to full equilibrium the pseudo-dimensionality can be reduced to two and the trigger is then located unevenly (e.g., after 32 points in a 128 points large dimension) in the pseudo-second dimension.

9. The equilibrium state as well as the photo-protected non-equilibrium state should be characterized by standard NMR technologies [29]. As the NMR signals of the imino protons are analyzed in the RT-NMR experiment special emphasis should be place on the elucidation of the base pairing pattern and secondary structure. Therefore ^{1}H-^{1}H NOESYs should be recorded for different temperatures within the temperature range of the kinetic analysis. If the RNA sequences are available in ^{15}N-labeled form, then ^{15}N HSQC and HNN-COSY experiments are adjuvant tools to describe the base pairing pattern.

10. Normally it is sufficient to describe the refolding behavior of bistable RNAs with a two state model. Nevertheless, bistable RNAs can also be understood as an ensemble of different RNA conformations containing different numbers of base pairing interactions. A distribution function that relates the relative population of RNA conformers with different numbers of base pairs to the free energy of each conformer describes the ensemble and is based on the free energy of each conformer on the basis of secondary structure prediction models. A forward refolding rate is assigned to each conformer based on the experimental finding that the rate-limiting step on the refolding trajectory is the unfolding of secondary structure. The changes in the concentration of the two conformations A and B involved in RNA refolding can then be mathematically described by:

$$\left|\begin{array}{l} \dfrac{dA}{dt} = \left(\sum_{i=1}^{l} -k_i \times a_i \times A[t]\right) + \left(\sum_{i=1}^{m} n_i \times b_i \times B[t]\right) \\ \dfrac{dB}{dt} = \left(\sum_{i=1}^{l} k_i \times a_i \times A[t]\right) + \left(\sum_{i=1}^{m} -n_i \times b_i \times B[t]\right) \end{array}\right| \quad (1)$$

Here a_i and b_i denote the population of a certain conformer in the ensemble of structures representing states A and B, respectively, and k_i and n_i the refolding rates according to the number of base pairs that are present in this state. l and m are the maximum numbers of base pairs in states A and B, respectively.

Acknowledgements

The work is supported by the state of Hesse (BMRZ), the DFG collaborative research center: Molecular principles of RNA-based regulation. H.S. is a member of the DFG center of excellence: macromolecular complexes.

References

1. Kleckner IR, Foster MP (2011) An introduction to NMR-based approaches for measuring protein dynamics. Biochim Biophys Acta 1814:942–968
2. Thirumalai D, Woodson SA (2000) Maximizing RNA folding rates: a balancing act. RNA 6:790–794
3. Webb AE, Weeks KM (2001) A collapsed state functions to self-chaperone RNA folding into a native ribonucleoprotein complex. Nat Struct Biol 8:135–140
4. Porschke D, Hoffman GW, Senear A (1973) Double helical complex formed from a polynucleotide and a complementary monomer. Nat New Biol 242:45–46
5. Schwalbe H, Buck J, Furtig B et al (2007) Structures of RNA switches: insight into molecular recognition and tertiary structure. Angew Chem Int Ed Engl 46:1212–1219
6. Dethoff EA, Chugh J, Mustoe AM et al (2012) Functional complexity and regulation through RNA dynamics. Nature 482:322–330
7. Treiber DK, Rook MS, Zarrinkar PP et al (1998) Kinetic intermediates trapped by native interactions in RNA folding. Science 279:1943–1946
8. Zarrinkar PP, Williamson JR (1994) Kinetic intermediates in RNA folding. Science 265:918–924
9. Hobartner C, Micura R (2003) Bistable secondary structures of small RNAs and their structural probing by comparative imino proton NMR spectroscopy. J Mol Biol 325:421–431
10. Wenter P, Furtig B, Hainard A et al (2005) Kinetics of photoinduced RNA refolding by real-time NMR spectroscopy. Angew Chem Int Ed Engl 44:2600–2603
11. Wenter P, Furtig B, Hainard A et al (2006) A caged uridine for the selective preparation of an RNA fold and determination of its refolding kinetics by real-time NMR. Chembiochem 7:417–420
12. Kühn T, Schwalbe H (2000) Monitoring the kinetics of ion-dependent protein folding by time-resolved NMR spectroscopy at atomic resolution. J Am Chem Soc 122:6169–6174
13. Wirmer J, Kühn T, Schwalbe H (2001) Millisecond time resolved photo-CIDNP NMR reveals a non-native folding intermediate on the ion-induced refolding pathway of bovine α-lactalbumin. Angew Chem 113:4378–4381
14. Manoharan V, Furtig B, Jaschke A et al (2009) Metal-induced folding of Diels-Alderase ribozymes studied by static and time-resolved NMR spectroscopy. J Am Chem Soc 131:6261–6270
15. Micura R, Hobartner C (2003) On secondary structure rearrangements and equilibria of small RNAs. Chembiochem 4:984–990
16. Micura R, Pils W, Hobartner C et al (2001) Methylation of the nucleobases in RNA oligonucleotides mediates duplex-hairpin conversion. Nucleic Acids Res 29:3997–4005
17. Reining A, Nozinovic S, Schlepckow K, Buhr F, Fürtig B, Schwalbe H (2013) Three-state mechanism couples ligand and temperature sensing in riboswitches. Nature 499: 355–359
18. Furtig B, Wenter P, Reymond L et al (2007) Conformational dynamics of bistable RNAs studied by time-resolved NMR spectroscopy. J Am Chem Soc 129:16222–16229
19. Gruber AR, Lorenz R, Bernhart SH et al (2008) The Vienna RNA websuite. Nucleic Acids Res 36:W70–W74
20. Zuker M (2003) Mfold web server for nucleic acid folding and hybridization prediction. Nucleic Acids Res 31:3406–3415
21. Furtig B, Wenter P, Pitsch S et al (2010) Probing mechanism and transition state of RNA refolding. ACS Chem Biol 5:753–765
22. Kawashima E, Kamaike K (2004) Synthesis of stable-isotope (C-13 and N-15) labeled nucleosides and their applications. Mini Rev Org Chem 1:309–332
23. Heckel A (2007) Nucleobase-caged phosphoramidites for oligonucleotide synthesis.

Curr Protoc Nucleic Acid Chem Chapter 1, Unit 1 17

24. Mayer G, Heckel A (2006) Biologically active molecules with a "light switch". Angew Chem Int Ed Engl 45:4900–4921
25. Kuprov I, Hore PJ (2004) Uniform illumination of optically dense NMR samples. J Magn Reson 171:171–175
26. Otting G, Wüthrich K (1989) Extended heteronuclear editing of 2D 1H NMR spectra of isotope-labeled proteins, using the X(w1, w2)-double-half-filter. J Magn Reson 85:586–594
27. Buck J, Furtig B, Noeske J et al (2007) Time-resolved NMR methods resolving ligand-induced RNA folding at atomic resolution. Proc Natl Acad Sci USA 104:15699–15704
28. Schanda P, Kupce E, Brutscher B (2005) SOFAST-HMQC experiments for recording two-dimensional heteronuclear correlation spectra of proteins within a few seconds. J Biomol NMR 33:199–211
29. Furtig B, Richter C, Wohnert J et al (2003) NMR spectroscopy of RNA. Chembiochem 4:936–962

Chapter 19

Exploring RNA Oligomerization and Ligand Binding by Fluorescence Correlation Spectroscopy and Small Angle X-Ray Scattering

Eileen Magbanua, Petr V. Konarev, Dmitri I. Svergun, and Ulrich Hahn

Abstract

RNA forms defined structures and binds specifically to target molecules. The combination of data which results from fluorescence correlation spectroscopy (FCS) and small angle X-ray scattering (SAXS) measurements can be used to determine intermolecular interactions between RNA and its binding partners. To define oligomerization states of free RNA and its complexes with bound target molecules, hydrodynamic radii, radii of gyration as well as the maximum sizes of the components have to be determined and compared. Furthermore, the program OLIGOMER allows calculating the portions of monomeric and dimeric RNA, for instance, within a mixture.

Key words FCS, FCCS, SAXS, RNA dimerization, RNA oligomerization, Aptamers

1 Introduction

Aptamers are one group of nucleotides that bind target molecules with high affinity and specificity due to their three-dimensional structure. The target molecules can range from low molecular weight components to complex proteins. Depending on the binding mode between target and aptamer, the aptamer can interact as monomer or dimer for instance. This oligomerization state might change dependent on the bound molecule [1]. To monitor the dynamics of free and/or complexed aptamers one can use fluorescence correlation spectroscopy (FCS) and small angle X-ray scattering (SAXS).

FCS is a fluorescence-based method to study molecular interactions based on changes in mobility [2]. FCS requires fluorescence labeling of at least one of the interacting components. In case of two associated molecules the lower molecular weight component is fluorescently labeled and the complex has to differ

Christina Waldsich (ed.), *RNA Folding: Methods and Protocols*, Methods in Molecular Biology, vol. 1086,
DOI 10.1007/978-1-62703-667-2_19, © Springer Science+Business Media, LLC 2014

significantly from the fluorescently labeled item regarding molar mass or rather diffusion time. To investigate the interaction between those two molecules, first the diffusion time of the fluorescently labeled component has to be characterized to calculate its hydrodynamic radius. Then the second unlabeled component is stepwise added yielding a significant increase of the diffusion time. The proportion of the different fluorescing components as well as their diffusion times can be calculated applying the two-component model. Using diffusion time of the complex, the hydrodynamic radius can be defined.

Fluorescence cross correlation spectroscopy (FCCS) is used to identify and characterize interactions between two fluorescing components [3]. The big advantage of FCCS over FCS is the fact that mass differences of the components investigated are irrelevant. Labeling the examined RNA with different fluorophores allows investigating its intermolecular dimerization.

SAXS is a powerful method for the analysis of biological macromolecules in solution [4]. Over the last decade, major advances in instrumentation and computational methods have led to new and exciting developments in the application of SAXS to structural biology including globular proteins, macromolecular complexes but also flexible systems [5–9]. SAXS can probe structure on an extremely broad range of macromolecular sizes, ranging from small proteins and polypeptides (a few kDa), to macromolecular complexes and large viral particles (to several hundred MDa). RNA molecules can be measured in solution under near native conditions, and the effect of changes in sample environment (like pH, temperature, ionic strength, ligand addition) can be easily followed. Still, the sample solutions must be well prepared, thoroughly purified and characterized before doing SAXS experiments. In such an experiment, samples containing dissolved macromolecules are exposed to an X-ray beam and the scattered intensity is recorded by a detector as a function of the scattering angle. Dilute aqueous solutions of nucleic acids give rise to an isotropic scattering intensity I, which depends on the modulus of the momentum transfer s ($s = 4\pi \sin(\theta)/\lambda$, where λ is the wavelength of the beam, and 2θ is the angle between the incident and scattered beam). The solvent scattering is subtracted and the background corrected intensity is presented as a radially averaged one-dimensional curve $I(s)$. For monodisperse solutions of identical randomly oriented molecule, the SAXS curve is proportional to the scattering of a single particle averaged over all orientations.

Several overall parameters can be directly obtained from an SAXS curve providing information about the size, oligomeric state, and overall shape of the molecule. Moreover, for folded nucleic acids or macromolecular complexes, low resolution three-dimensional structures can be determined from the scattering data either ab initio or through the refinement using available high resolution structures and/or homology models [10]. SAXS is routinely employed for the validation of structural models, analysis of oligomeric states and the estimation of volume fractions of components in mixtures. SAXS has been actively used also for flexible systems including solutions of intrinsically disordered proteins (IDPs) [11]. In the present chapter the experimental and methodological peculiarities of the use of FCS and SAXS for exploring RNA oligomerization and ligand binding are presented.

2 Materials

Prepare all solutions RNase-free (*see* **Note 1**).

2.1 FCS and FCCS

1. ConforCor2 (Carl Zeiss) and software package.
2. Cover slip 24 × 50 mm.
3. Fluorophore or fluorescently labeled RNA (store at −20 °C).
4. FCS buffer: 10 mM Na-HEPES, pH 7.4, 100 mM KCl, 5 mM $MgCl_2$.
5. 100 mM $MgCl_2$ stock solution.
6. For FCCS measurements: Fluorescently labeled RNA (*see* **Note 2**) with two different fluorophores, respectively (e.g., ATTO 488 and ATTO 647 N). Store at −20 °C.
7. RNase-free water.

2.2 SAXS

1. RNA solution: Precise knowledge of the RNA concentration is important for the assessment of the MW and oligomeric state of the RNA. The SAXS typical solute concentrations range from about 0.5–1 mg/mL to about 5–10 mg/mL (*see* **Note 3**), such that about 1–2 mg of purified material is usually required for a complete SAXS experiment on the given construct at the given conditions (e.g., buffer composition, temperature). RNA is stored at −20 °C.
2. SAXS buffer: 20 mM Na-HEPES, pH 7.4, 180 mM KCl, 50 mM $MgCl_2$, and 2 mM dithiothreitol (DTT); stored at −20 °C. This buffer composition was used for SAXS experiments of fluorophore binding RNA aptamer in ref. [1]. Sufficient amount (>10 mL) of matching buffer(s) must be brought to the SAXS station to make the buffer collections and to dilute the samples if necessary. The buffer composition must precisely match the composition of the sample: even small mismatches in the chemical composition of the solvent between the buffer and the sample may lead to difficulties during background subtraction. At best, the last dialysis buffer should be used for the background measurements (*see* **Note 4**).

2.3 SAXS Data Analysis

We present here the protocols based on the use of the program package ATSAS [12], which allows one to perform major processing and analysis steps. This package and all the programs mentioned below are publicly available for download from http://www.embl-hamburg.de/biosaxs/software.html.

3 Methods

3.1 RNA

1. Produce RNA by standard in vitro T7 transcription followed by gel-purification, extraction, and alcohol precipitation.
2. Alternatively, RNA can be chemically synthesized and purchased elsewhere.
3. For FCS titration experiments dilute RNA to a final concentration of 70 μM in FCS buffer without $MgCl_2$ (*see* **Note 5**).
4. For SAXS measurements prepare RNA in SAXS buffer with concentrations up to 45 μM.

3.2 RNA Denaturation and Renaturation

1. Prior to the experiments heat the RNA to 70 °C for 10 min and transfer it directly on ice.
2. Add $MgCl_2$ to a final concentration of 5 mM to the RNA (*see* **Note 6**).
3. Incubate RNA for 1 h at room temperature to allow RNA to fold properly.

3.3 FCS Measurement

FCS titration experiments are carried out at room temperature (22 °C).

1. Adjust pinhole with a standard fluorophore in *x*, *y*, and *z* direction.
2. Define the diffusion time $\tau_{\text{Diff};\,i}$ of the fluorescent component.
3. Add non-fluorescent binding component to the fluorescent component and diffusion time increases significantly (*see* **Note 7**).
4. Titrate the non-fluorescent component stepwise to fluorescent component until diffusion time remains constant (*see* **Note 8**).
5. The diffusion time of the complex can be determined easily by fitting FCS data with Zeiss software. Therefore, select the two-component model and set the diffusion time of the unbound fluorescent component as fixed value.

   The two-component model describes free three-dimensional diffusion and triplet excited state with Eq. 1

$$G(\tau) = 1 + \frac{1}{N} \cdot \left(\frac{1 - T + T \cdot e^{-\frac{\tau}{\tau_T}}}{1 - T} \right) \cdot \left\{ \sum_{i=1}^{2} \frac{f_i}{\left(1 + \frac{\tau}{\tau_{Diff;\,i}}\right)\sqrt{1 + \frac{\tau}{\tau_{Diff;\,i}} S^{-2}}} \right\} \quad (1)$$

and uses number of particles N, diffusion time $\tau_{\mathrm{Diff};\,i}$, fractional population f_1 of two different diffusion species ($n=2$) with triplet fraction T, lag time τ, and structure parameter S.

6. Calculate the complex fraction Υ using the concentration of complex $[C]$ and free fluorescent component $[F]$:

$$\Upsilon = \frac{[C]}{[F]+[C]}. \tag{2}$$

7. To calculate the dissociation constant K_{d} plot the complex fraction Υ against concentration of non-fluorescent species. K_{d} equates to the non-fluorescent species concentration at the half-maximum of complex fraction Υ.
8. Use diffusion time $\tau_{\mathrm{Diff};i}$ to calculate diffusion coefficient D_i by using radial diameter r_0 of standard fluorophores (Alexa 488 = 169 nm, R6G = 189 nm, and Cy5 = 250 nm) and

$$D_i = \frac{r_0^2}{4\,\tau_{Diff;\,i}}. \tag{3}$$

9. Calculate the hydrodynamic radius r_H by

$$r_H = \frac{\kappa T}{(6\pi\eta D_i)} \tag{4}$$

and applying the Boltzmann constant κ (138×10^{-23}kg m^2 s^{-2} K^{-1}), the viscosity η (1.002×10^{-3} kg m^{-1}s^{-1}), temperature T (22 °C = 295.15 K), and calculated diffusion coefficient D_i.

3.4 SAXS Measurement

A typical sample volume required for a single measurement is about 20–60 μL depending on the SAXS station used (*see* **Note 9**). The measurements of the SAXS data are usually made below room temperature (typically, at 5–10 °C) to reduce the possible radiation damage effects (*see* **Note 10**).

3.4.1 General Preparative Steps

1. Check for radiation damages during X-ray measurements. If the effect is observed, measures should be taken (reduction of the data takes time, flow of the sample during exposure, addition of glycerol or DTT as scavengers).
2. Record a series of scattering curves at varying concentrations to ensure that the conditions of a 'dilute' solution are fulfilled. For fluorophore binding RNA aptamer [1] concentrations ranging from 0.2 to 0.8 mg/mL were used.
3. Measure corresponding buffer before and after the sample (*see* **Notes 11** and **12**).

3.4.2 The SAXS Experiment

1. Perform SAXS measurements of RNA alone and of the corresponding buffer.
2. Add binding component in excess to RNA to ensure that all RNA molecules are bound.
3. Perform SAXS measurements of RNA in complex and of the corresponding buffer.

3.5 SAXS Data Analysis

3.5.1 Primary Analysis of SAXS Data

The scattering data are usually collected by two-dimensional detectors, and the images are appropriately processed, corrected, and reduced to one-dimensional scattering profiles by the local software of the SAXS stations. The output is typically an ASCII file containing the subtracted data in columnar format: $s, I(s), \sigma(s)$, where the latter column represents the standard deviation of the processed intensity. The normalization and buffer subtraction are performed with following Eq. 5

$$I(s) = \frac{\left(\frac{I_{sample}(s)}{T_{sample}} - \frac{I_{buffer}(s)}{T_{buffer}} \right)}{c} \tag{5}$$

where $I_{\text{sample}}(s)$ and $I_{\text{buffer}}(s)$, respectively, are the measured scattering data from the sample and the buffer, T_{sample} and T_{buffer} are the intensities of the transmitted X-ray beam, and c is the solute concentration. This basic operation is often performed by the station-specific software, scripts or completely automatically (the latter is the case, e.g., at the synchrotron P12 station of the EMBL in Hamburg). In the following we assume that the radially averaged and subtracted patterns are available and describe the sequence of actions using the interactive data analysis program PRIMUS from the ATSAS package [12].

1. Load the SAXS curve in PRIMUS.
2. Do a Kratky representation of the SAXS curve, which can be obtained with the SASPLOT application of PRIMUS that allows several data representations.

   A Kratky representation of the SAXS curve ($I(s) \times s^2$ vs s) is an excellent tool to qualitatively identify conformational disorder in biomacromolecules. Unstructured RNA presents a continuous raise of $I(s) \times s^2$, whereas globular constructs display a peak. Partly disordered systems present a mixed behavior depending on the relative proportion of each of the parts. An example of Kratky plots for the folded tRNA (PDB code: 4TRA.pdb), unfolded DsrA_{34} [13] and folded fluorophore-binding RNA aptamer [6] is given in Fig. 1.
3. Use Guinier approximation, R_g and the forward scattering, $I(0)$ that derived from the initial part of the scattering profile [14]. At very small angles the intensity is represented as:

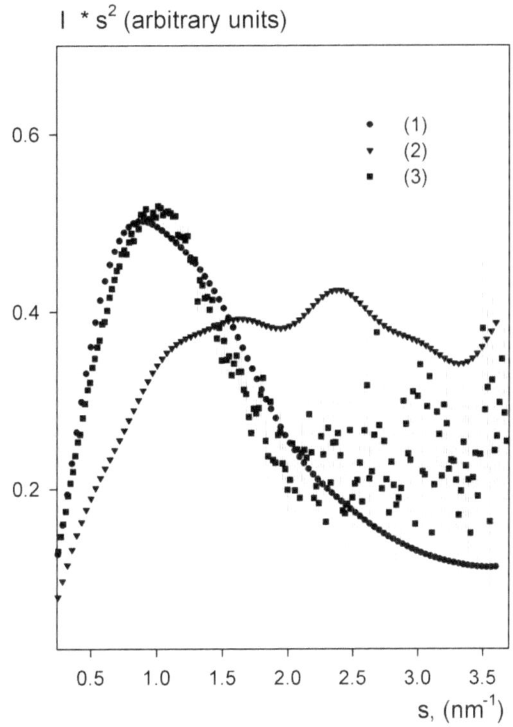

Fig. 1 SAXS Kratky plots of a folded tRNA (*curve 1*, *circles*), unfolded $DsrA_{34}$ (*curve 2*, *triangles*), and folded SRB2m (*curve 3*, *squares*). The plot displays the weighted scattering intensity ($I \times s^2$) as a function of momentum transfer $s = 4\pi \sin(\theta)/\lambda$, where 2θ is the scattering angle and $\lambda = 0.15$ nm is the X-ray wavelength. The theoretical scattering from tRNA and $DsrA_{34}$ models are computed with CRYSOL

$$I(s) = I(0)\exp\left(\frac{-\left(sR_g\right)^2}{3}\right) \tag{6}$$

The R_g and $I(0)$ are obtained by a simple linear fit in logarithmic scale using the *Guinier* option of PRIMUS where the optimal range of points for the analysis can be selected (*see* **Note 13**). Note that there is an option *Autorg* available, which selects the appropriate range and computes the R_g automatically. In some cases, however, it might be useful to cross-validate the results by interactive analysis.

4. Divide $I(0)$ values by the concentration of the sample in mg/mL to apply the equation.

$$MW_{RNA} = \frac{I(0)_{RNA} \cdot MW_{stand}}{I(0)_{stand}} \tag{7}$$

5. Compare $I(0)$ value of the molecule studied with that obtained for a standard RNA molecule of known molecular weight (MW) and concentration (*see* **Note 14**). The forward intensity $I(0)$ provides the estimate of the MW of the molecule and hence suggests its oligomeric state.
6. Obtain the distance distribution function $P(r)$ from the SAXS curve with the program GNOM [15] which is called directly from PRIMUS or its automated version AUTOGNOM [16]. $P(r)$ is defined as the distribution of distances between volume elements insight the particle. $P(r)$ defines the maximum particle distance, D_{max} beyond which $P(r)$ is equal to zero. Porod's law [17] defines the asymptotic behavior of $I(s)$ for homogeneous particles at large angles which decays as s^{-4}. Using the Porod invariant, the values of the excluded particle volume can be obtained from the scattering data using the Porod option of PRIMUS where the optimal range of points for the analysis can be selected (*see* **Note 15**).

3.5.2 Ab Initio Shape Determination

Distance distribution within the particle depends on its shape; therefore the scattering intensity can also be considered a shape's footprint. It is clear, however, that the reconstruction of a three-dimensional model of an object from its one-dimensional scattering pattern is an ill-posed problem, as multiple structures can provide one and the same scattering profile. A simplification reducing the ambiguity of such reconstruction consists in representing the particle at low resolution as a homogeneous body. The model could be a complex shape built from densely packed beads on a regular grid. In a generic modeling approach the search of an optimal model is performed to minimize discrepancy between the experimental data $I_{exp}(s)$ and a scattering profile computed from the model $I_{calc}(s)$:

$$\chi^2 = \frac{1}{N-1}\sum_{N}^{j=1}\left[\frac{I_{exp}\left(s_j\right)-cI_{calc}\left(s_j\right)}{\sigma\left(s_j\right)}\right]^2, \quad (8)$$

where c is a scaling factor, N is the number of points, and σ denotes experimental errors. In practice, the target function to be minimized $TF = \chi^2 + \sum \alpha_i P_i$ contains not only the discrepancy term χ^2 but also a set of weighted penalties $\alpha_i P_i$. The most wide-spread low resolution shape determination approach (implemented in programs DAMMIN/DAMMIF) employs simulated annealing minimization to build a compact interconnected ensemble of beads within a search volume (typically a sphere with the diameter D_{max}) that fits the experimental data [18, 19]. The natural restraints in this case are requirement of interconnectivity (a molecule cannot consist of separate fragments) and compactness (modeling is done at low resolution using homogeneous approximation; no fine details of the structure are expected).

The following steps should be made for ab initio shape determination analysis:

1. Run DAMMIN/DAMMIF for RNA molecule alone and with a ligand. Provide GNOM files as input data.
2. The output pdb files will contain ab initio shapes of the RNA molecule. Check the fit quality χ obtained with DAMMIN/DAMMIF, it should be around unity and the **.fir* file should not display systematic deviation between the experimental and calculated data. An example of ab initio model of SRB2m with PBV (patent blue V) computed by DAMMIN [6] is presented in Fig. 2.
3. Compare the volumes of the ab initio models for the RNA molecule alone and with a ligand in order to check whether the oligomerization state of RNA was changed (*see* **Note 16**).

3.6 Identification of Oligomerization State

In case of a polydisperse system without interparticle interactions, the scattering profile is a linear combination of the scattering

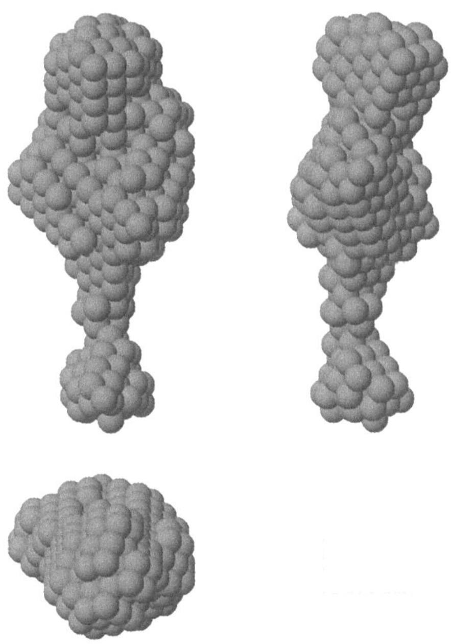

Fig. 2 The typical ab initio monomer model of SRB2m with PBV restored by DAMMIN. The *right* and *bottom panels* show the models rotated by 90 around the *y* and *x* axes, respectively. The scale bar is equal to 2 nm. Reprinted with permission from: Werner, A., Konarev, P.V., Svergun, D.I., and Hahn, U. (2009) Characterization of a fluorophore binding RNA aptamer by fluorescence correlation spectroscopy and small angle X-ray scattering. *Anal Biochem* 389, 52–62 (Elsevier)

intensities of individual components, weighted by their volume fractions [6] and is described by Eq. 9.

$$I(s) = \frac{1}{N} \sum_{N}^{n=1} I_n(s) \quad (9)$$

If the scattering patterns of all the mixture components are known (e.g., a mixture of monomers and dimers with known structure), the volume fractions are directly computed from SAXS data using the program OLIGOMER [20]. This approach is widely used to study oligomeric equilibria, complex formation, and structural transitions [21].

To identify the oligomerization state of the RNA the following steps should be performed:

1. The excluded volume in nm^3 of RNA calculated by SAXS data is numerically equal to the molecular weight in kDa [22]. Compare V_p and MW of RNA. If values differ this provides a first indication for an oligomerization state.
2. Compare V_p, D_{max}, R_g (from SAXS data), and r_H (from FCS data) RNA alone and in complex with bound component (*see* Table 1 with overall parameters obtained for fluorophore-binding RNA aptamer in ref. [6]). The bound component can influence oligomerization state of RNA, e.g., either to form monomers or dimers.

Table 1
Overall structural parameters of SRB2m RNA

Sample	Method	r_H or R_g (nm)	D_{max} (nm)	V_p (nm^3)	oligom. state
SRB2m	DLS	3.1 ± 0.4			
	SAXS	2.8 ± 0.1	10.0 ± 1.0	38.0 ± 4.0	++ and +
SRB2m + Sulforhodamine B	FCS	2.9 ± 0.1			
	SAXS	2.5 ± 0.1	8.5 ± 1.0	30.5 ± 3.0	++ and +
SRB2m + PBV	FCS	1.9 ± 0.1			
	SAXS	2.3 ± 0.1	8.0 ± 1.0	23.5 ± 3.0	+

Hydrodynamic radii r_H determined by FCS or DLS and radii of gyration R_g determined by SAXS, maximum sizes D_{max}, excluded volumes V_p from SAXS studies of SRB2m, SRB2m + Sulforhodamine B and SRB2m + PBV (patent blue V). The oligomerization state is presented by ++ in case of dimeric molecules and by + in the case of monomeric molecules. The errors represent standard deviations. The data were derived from at least two independent experiments.
Reprinted with permission from: Werner, A., Konarev, P.V., Svergun, D.I., and Hahn, U. (2009) Characterization of a fluorophore binding RNA aptamer by fluorescence correlation spectroscopy and small angle X-ray scattering. *Anal Biochem* 389, 52–62 (Elsevier)

3. In case of oligomerization use the program OLIGOMER to calculate proportion of respective monomer and dimer, for instance. For calculation the results of SAXS measurements are required as well as molecular weight of RNA.

3.7 FCCS Measurement

FCCS can also be used to identify oligomerization state of RNA molecules. Therefore the same RNA has to be fluorescently labeled with different fluorophores (*see* **Note 17**).

1. Mix both fluorescently labeled RNAs in equal amounts.
2. Denature and refold RNA following **steps 1–3** in Subheading 3.2.
3. Measure the cross-correlation signal. Cross-correlation signal should occur in case of oligomerization of RNA.

4 Notes

1. Aqueous solutions can be treated with diethylpyrocarbonate (DEPC) to inactivate RNases due to acetylation of histidin, lysine, cysteine, and tyrosine residues. Therefore solution is treated with 0.1 % (v/v) DEPC and stirred overnight. Autoclaving afterwards leads to destruction of DEPC. Tris-buffered solutions should not be treated with DEPC directly because it inactivates DEPC.
2. In case of precipitants or inhomogeneity RNA solution can be centrifuged for 2 h at maximum speed.
3. Typically, for each sample a concentration series (e.g., 1, 2, 5, 10 mg/mL) has to be measured. If the sample tends to (or may) aggregate at higher concentrations, it is highly recommended to bring the low concentration stocks to the SAXS instrument and concentrate the samples prior to the X-ray measurements. If the sample is well-behaved, one may bring high concentration stocks to dilute them before the measurements.
4. To achieve the best SAXS results, the RNA molecules should normally be dissolved in the buffers yielding close to ideal solutions. The ionic strength and pH should be selected in such a way that the RNA molecules minimally interact with each other unless it is required by the nature of the experiment. Any additives (salts, small molecules, co-solutes, etc.) diminish the contrast of the RNA molecule and increase the X-ray absorption; therefore buffers with excessive amounts of additives (typically, not more than 0.5 M NaCl, not more than 5 % glycerol, not more than 5 mM ATP, etc.) should be avoided unless this is not dictated by the biological or chemical considerations.

5. Concentration of RNA refers to titration experiments in which RNA binds to a fluorescent component. If RNA itself is fluorescently labeled and the interacting component is non-fluorescent, lower concentrations of RNA are used.
6. Heating RNA in presence of $MgCl_2$ might lead to hydrolysis of RNA.
7. Molar mass of a molecule determines diffusion time with $\tau_{Diff} \approx \sqrt[3]{m}$. Therefore the molar masses of investigated molecules have to differ by a factor of 8 or higher for distinguishing between different diffusion species.
8. Volume differentials due to evaporation can be adjusted by adding RNase-free water.
9. On second generation synchrotrons and on older X-ray laboratory sources, one needs about 50 μL and more of the sample; on modern high brilliance synchrotrons and on new laboratory cameras, about 10 μL of the sample are required.
10. SAXS stations normally allow for controlled measurement of the temperature series, in the range from ice melting to water boiling points. The temperature measurements are often useful for the studies of the thermodynamic folding/unfolding processes.
11. Note that typically all the SAXS measurements for the given experimental session are performed in one and the same measuring cell to ensure that the background scattering remains unchanged allowing for reliable buffer subtraction. The sample compartment (which is either a cell with flat windows or a capillary) is cleaned and refilled after each measurement. In the past, the cleaning and filling procedure was done manually; nowadays, liquid handling robots are becoming more and more popular to facilitate the automatic data collection [23]. On some stations (e.g., a synchrotron P12 station of the EMBL in Hamburg) remote operation is possible where the users send the samples and perform the data collection from their laboratory.
12. On high brilliance synchrotrons, the samples can be damaged by the X-ray radiation even during the seconds and sub-seconds collection time. A reducing agent (e.g., 2 mM DTT or TCEP) or an additive slowing down the formation of free radicals (e.g., up to 5 % glycerol) may be added to the sample before the experiment to diminish the damage/aggregation during the data collection. In any case, synchrotron data should be collected in short individual time frames, which have to be compared to each other to detect the radiation damage effects.
13. The momentum transfer range used for Guinier's approach must not exceed $s < 1.3 \times R_g$. The first points of the curve, if

they are influenced by the primary beam and show nonlinear behavior, must be suppressed from Guinier's analysis.

14. If one uses a standard protein molecule in Eq. 7, the experimental molecular weight of the RNA molecule from Eq. 7 will be two times overestimated due to the two times higher density of the RNA molecule compared to the protein.
15. There is a program AUTOPOROD available, which selects the appropriate range and computes the V_p automatically. In some cases, however, it might be useful to cross-validate the results by interactive analysis.
16. Ab initio shape determination can be applied only for well-behaved monodisperse systems.
17. A fluorophore can be inserted into RNA at the 5′ or 3′ end as well as intramolecularly. Location of the label is dependent on RNA binding site and should not interfere with binding component.

References

1. Werner A, Konarev PV, Svergun DI et al (2009) Characterization of a fluorophore binding RNA aptamer by fluorescence correlation spectroscopy and small angle X-ray scattering. Anal Biochem 389:52–62
2. Haustein E, Schwille P (2007) Fluorescence correlation spectroscopy: novel variations of an established technique. Annu Rev Biophys Biomol Struct 36:151–169
3. Weisshart K, Jungel V, Briddon SJ (2004) The LSM 510 META - ConfoCor 2 system: an integrated imaging and spectroscopic platform for single-molecule detection. Curr Pharm Biotechnol 5:135–154
4. Feigin LA, Svergun DI (1989) Structure analysis by small-angle X-ray and neutron scattering. In: Taylor GW (ed) Acta polymerica, vol 40. Plenum Press, New York/London, p 224
5. Svergun DI, Koch MH (2002) Advances in structure analysis using small-angle scattering in solution. Curr Opin Struct Biol 12:654–660
6. Koch MH, Vachette P, Svergun DI (2003) Small-angle scattering: a view on the properties, structures and structural changes of biological macromolecules in solution. Q Rev Biophys 36:147–227
7. Petoukhov MV, Svergun DI (2005) Global rigid body modeling of macromolecular complexes against small-angle scattering data. Biophys J 89:1237–1250
8. Petoukhov MV, Svergun DI (2007) Analysis of X-ray and neutron scattering from biomacromolecular solutions. Curr Opin Struct Biol 17:562–571
9. Jacques DA, Trewhella J (2010) Small-angle scattering for structural biology–expanding the frontier while avoiding the pitfalls. Protein Sci 19:642–657
10. Mertens HD, Svergun DI (2010) Structural characterization of proteins and complexes using small-angle X-ray solution scattering. J Struct Biol 172:128–141
11. Bernado P, Mylonas E, Petoukhov MV et al (2007) Structural characterization of flexible proteins using small-angle X-ray scattering. J Am Chem Soc 129:5656–5664
12. Konarev PV, Petoukhov MV, Volkov VV et al (2006) ATSAS 2.1, a program package for small-angle scattering data analysis. J Appl Cryst 39:277–286
13. de Almeida Ribeiro E Jr, Beich-Frandsen M, Konarev PV et al (2012) Structural flexibility of RNA as molecular basis for Hfq chaperone function. Nucleic Acids Res 40:8072–8084
14. Guinier A (1939) La diffraction des rayons x aux tres petits angles: application a l'etude de phenomenes ultramicroscopiques. Ann Phys 12:161–237
15. Svergun DI (1992) Determination of the regularization parameter in indirect-transform

methods using perceptual criteria. J Appl Cryst 25:495–503
16. Petoukhov MV, Konarev PV, Kikhney AG et al (2007) ATSAS 2.1 - towards automated and web-supported small-angle scattering data analysis. J Appl Cryst 40:s223–s228
17. Porod G (1982) General theory. In: Kratky O, Glatter O (eds) Small angle X-ray scattering. Academic Press, London, pp 17–51
18. Svergun DI (1999) Restoring low resolution structure of biological macromolecules from solution scattering using simulated annealing. Biophys J 76:2879–2886
19. Franke D, Svergun DI (2009) DAMMIF, a program for rapid ab-initio shape determination in small-angle scattering. J Appl Cryst 42:342–346
20. Konarev PV, Volkov VV, Sokolova AV et al (2003) PRIMUS: a Windows PC-based system for small-angle scattering data analysis. J Appl Cryst 36:1277–1282
21. Xu X, Reinle W, Hannemann F et al (2008) Dynamics in a pure encounter complex of two proteins studied by solution scattering and paramagnetic NMR spectroscopy. J Am Chem Soc 130:6395–6403
22. Funari E, Cavalieri M, Ade P et al (2000) Environmental and health problems of cyanobacteria blooms in surface waters in reference to the Italian situation. Ann Ig 12:381–400
23. Round AR, Franke D, Moritz S et al (2008) Automated sample-changing robot for solution scattering experiments at the EMBL Hamburg SAXS station X33. J Appl Cryst 41:913–917

INDEX

Christina Waldsich (ed.), *RNA Folding: Methods and Protocols*, Methods in Molecular Biology, vol. 1086, DOI 10.1007/978-1-62703-667-2, © Springer Science+Business Media, LLC 2014

T

MIX
Papier aus verantwortungsvollen Quellen
Paper from responsible sources
FSC® C105338

If you have any concerns about our products,
you can contact us on
ProductSafety@springernature.com

In case Publisher is established outside the EU,
the EU authorized representative is:
Springer Nature Customer Service Center GmbH
Europaplatz 3, 69115 Heidelberg, Germany

Printed by Libri Plureos GmbH
in Hamburg, Germany